AUTOPHAGY

AND

SIGNALING

METHODS IN SIGNAL TRANSDUCTION SERIES

Joseph Eichberg, Jr. and Michael X. Zhu
Series Editors

Published Titles

Autophagy and Signaling, Esther Wong

Calcium Entry Channels in Non-Excitable Cells, Juliusz Ashot Kozak and James W. Putney, Jr.

Lipid-Mediated Signaling Transduction, Second Edition, Eric Murphy, Thad Rosenberger, and Mikhail Golovko

Signaling Mechanisms Regulating T Cell Diversity and Function, Jonathan Soboloff and Dietmar J. Kappes

Gap Junction Channels and Hemichannels, Donglin Bai and Juan C. Sáez

Cyclic Nucleotide Signaling, Xiaodong Cheng

TRP Channels, Michael Xi Zhu

Lipid-Mediated Signaling, Eric J. Murphy and Thad A. Rosenberger

Signaling by Toll-Like Receptors, Gregory W. Konat

Signal Transduction in the Retina, Steven J. Fliesler and Oleg G. Kisselev

Analysis of Growth Factor Signaling in Embryos, Malcolm Whitman and Amy K. Sater

Calcium Signaling, Second Edition, James W. Putney, Jr.

G Protein-Coupled Receptors: Structure, Function, and Ligand Screening, Tatsuya Haga and Shigeki Takeda

G Protein-Coupled Receptors, Tatsuya Haga and Gabriel Berstein

Signaling Through Cell Adhesion Molecules, Jun-Lin Guan

G Proteins: Techniques of Analysis, David R. Manning

Lipid Second Messengers, Suzanne G. Laychock and Ronald P. Rubin

AUTOPHAGY
AND
SIGNALING

Edited by

Esther Wong

Nanyang Technological University
Singapore

CRC Press is an imprint of the
Taylor & Francis Group, an **informa** business

CRC Press
Taylor & Francis Group
6000 Broken Sound Parkway NW, Suite 300
Boca Raton, FL 33487-2742

First issued in paperback 2020

© 2018 by Taylor & Francis Group, LLC
CRC Press is an imprint of Taylor & Francis Group, an Informa business

No claim to original U.S. Government works

ISBN-13: 978-1-4987-3189-8 (hbk)
ISBN-13: 978-0-367-65772-7 (pbk)

Library of Congress Cataloging-in-Publication Data

Names: Wong, Esther, 1974- editor.
Title: Autophagy and signaling / [edited by] Esther Wong.
Other titles: Methods in signal transduction.
Description: Boca Raton, FL : CRC Press, 2018. | Series: Methods in signal transduction series
Identifiers: LCCN 2017014653 | ISBN 9781498731898 (hardback : alk. paper)
Subjects: | MESH: Autophagy--physiology | Signal Transduction--physiology
Classification: LCC QR181.7 | NLM QU 375 | DDC 616.07/9--dc23
LC record available at https://lccn.loc.gov/2017014653

**Visit the Taylor & Francis Web site at
http://www.taylorandfrancis.com**

**and the CRC Press Web site at
http://www.crcpress.com**

Contents

SECTION I   Signaling Pathways Regulating Autophagy

SECTION II   Autophagy and Cell Fate

SECTION III   Autophagy in Immunity and Metabolism

SECTION IV   *Autophagy in Neural Homeostasis and Neurodegeneration*

Series Preface

The concept of signal transduction is now long established as a central tenet of biological sciences. Since the inception of the field close to 50 years ago, the number and variety of signal transduction pathways, cascades, and networks have steadily increased and now constitute what is often regarded as a bewildering array of mechanisms by which cells sense and respond to extracellular and intracellular environmental stimuli. It is not an exaggeration to state that virtually every cell function is dependent on the detection, amplification, and integration of these signals. Moreover, there is increasing appreciation that in many disease states, aspects of signal transduction are critically perturbed.

Our knowledge of how information is conveyed and processed through these cellular molecular circuits and biochemical switches has increased enormously in scope and complexity since this series was initiated 15 years ago. Such advances would not have been possible without the supplementation of older technologies, drawn chiefly from cell and molecular biology, biochemistry, physiology, pharmacology, with newer methods that make use of sophisticated genetic approaches as well as structural biology, imaging, bioinformatics, and systems biology analysis.

The overall theme of this series continues to be the presentation of the wealth of up-to-date research methods applied to the many facets of signal transduction. Each volume is assembled by one or more editors who are preeminent in their specialty. In turn, the guiding principle for editors is to recruit chapter authors who will describe procedures and protocols with which they are intimately familiar in a reader-friendly format. The intent is to assure that each volume will be of maximum practical value to a broad audience, including students and researchers just entering an area, as well as seasoned investigators.

As a common pathway important for cell survival, autophagy has received great attention in recent years, encompassing such broad areas as normal physiology of all tissues and organ systems as well as disease conditions, including muscle atrophy, neurodegeneration, cancer, etc. This fast growing area of signal transduction is constantly evolving, with new concepts and methodology frequently developed, many of which are covered in the current volume. It is hoped that the information contained in this volume, as well as other books of this series, will constitute a useful resource to the life sciences research community well into the future.

Joseph Eichberg
Michael Xi Zhu
Series Editors

Preface

Autophagy is a self-eating pathway that the cell uses to recycle its cellular components in lysosomes. Over the years, our understanding of autophagy has evolved from a mere "garbage disposal" pathway to a highly sophisticated system that plays integral roles in multiple cell survival processes. There is a paradigm shift in the recognition of autophagy as a significant pathway that triages health and diseases. This was recently affirmed by the 2016 Nobel Prize in Physiology or Medicine to Professor Yoshinori Ohsumi for his pioneering work on elucidating how autophagy works in yeast.

Besides being a quality control process, autophagy participates in a myriad of physiological functions that directly impact cellular and organismal health. Healthy functional autophagy is associated with longevity while dysfunction in this pathway contributes to the aging process and an array of diseases. Autophagy maintains the health of cells and tissues by regulating cellular homeostasis, cell fate and remodeling, metabolism, immune defense, stress responses and repairs. In recent years, different forms of autophagy have been identified that endow cells with multifaceted mechanisms to better counteract different stressors and toxic cargoes. Further findings support a central role of autophagy in many molecular networks, where autophagy is interconnected with other signaling pathways, serving either as a regulator or an effector to coordinate and orchestrate specific or systematic cellular responses and adaptation. These different autophagic variants and multiplicity of connections together confer cells with great versatility for survival. Thus, understanding autophagy signaling has direct relevance for promoting health and mitigating disease progression. This sets the backdrop for this book, which aims to provide an up-to-date overview of the plethora of signaling pathways that converge on regulating autophagy in response to different cellular needs. The various chapters cover the current perspective of autophagy in regulating cell fate, immune response, nutrient sensing and metabolism, neural functions, and homeostasis. Some chapters also include exploration of the mechanisms and significance of cross talk between autophagy and other cellular processes. Alterations in autophagy observed in aging and age-related pathologies are also discussed. I hope each reader will gain new insights on the different flavors of autophagic mechanisms and come to appreciate the diverse functions and importance of autophagy from the different chapters. Finally, special thanks to all the authors for contributing their time and talent for this book.

Esther Wong
Nanyang Technological University
Singapore

Contributors

Cheng Bing
Lee Kong Chian School of Medicine
Nanyang Technological University
Singapore

Hui-Ying Chan
Neurodegeneration Research
    Laboratory
National Neuroscience Institute
Singapore

Patrice Codogno
Institut Necker Enfants-Malades
    (INEM)
Université Paris Descartes-Sorbonne
    Paris
Paris, France

Karen Crasta
Lee Kong Chian School of Medicine
Nanyang Technological University
and
Agency of Science, Technology
    and Research
Institute of Molecular and Cell Biology
Singapore
and
Department of Medicine
Imperial College London
London, United Kingdom

Andre Du Toit
Department of Physiological Sciences
Faculty of Natural Sciences, University
    of Stellenbosch
Stellenbosch, South Africa

Sheng-Han Kuo
Department of Neurology
New York State Psychiatric Center
Columbia University Medical Center
New York, NY

Grace G.Y. Lim
Neurodegeneration Research
    Laboratory
National Neuroscience Institute
Singapore

Kah-Leong Lim
Neurodegeneration Research
    Laboratory
National Neuroscience Institute
and
Department of Physiology
National University of Singapore
Duke-NUS Medical School
Singapore

Ben Loos
Department of Physiological Sciences
Faculty of Natural Sciences
University of Stellenbosch
Stellenbosch, South Africa

Nuria Martinez-Lopez
Department of Medicine
Albert Einstein College of Medicine
Bronx, NY

Olatz Pampliega
Institut des Maladies Neurodégénératives,
    UMR 5293
CNRS, Université de Bordeaux
Bordeaux, France

Tassula Proikas-Cezanne
Department of Molecular Biology
Interfaculty Institute of Cell Biology
Eberhard Karls University Tuebingen
and
International Max Planck Research
    School "From Molecules to
    Organisms"
Max Planck Institute for Developmental
    Biology,
Tuebingen, Germany

Moumita Rakshit
School of Biological Sciences
Nanyang Technological University
Singapore

Yi Ren
Department of Physiology
Yong Loo Lin School of Medicine
National University of Singapore
Singapore

Han-Ming Shen
Department of Physiology
Yong Loo Lin School of Medicine
National University of Singapore
Singapore

David Sulzer
Department of Neurology
New York State Psychiatric Center
Columbia University Medical Center
and
Departments of Psychiatry, Pharmacology
New York State Psychiatric Center
Columbia University Medical Center
New York, NY

Chrisna Swart
Department of Physiological Sciences
Faculty of Natural Sciences, University
    of Stellenbosch
Stellenbosch, South Africa

Sijie Tan
School of Biological Sciences
Nanyang Technological University
Singapore

Guomei Tang
Department of Neurology
New York State Psychiatric Center
Columbia University Medical Center
New York, NY

Rut Valdor
Department of Internal Medicine
Medicine School, University of Murcia
Biomedical Research Institute of
    Murcia (IMIB-Arrixaca)
Murcia, Spain

Esther Wong
School of Biological Sciences
Nanyang Technological University
Singapore

Han Xie
Faculty of Dentistry Research
    Laboratories
National University of Singapore
Singapore

Tso-Pang Yao
Department of Pharmacology and
    Cancer Biology
Duke University School of Medicine
Durham, NC

Theresia Zuleger
Department of Molecular Biology
Interfaculty Institute of Cell Biology
Eberhard Karls University Tuebingen
Tuebingen, Germany

Section I

Signaling Pathways Regulating Autophagy

1 Regulation of Autophagy by AMPK

Yi Ren, Han Xie, and Han-Ming Shen
National University of Singapore
Singapore

CONTENTS

1.1 INTRODUCTION

Metabolism is the most important cellular process that comprises different biochemical reactions. Two opposite metabolic pathways, that is, catabolic and anabolic pathways, are exquisitely regulated in order to generate enough energy and building blocks for diverse biological functions and syntheses of new cellular components. The central regulator of cellular metabolism is the 5'-adenosine monophosphate (AMP)-activated protein kinase (AMPK), an evolutionarily conserved serine/threonine kinase complex that, when activated by metabolic stress, acts to maintain cellular energy homeostasis through suppressing anabolic pathways and enhancing catabolic pathways (Hardie et al. 2015). Macroautophagy (hereafter referred to as autophagy in this chapter) is one of the critical catabolic pathways that can be promoted by AMPK under various stress conditions. Autophagy is a dynamic cellular recycling process in which proteins and organelles are sequestered within autophagosomes, delivered to lysosomes, and eventually degraded by hydrolases to produce energy and building blocks for cellular renovation (Feng et al. 2015; Mizushima and Komatsu 2011; Shen and Mizushima 2014). Given its role in maintaining cellular homeostasis, dysregulation of autophagy is widely implicated in human diseases, including cancer, neurodegenerative disorders, and microbial infection (Choi et al. 2013; Mizushima et al. 2008). Two major upstream regulators of autophagy are the mechanistic target of rapamycin (mTOR), a key controller of cellular metabolism which inhibits autophagy when nutrients are rich, and AMPK that, the other way around, activates autophagy when nutrients are scarce (Hardie 2011; Jung et al. 2010). To date, numerous studies have reported that under stress conditions, for example, glucose starvation, hypoxia, and drug treatment, autophagy is activated in an AMPK-dependent manner (Kim et al. 2011; Papandreou et al. 2008; Yu et al. 2013). Three major ways have been reported in which AMPK activates autophagy: (1) through phosphorylation and activation of tuberous sclerosis complex 1/2 (TSC1/TSC2) (Inoki et al. 2003b), or phosphorylation of regulatory-associated protein of mammalian target of rapamycin (raptor) and its subsequent inhibition by 14-3-3 (Gwinn et al. 2008), thus suppressing mTOR complex 1 (mTORC1) activity, (2) through activation of UNC-51-like kinase 1(Ulk1) (Egan et al. 2011; Kim et al. 2011), and (3) through phosphorylation of Beclin1 in the presence of Atg14L, and therefore activation of the Vps34 complex (Kim et al. 2013). Besides, cyclin-dependent kinase (Cdk) inhibitor p27[kip1], eukaryotic elongation factor 2 (eEF2) kinase, and Forkhead Box O3a (FoxO3a) have been identified as direct AMPK targets to regulate autophagy (Browne et al. 2004; Liang et al. 2007; Sanchez et al. 2012; Wu et al. 2006). And recently, it has been reported that Sirtuin1 (Sirt1) activation mediates autophagy via an AMPK-dependent manner (Chang et al. 2015; Huang et al. 2015). In this chapter, we are going to give a systematic introduction about the basics of AMPK and the above mentioned regulatory pathways of autophagy by AMPK.

1.2 AMPK

In 1973, two groups described protein fractions that, in the presence of ATP, inactivated two regulatory enzymes in lipid synthesis: acetyl-CoA carboxylase (ACC) in fatty acid synthesis (Carlson and Kim 1973) and 3-hydroxy-3-methylglutaryl-CoA (HMG-CoA) reductase in cholesterol synthesis (Beg et al. 1973). Researchers failed to realize that observations from these two groups were actually two different functions of one protein kinase until 1987, when Hardie's group provided evidence that both ACC and HMG-CoA reductase can be phosphorylated and inactivated in an identical manner by AMP (Carling et al. 1987). Since then, this protein kinase is named as AMPK.

1.2.1 STRUCTURE OF AMPK

The heterotrimeric AMPK complex consists of catalytic α subunit and regulatory β and γ subunits (Figure 1.1a). Multiple isoforms of these subunits (α1/α2, β1/β2, and γ1/γ2/γ3) are encoded by distinct genes (*PRKAA1/2*, *PRKAB1/2*, and *PRKAG1/2/3*) in mammals, which can form 12 heterotrimeric combinations (Carling 2004; Hardie 2014). Given its role in sensing changes in energy level and controlling activities of enzymes involved in cellular metabolism, the structure of AMPK heterotrimer can be divided into two modules, that is, the catalytic module and the nucleotide-binding module (Hardie 2014).

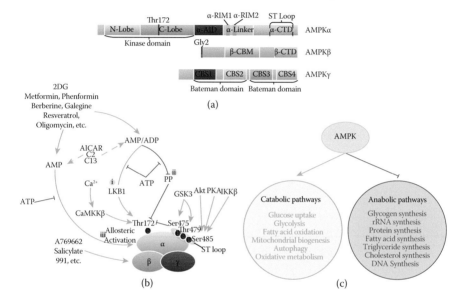

FIGURE 1.1 Structure, regulation and biological functions of AMPK. (a) The domain structure of typical mammalian AMPK subunits. (b) Canonical and noncanonical regulation of AMPK. AMP activates AMPK via three canonical ways: (i) promoting Thr172 phosphorylation by LKB1; (ii) inhibiting Thr172 dephosphorylation by protein phosphatases (PP); and (iii) allosteric activation. Only effects (1) and (2) but not (3) can be mimicked by ADP. All of these can be antagonized by ATP. In addition, CaMKKβ also phosphorylates AMPK at Thr172 in response to Ca^{2+}, but independent of changes in AMP/ADP. Moreover, three classes of compounds activate AMPK via noncanonical ways: (1) inhibiting ATP synthesis and causing increase of AMP/ADP levels, for example, 2DG, metformin, berberine, and resveratrol; (2) generating AMP analogs, for example, AICAR, C2, and C13; (3) directly binding to the β sunbunit thus causing allosteric activation and inhibition of Thr172 dephosphorylation, for example, A769662, salicylate, and 991. On the other hand, the phosphorylation of the ST Loop of the α-subunit by Akt, PKA, or GSK3 suppressing Thr172 phosphorylation and AMPK activity. (c) The biological functions of AMPK. Once activated, AMPK promotes catabolic pathways to generate ATP, while inhibits anabolic pathways to decrease ATP consumption. N-Lobe, N-terminal lobe of kinase domain; C-Lobe, C-terminal lobe of kinase domain; α-AID, α-autoinhibitory domain; α-RIM1/2, α-regulatory subunit interacting motif 1/2; ST Loop, serine/threonine-rich loop; α-CTD, α-C-terminal domain; β-CBM, β-carbohydrate-binding module; β-CTD, β-C-terminal domain; CBS1-CBS4, cystathionine β-synthase repeats.

1.2.1.1   The α Subunit

The serine/threonine kinase domain (α-KD) of AMPK is located at the N-terminus of the α subunit. When a conserved threonine residue (Thr172 in α2, equivalent to Thr183 in α1) within the kinase domain is phosphorylated by upstream kinases, AMPK is activated >100-fold (Hardie 2015a). Following the kinase domain is the autoinhibitory domain (α-AID), which inhibits the kinase activity through its binding to the α-KD (Moreira et al. 2015). Without α-AID, the α-KD alone containing constructs are 10-fold more active than α-KD-AID constructs, regardless of the Thr172 phosphorylation status (Goransson et al. 2007; Pang et al. 2007). A polypeptide, termed as α-linker, connects the α-AID to the C-terminal domain (α-CTD). The α-linker contains two conserved α-regulatory subunit interacting motifs (α-RIM1 and α-RIM2) (Xin et al. 2013). Both of them can interact with the γ subunit, which will be discussed later. In addition, the α-linker and α-AID form the hinge between the catalytic module and the nucleotide-binding module (Hardie 2014). The α-CTD contains a long serine/threonine-rich loop (ST loop). A few studies have reported that the phosphorylation at S485 by Akt (also known as protein kinase B, PKB), protein kinase A (PKA) or inhibitor of nuclear factor kappa-B kinase subunit β (IKKβ), as well as the phosphorylation at S475 and T479 by glycogen synthase kinase 3 (GSK3) within the ST loop, inhibits the AMPK activity (Hawley et al. 2014; Horman et al. 2006; Hurley et al. 2006; Park et al. 2014; Suzuki et al. 2013).

1.2.1.2   The β Subunit

The β subunit contains two conserved domains, that is, the carbohydrate-binding module (β-CBM) and the C-terminal domain (β-CTD). The β-CBM mediates the binding of AMPK to the glycogen particles, which may serve to colocalize AMPK with its downstream target glycogen synthase (Hudson et al. 2003; McBride et al. 2009; Polekhina et al. 2003, 2005). An autophosphorylation site at Ser108 within the β-CBM is found to be critical for the allosteric activation of AMPK. The β-CTD functions as a bridge between both α-CTD and the γ subunit, forming the core of the AMPK complex. In addition, Gly2 myristoylation at the N-terminus of the β subunit is essential for AMP-induced Thr172 phosphorylation and membrane association of AMPK (Liang et al. 2015; Oakhill et al. 2010).

1.2.1.3   The γ Subunit

The γ subunit is responsible for sensing changes in AMP through the direct binding of adenine nucleotides (Cheung et al. 2000; Scott et al. 2004). Although the N-terminal regions vary greatly in length and sequence among different γ isoforms, they all contain four tandem cystathionine β-synthases (CBS) repeats (CBS1-4), with each two forming a Bateman domain that provides a binding site in the cleft between two repeats for adenosine-containing ligands (Scott et al. 2004). The four CBS repeats assemble into a flattened disk with one repeat in each quadrant, which generates two Bateman domains, and thus four potential binding sites. These sites are numbered 1–4 according to the number of the CBS repeats with a conserved aspartate residue that is involved in ligand binding (Kemp et al. 2007). However, in mammalian AMPK γ subunits, site 2, with the corresponding aspartate in CBS2 replaced by an arginine, is always vacant (Oakhill et al. 2012). On the contrary, site 4 was originally suggested to be an AMP nonexchangeable site (Xiao et al. 2007, 2011), which has been challenged that ATP can also be found in site 4 (Chen et al. 2012). In addition, although sites 1 and 3 are able to bind AMP, ADP, or ATP in competition, site 3 appears to be the only site determining AMPK activity (Xiao et al. 2011, 2013; Xin et al. 2013).

1.2.2 REGULATION OF AMPK

1.2.2.1 Canonical Regulation by Adenine Nucleotides

As its name suggests, AMPK is activated by AMP. Subsequent studies showed that ADP also activates AMPK (Oakhill et al. 2010, 2011; Xiao et al. 2011). The regulation of AMPK by AMP, ADP, and ATP is referred to as the canonical pathway (Figure 1.1b).

AMPK is activated by AMP via three complementary ways: (i) promoting Thr172 phosphorylation by upstream kinases; (ii) inhibiting Thr172 dephosphorylation by protein phosphatases; and (iii) causing allosteric activation. Effects (i) and (ii) can be mimicked by ADP, while effect (iii) is triggered only by AMP. All of these can be antagonized by ATP.

The major upstream kinases that phosphorylate Thr172 are the tumor suppressor LKB1-STRAD-MO25 complex (Hawley et al. 2003; Shaw et al. 2004; Woods et al. 2003) and the Ca^{2+}/calmodulin-dependent kinase kinases (CaMKKs), especially CaMKKβ (Hawley et al. 2005; Hurley et al. 2005; Woods et al. 2005). It is generally believed that the effect (1) is only achieved by AMP through LKB1, whereas activation of AMPK by CaMKKβ is dependent on the increase of intracellular Ca^{2+} but independent of changes in adenine nucleotides. Intriguingly, there were studies reporting that binding of ADP promotes Thr172 phosphorylation by CaMKKβ (Oakhill et al. 2010, 2011), while Ca^{2+} was found to activate AMPK synergistically with AMP/ADP, via suppression of Thr172 dephosphorylation (Fogarty et al. 2010). Remarkable progresses have been made recently for revealing a new mechanism of AMPK activation; nutrient deprivation causes the regulator to recruit the axin/LKB1 complex to lysosomes, while increased AMP causes translocation of AMPK to this complex, where it is phosphorylated and activated by LKB1 (Zhang et al. 2013, 2014).

The effect (2) is due to the conformational change caused by AMP (Davies et al. 1995). When ATP replaces AMP at site 3, the α-linker that connects the catalytic and nucleotide-binding modules is released from the γ subunit, which causes two modules to move apart, increasing the accessibility of Thr172 to phosphatases and thus dephosphorylation (Xiao et al. 2011). In addition to AMP, binding of ADP can also inhibit Thr172 dephosphorylation, although AMP is more potent and may be more physiologically important as the ratio changes in AMP are always larger than those in ADP in cells under stress conditions (Gowans et al. 2013; Xiao et al. 2011).

The effect (3) can only be caused by AMP but not ADP through its binding to the site 3 in the γ subunit. The α-RIM1 in the α-linker binds to the vacant site 2, while α-RIM2 associates with site 3 bound with AMP. This association is proposed to release the α-AID from the α-KD, thus causing allosteric activation (Xiao et al. 2013; Xin et al. 2013).

1.2.2.2   Noncanonical Activation by Natural or Synthetic Compounds

Many high-throughput screens have been conducted to identify AMPK activators. To date, numerous activators have been found that can be grouped into the following three classes according to their mechanisms of activation (Figure 1.1b):

1. Ligands that bind to the β subunit; the first activator belonging to this class is A769662 (Cool et al. 2006), which activates AMPK through its direct binding to the β-CBM, thus causing both allosteric activation and inhibition of Thr172 dephosphorylation (Goransson et al. 2007; Sanders et al. 2007). Activation of AMPK by A769662 is independent of AMP but requires β1 Ser108 autophosphorylation, as it activates AMPK in AMP-insensitive γ2 R531G-expressing cells (Hawley et al. 2010), whereas β1 S108A mutation totally abolishes the activation (Sanders et al. 2007). In addition, A769662 is more effective in β1-containing complexes than β2, indicating that this activation involves the β subunit (Scott et al. 2008). Recently, another two new compounds, 991 and MT-63-78, were identified as AMPK allosteric activators. The compound 991 binds to β subunit in the same way as A769662 (Xiao et al. 2013), while MT-63-78 shows stronger activation in β1-containing complexes than β2 (Zadra et al. 2014), which is thought to bind to the same site as A769662. In addition to synthetic compounds, salicylate, a natural product that has been used as a medicine with a long history, was found to activate AMPK through its binding to the same site and is dependent on β Ser108 phosphorylation (Calabrese et al. 2014; Hawley et al. 2012). The finding of synthetic compounds as well as natural plant product that regulate AMPK through their binding at the specific site raises the question of whether there is any metabolite in mammals that functions in the same way (Hardie et al. 2015). Although such a metabolite has not yet been identified, the specific binding site has been referred to as the allosteric drug and metabolite (ADaM)-binding pocket (Langendorf and Kemp 2015).
2. Indirect activators that increase AMP/ADP through inhibiting ATP production; the majority of these compounds activate AMPK by inhibiting mitochondrial respiratory chain, for example, the antidiabetic drugs metformin and phenformin (Hawley et al. 2003; Zhou et al. 2001), as well as the natural plant products like berberine and galegine (Lee et al. 2006; Mooney et al. 2008). In addition, some compounds inhibit mitochondrial ATPase, thus increasing AMP/ADP, for example, resveratrol and oligomycin (Gledhill et al. 2007; Hawley et al. 2002). Another way to lower ATP production is the glycolytic inhibitor 2-deoxyglucose (2DG), which blocks glycolysis, thus activating AMPK (Inoki et al. 2003b).

3. Compounds that generate AMP analogs; for many years, 5-aminoimidazole-4-carboxamide ribonucleoside (AICAR) is the only member belonging to this class. When taken up into cells by adenosine transporters, AICAR is phosphorylated to ZMP, an AMP analog which can mimic all the effects of AMP, leading to AMPK activation (Corton et al. 1995; Gadalla et al. 2004). Recently, a new AMP analog, 5-(5-hydroxyl-isoxazol-3-yl)-furan-2-phosphonic acid (C2), and its cell-permeable phosphonate diester C13, were described to activate AMPK more potently than ZMP and AMP (Gomez-Galeno et al. 2010).

1.2.3 Biological Functions of AMPK

Once activated, AMPK has two principal functions: promoting catabolic pathways to generate ATP, while suppressing anabolic pathways to decrease ATP consumption. The functions of AMPK are involved in many important aspects of cell metabolism (Figure 1.1c): (1) glucose metabolism, including promoting glucose uptake and glycolysis, whereas inhibiting glycogen synthesis; (2) protein metabolism, including inhibiting rRNA and protein syntheses; (3) increasing mitochondrial biogenesis; (4) lipid metabolism, including promoting fatty acid oxidation, whereas inhibiting fatty acid, triglyceride, and cholesterol syntheses; (5) inhibiting DNA synthesis and promoting cell-cycle arrest; and (6) promoting autophagy (Hardie 2015b; Hardie et al. 2012). Here, we provide a detailed introduction on the regulatory function of AMPK in autophagy.

1.3 AUTOPHAGY

Autophagy is an evolutionarily conserved cellular process in which cellular components such as proteins and organelles are degraded and recycled through lysosomal machinery (Boya et al. 2013; Mizushima and Komatsu 2011; Yang and Klionsky 2010; Yu et al. 2015). In the case of nutrient starvation or overwhelming energy demands, autophagy will be promoted to produce sufficient energy and building blocks for critical cellular activities (Mizushima 2009). The activated AMPK targets the Ulk1/Atg13/FIP200 (focal adhesion kinase family–interacting protein of 200 kDa)/Atg101 complex and promotes its translocation to endoplasmic reticulum (ER), where it regulates the Vps34 complex (Beclin1/Vps15/Vps34) to produce phosphatidylinositol (PI)-3-phosphate (PI3P), which facilitates the maturation of the phagophore (Yu et al. 2015). Two ubiquitin-like protein conjugation systems, that is, Atg12 system and LC3 system are then involved to form Atg12/Atg5/Atg16L1 complex and LC3-phosphatidylethanolamine (PE) conjugate, which play important roles in the elongation of the phagophore and formation of the autophagosome. The autophagosome eventually fuses with lysosome for the degradation of the cargoes (Boya et al. 2013; Marino et al. 2014; Mizushima and Komatsu 2011). To date, autophagy has been well established to play important roles in maintaining cellular homeostasis and renovation, as well as in many other cellular processes related to aging, development, and immune response (Mizushima and Komatsu 2011). Accordingly, autophagy is closely implicated in many important human diseases such as cancer, infection, diabetes, and neurodegenerative diseases (Choi et al. 2013; Mizushima et al. 2008).

1.4 THE REGULATION OF AUTOPHAGY BY AMPK

At present, there is substantial evidence demonstrating the important regulatory function of AMPK in autophagy. As summarized in Figure 1.2, there are multiple pathways that have been implicated, including (1) through activation of TSC2 or inhibition of raptor, thus suppressing mTORC1; (2) through activation of Ulk1; and (3) through activation of the Vps34 complex.

1.4.1 THROUGH MTORC1

mTOR is a serine/threonine kinase that forms two distinct complexes, namely mTOR complex 1 (mTORC1) and mTOR complex 2 (mTORC2). mTORC1 and mTORC2 share mTOR as the catalytic subunit, and other subunits including mammalian lethal with sec-13 protein 8 (mLST8, also known as GβL) (Jacinto et al. 2004; Kim et al. 2003), mTOR inhibitor DEP domain containing mTOR-interacting protein (DEPTOR) (Peterson et al. 2009), and Tti1/Tel2 complex (Kaizuka et al. 2010). Although regulatory-associated protein of mammalian target of rapamycin (raptor) (Hara et al. 2002; Kim et al. 2002) and proline-rich Akt substrate 40 kDa (PRAS40) (Sancak et al. 2007; Thedieck et al. 2007; Vander Haar et al. 2007; Wang et al. 2007) only exist in mTORC1, rapamycin-insensitive companion of mTOR (rictor) (Jacinto et al. 2004; Sarbassov et al. 2004), mammalian stress-activated map kinase-interacting protein 1 (mSin1) (Frias et al. 2006; Jacinto et al. 2006), and protein observed with rictor 1 and 2 (protor1/2) (Peterson et al. 2009; Thedieck et al. 2007) are specific

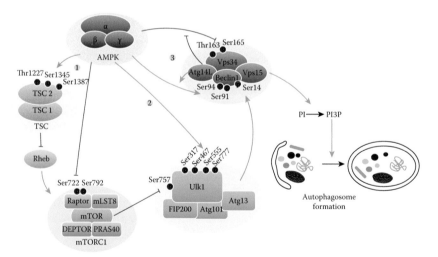

FIGURE 1.2 Regulation of autophagy by AMPK. Activated AMPK promotes autophagy via three major mechanisms: (1) through phosphorylation and activation of TSC2, which inhibits Rheb GTPase, or through phosphorylation and inhibition of raptor, thus suppressing mTORC1; (2) through direct phosphorylation and activation of Ulk1; and (3) through phosphorylation of Beclin1 and thus activation of the Vps34 complex, which generates PI3P and thus is important for autophagosome formation. PI, phosphatidylinositol; PI3P, phosphatidylinositol-3-phosphate.

to mTORC2. Besides their component difference, mTORC1 and mTORC2 are also distinct in the sensitivity to rapamycin as well as upstream regulators and downstream targets. mTORC1 is activated by oxygen, amino acids, growth factors, and energy, while inhibited by stress and rapamycin, whereas mTORC2 is insensitive to rapamycin and only responds to growth factors (Laplante and Sabatini 2012). When activated in nutrient-rich conditions, mTORC1 promotes cell growth through initiating protein synthesis, lipogenesis, and energy metabolism, and inhibiting autophagy and lysosome biogenesis, which functions in an opposite manner to AMPK. mTORC2, although less studied, is reported to regulate cytoskeletal organization and to act as an upstream kinase for Akt (Laplante and Sabatini 2012). mTORC1 suppresses autophagy through direct phosphorylation of Ulk1 and Atg13, thus inhibiting Ulk1 complex and blocking autophagy at the early stage (Ganley et al. 2009; Hosokawa et al. 2009; Jung et al. 2009), which makes mTORC1 the gatekeeper of autophagy signaling.

Tuberous sclerosis complex (TSC) is an autosomal dominant disease resulting from mutations in either *TSC1* or *TSC2* tumor suppressor genes (Young and Povey 1998). Harmatin and tuberin are the products of *TSC1* gene and *TSC2* gene, respectively (van Slegtenhorst et al. 1998). They form a stable complex that negatively regulates mTORC1, cell growth, and cell proliferation (Gao and Pan 2001; Ito and Rubin 1999; Potter et al. 2001; Tapon et al. 2001). In TSC1/TSC2 complex, TSC1 is responsible for regulating the cellular localization of the complex (van Slegtenhorst et al. 1998) and stabilizing TSC2 by preventing the HERC1 ubiquitin ligase–mediated ubiquitination and degradation of TSC2 (Benvenuto et al. 2000; Chong-Kopera et al. 2006). TSC2 contains a GTPase-activating protein (GAP) domain, which inhibits its downstream GTPase. It has been reported that TSC1/TSC2 complex inhibits S6K and activates eIF4E-binding protein 1 (4E-BP1), two well-established downstream targets of mTOR (Inoki et al. 2002). The functions of TSC1/TSC2 are indeed mediated through inhibition of mTOR, as Ras homolog enriched in brain (Rheb), a small GTPase and the activator of mTORC1, was shown to be the direct downstream target of TSC2 (Inoki et al. 2003a; Tee et al. 2003; Zhang et al. 2003). Thus, TSC1/TSC2 complex negatively regulates mTOR (Inoki et al. 2002).

It was originally reported that mTOR activity was linked to intracellular ATP level (Dennis et al. 2001). Subsequent studies found that AMPK suppresses mTOR and its downstream pathways (Bolster et al. 2002; Kimura et al. 2003). To date, two different ways in which mTOR is inhibited by AMPK have been described. First, AMPK phosphorylates TSC2 and activates TSC1/TSC2 complex, which plays an inhibitory role in controlling mTOR pathway. Second, AMPK directly phosphorylates raptor at Ser722 and Ser792, which induces the binding of 14-3-3 to raptor, thus preventing the activation of mTORC1 (Gwinn et al. 2008) and initiating cellular stress response, including autophagy.

It is known that 2DG and AICAR inhibit S6K activity, and this inhibition is mediated by AMPK through suppression of mTOR (Bolster et al. 2002; Kimura et al. 2003). Subsequent studies revealed that TSC2 is phosphorylated at Thr1227 and Ser1345, and directly activated by AMPK both *in vitro* and *in vivo*, therefore enhancing its inhibition of mTOR (Inoki et al. 2003b). Among the two phosphorylation sites, Ser1345 was a crucial priming site for the successive phosphorylation of Ser1337, Ser1333, and Thr1329 by GSK3 (Inoki et al. 2006). This coordinated phosphorylation of TSC2 by AMPK and GSK3 is required for activation of TSC2 and subsequent inhibition of mTOR pathway (Inoki et al. 2006). Recently, another AMPK phosphorylation site on TSC2 was identified. Ethanol, which suppresses insulin-stimulated protein synthesis, was shown to enhance AMPK activity, resulting in increased Ser1387 phosphorylation on TSC2 and thus inhibition of mTOR (Hong-Brown et al. 2012). Furthermore, mitogen-activated protein kinase (MEK)/extracellular signal–regulated kinases (ERK) were positioned downstream of AMPK and upstream of TSC1/TSC2 to regulate autophagy (Wang et al. 2009). Upon autophagy stimuli, AMPK activates MEK/ERK, leading to its binding to and activating TSC2, which disassembles mTORC1 and enhances Beclin1 expression to regulate autophagy (Wang et al. 2009). From these studies, it appears that activation of TSC2 involves multiple phosphorylation sites, while it remains unclear whether they act separately or are coordinated to activate TSC2. Interestingly, besides Rheb, TSC1/TSC2 complex was recently proved to suppress mTOR through another GTPase RalA/B. Loss of TSC1/TSC2 activates RalA/B, therefore activating mTOR while suppressing autophagy even in the absence of Rheb (Martin et al. 2014). Nevertheless, although many lines of evidence suggest that TSC1/TSC2 complex plays a pivotal role in promoting AMPK-mediated autophagy via inhibition of mTOR (Suh et al. 2010; Wang et al. 2009; Zhu et al. 2012), autophagy is still inducible in TSC1/TSC2-deficient cells (Di Nardo et al. 2014; Lee et al. 2010), suggesting that TSC1/TSC2 complex is at the upstream of signaling pathway controlling mTOR and autophagy, or alternative pathways that regulate autophagy exist.

On the other hand, the mTORC1 subunit raptor has been reported to be a direct AMPK substrate (Gwinn et al. 2008). Once phosphorylated by AMPK at Ser722 and Ser792, raptor is sequestered by 14-3-3, leading to inhibition of mTORC1 (Gwinn et al. 2008). In TSC2-deficient cells, AICAR induces autophagy in the presence of wild-type raptor but not the mutant raptor that lacks the AMPK phosphorylation sites (S722A and S792A) (Lee et al. 2010), suggesting that phosphorylation of raptor by AMPK negatively regulates mTORC1 and is required for autophagy induction. Moreover, many recent studies described that autophagy is triggered by different stimuli in an AMPK-dependent manner, which coincides with raptor phosphorylation and mTORC1 inhibition (Aryal et al. 2014; Chao et al. 2011; Ethier et al. 2012; Zhu et al. 2013).

1.4.2 THROUGH ULK1

The Ulk1/Atg13/FIP200/Atg101 complex is essential for initiation of autophagy, leading to autophagosome biogenesis (Kim and Guan 2011; Suzuki et al. 2015). mTORC1 and AMPK have been reported to regulate Ulk1 complex in opposite ways. Under nutrients-rich conditions, the active mTORC1 phosphorylates Ulk1 at Ser757, which prevents its interaction with AMPK (Kim et al. 2011). While under energy stress conditions, AMPK is activated and thus mTORC1 is inhibited through phosphorylation of TSC2 and raptor by AMPK. Then, Ser757 is dephosphorylated and concomitantly Ulk1 is released to interact with AMPK, leading to phosphorylations at Ser317, Ser777 (Kim et al. 2011), Ser467, and Ser555 (Egan et al. 2011), and eventually activation of Ulk1 and initiation of autophagy. The loss of either AMPK or Ulk1, and serine to alanine mutation of these phosphorylation sites, all resulted in decreased lipidation of LC3 and formation of GFP-LC3 puncta and accumulation of autophagy adaptor p62 (Egan et al. 2011; Kim et al. 2011). Such evidence suggests that Ulk1 is a direct AMPK target and its phosphoryla-tion by AMPK is required for autophagy. Given that AMPK inhibits mTORC1 and mTORC1 inhibits Ulk1, and that AMPK directly activates Ulk1, Ulk1 is a key player in controlling autophagy downstream of AMPK. Interestingly, it was reported that Ulk1 phosphorylates all three AMPK subunits *in vitro*, thus decreas-ing starvation-induced AMPK activity (Loffler et al. 2011). This study established a direct negative feedback loop on the AMPK-mTOR and AMPK-Ulk1 signaling axes. However, further studies are required to elucidate the exact mechanisms of how AMPK and Ulk1 are involved in the signal transduction during autophagy process.

1.4.3 THROUGH THE VPS34 COMPLEX

Vps34 is a class-III phosphatidylinositol-3 kinase (PI3K) that phosphorylates PI to PI3P, which recruits PI3P-binding proteins to modulate their activities (Obara and Ohsumi 2011; Yu et al. 2015). Vps34 forms different complexes that are involved in various cellular functions, including endocytic sorting, mTOR signaling, and autophagy (Backer 2008). The pro-autophagy Vps34 complex consists of the core components Vps34 and Vps15, as well as Beclin1, which recruits additional proteins to regulate autophagy, such as Atg14L, Ambra1, and UVRAG (Itakura et al. 2008; Liang et al. 2006; Sun et al. 2008). Among them, Atg14L has been reported to be essential for the activity of Vps34 complex and autophagy initiation (Itakura et al. 2008; Itakura and Mizushima 2010; Matsunaga et al. 2010; Sun et al. 2008; Zhong et al. 2009). Recently, AMPK was described to regulate Vps34 complex through phosphorylation of Beclin1 or Vps34, depending on whether Atg14L is bound or not, respectively (Kim et al. 2013). In that study, it was found that under glucose starvation, AMPK phosphorylates Vps34 at Thr163 and Ser165 in the absence of Atg14L to inhibit its PI3K activity. In contrast, if Atg14L is present in the Vps34 complex, AMPK phosphorylates Beclin1, instead of Vps34, at Ser91 and Ser94, a process essential for autophagy (Kim et al. 2013). Importantly, this effect is specific for autophagy induced by glucose starvation, but not by rapamycin treatment (Kim et al. 2013). Interestingly, under amino acid starvation or mTOR inhibition, Beclin1 is phosphorylated by Ulk1 at Ser14, which also requires Atg14L to be present in the Vps34 complex (Russell et al. 2013). In addition, in both studies, UVRAG may regulate Vps34 complex in a manner similar to Atg14L except that Atg14L func-tions at the autophagy initiation stage, while UVRAG functions at the autophago-some maturation stage (Kim et al. 2013; Russell et al. 2013). As AMPK activates Ulk1 when mTORC1 is inhibited, these two studies thus reveal a new mechanism of how AMPK regulates Vps34 complex under different energy stress conditions. Nevertheless, given the complexity of Vps34 complexes, further studies are required to depict a more comprehensive network of Vps34 regulation of autophagy.

1.4.4 Through Other Pathways

Besides the three major ways as described above, several other mechanisms have been reported to be implicated in the regulatory function of AMPK in autophagy. For instance, under energy stress conditions, the cyclin-dependent kinase (Cdk) inhibitor p27[kip1] is phosphorylated at Thr198 by AMPK, therefore stabilizing p27[kip1] and leading cells to enter autophagy instead of apoptosis (Liang et al. 2007).

Another potential link between AMPK and autophagy is the eukaryotic elongation factor 2 (eEF2) kinase. It is known that eEF2 kinase is negatively regulated by mTOR signaling (Averous and Proud 2006). At least three phosphorylation sites are mediated by mTOR, that is, Ser78, Ser359, and Ser366 (Browne and Proud 2004; Knebel et al. 2001; Wang et al. 2001). When phosphorylated at these sites, eEF2 kinase activity is decreased and thus eEF2 is released to regulate protein synthesis (Browne and Proud 2004; Knebel et al. 2001; Wang et al. 2001). However, it has been reported that once activated by different stimuli, AMPK directly phosphorylates eEF2 kinase at Ser398, enhancing its kinase activity and thus inhibition of eEF2 (Browne et al. 2004). Furthermore, knockdown of eEF2 kinase inhibits autophagy, while overexpression increases autophagy (Browne and Proud 2004), indicating that AMPK and mTOR may regulate eEF2 kinase in opposite ways to modulate autophagy, similar to Ulk1.

In addition to phosphorylation, emerging evidence has suggested that the acetylation/deacetylation of Atg proteins is essential for autophagy (Lee et al. 2008; Lee and Finkel 2009; Lin et al. 2012a, 2012b; Yi et al. 2012). Sirt1, a highly conserved nicotinamide adenine dinucleotide (NAD$^+$)-dependent deacetylase, is activated under starvation, which deacetylates Atg5, Atg7, and LC3 and promotes autophagy (Huang et al. 2015; Lee et al. 2008). Recently, Sirt1 was reported to be activated in an AMPK-dependent manner (Chang et al. 2015). Under glucose starvation, activated AMPK phosphorylates glyceraldehyde-3-phosphate dehydrogenase (GAPDH) at Ser122, leading to its translocation into the nucleus. Nuclear-localized GAPDH interacts with Sirt1, dissociating Sirt1 from its inhibitor DBC1 (deleted in breast cancer 1), and therefore activating Sirt1 to regulate autophagy (Chang et al. 2015).

Moreover, AMPK also promotes autophagy from transcription level through activating transcription factor Forkhead Box O3a (FoxO3a) (Sanchez et al. 2012). Several sites of FoxO3a have been reported to be phosphorylated by AMPK, leading to its activation (Greer et al. 2007). It was shown later that AMPK-mediated activation of FoxO3a enhances the expression of LC3, Gabarapl, and Beclin1, therefore promoting autophagy (Sanchez et al. 2012).

1.5 CONCLUSIONS AND FUTURE DIRECTIONS

As discussed above, the regulation of autophagy by AMPK is known to be achieved via multiple mechanisms, including the following three main major pathways: (1) through inhibition of mTORC1, (2) through activation of Ulk1, and (3) through activation of the Vps34 complex. Among them, the inhibitory effect of AMPK on mTOR appears to be particularly important. There are several important directions to further understand the molecular mechanisms and biological importance of AMPK in autophagy. First, the cross talk between AMPK and mTOR is of greater interest and importance to elucidate the mechanisms of how autophagy is regulated under various stress conditions. Second, in addition to the known AMPK downstream substrates, such as Ulk1, Vps34, and Beclin1, further studies are needed to identify other potential targets in autophagy. Third, there is ongoing work for identification of novel AMPK pharmacological activators and inhibitors and exploration of such agents in modulation of autophagy for the therapeutic purpose of some autophagy-related human diseases such as cancer. Taken together, further investigations of the AMPK network in autophagy will be an important aspect of research in understanding of the physiological roles of this critical energy-sensing kinase in health and disease.

REFERENCES

Aryal, P., Kim, K., Park, P.H., Ham, S., Cho, J., and Song, K. (2014). Baicalein induces autophagic cell death through AMPK/ULK1 activation and downregulation of mTORC1 complex components in human cancer cells. *FEBS J* 281, 4644–4658.

Averous, J., and Proud, C.G. (2006). When translation meets transformation: The mTOR story. *Oncogene* 25, 6423–6435.

Backer, J.M. (2008). The regulation and function of Class III PI3Ks: Novel roles for Vps34. *Biochem J* 410, 1–17.

Beg, Z.H., Allmann, D.W., and Gibson, D.M. (1973). Modulation of 3-hydroxy-3-methylglutaryl coenzyme A reductase activity with cAMP and wth protein fractions of rat liver cytosol. *Biochem Biophys Res Commun* 54, 1362–1369.

Benvenuto, G., Li, S., Brown, S.J., Braverman, R., Vass, W.C., Cheadle, J.P., Halley, D.J., Sampson, J.R., Wienecke, R., and DeClue, J.E. (2000). The tuberous sclerosis-1 (TSC1) gene product hamartin suppresses cell growth and augments the expression of the TSC2 product tuberin by inhibiting its ubiquitination. *Oncogene* 19, 6306–6316.

Bolster, D.R., Crozier, S.J., Kimball, S.R., and Jefferson, L.S. (2002). AMP-activated protein kinase suppresses protein synthesis in rat skeletal muscle through down-regulated mammalian target of rapamycin (mTOR) signaling. *J Biol Chem* 277, 23977–23980.

Boya, P., Reggiori, F., and Codogno, P. (2013). Emerging regulation and functions of autophagy. *Nat Cell Biol* 15, 713–720.

Browne, G.J., Finn, S.G., and Proud, C.G. (2004). Stimulation of the AMP-activated protein kinase leads to activation of eukaryotic elongation factor 2 kinase and to its phosphorylation at a novel site, serine 398. *J Biol Chem* 279, 12220–12231.

Browne, G.J., and Proud, C.G. (2004). A novel mTOR-regulated phosphorylation site in elongation factor 2 kinase modulates the activity of the kinase and its binding to calmodulin. *Mol Cell Biol* 24, 2986–2997.

Calabrese, M.F., Rajamohan, F., Harris, M.S., Caspers, N.L., Magyar, R., Withka, J.M., Wang, H., et al. (2014). Structural basis for AMPK activation: Natural and synthetic ligands regulate kinase activity from opposite poles by different molecular mechanisms. *Structure* 22, 1161–1172.

Carling, D. (2004). The AMP-activated protein kinase cascade—A unifying system for energy control. *Trends Biochem Sci* 29, 18–24.

Carling, D., Zammit, V.A., and Hardie, D.G. (1987). A common bicyclic protein kinase cascade inactivates the regulatory enzymes of fatty acid and cholesterol biosynthesis. *FEBS Lett* 223, 217–222.

Carlson, C.A., and Kim, K.H. (1973). Regulation of hepatic acetyl coenzyme A carboxylase by phosphorylation and dephosphorylation. *J Biol Chem* 248, 378–380.

Chang, C., Su, H., Zhang, D., Wang, Y., Shen, Q., Liu, B., Huang, R., et al. (2015). AMPK-dependent phosphorylation of GAPDH triggers sirt1 activation and is necessary for autophagy upon glucose starvation. *Mol Cell* 60, 930–940.

Chao, A.C., Hsu, Y.L., Liu, C.K., and Kuo, P.L. (2011). α-Mangostin, a dietary xanthone, induces autophagic cell death by activating the AMP-activated protein kinase pathway in glioblastoma cells. *J Agric Food Chem* 59, 2086–2096.

Chen, L., Wang, J., Zhang, Y.Y., Yan, S.F., Neumann, D., Schlattner, U., Wang, Z.X., and Wu, J.W. (2012). AMP-activated protein kinase undergoes nucleotide-dependent conformational changes. *Nat Struct Mol Biol* 19, 716–718.

Cheung, P.C., Salt, I.P., Davies, S.P., Hardie, D.G., and Carling, D. (2000). Characterization of AMP-activated protein kinase gamma-subunit isoforms and their role in AMP binding. *Biochem J* 346 Pt 3, 659–669.

Choi, A.M., Ryter, S.W., and Levine, B. (2013). Autophagy in human health and disease. *N Engl J Med* 368, 1845–1846.

Chong-Kopera, H., Inoki, K., Li, Y., Zhu, T., Garcia-Gonzalo, F.R., Rosa, J.L., and Guan, K.L. (2006). TSC1 stabilizes TSC2 by inhibiting the interaction between TSC2 and the HERC1 ubiquitin ligase. *J Biol Chem* 281, 8313–8316.

Cool, B., Zinker, B., Chiou, W., Kifle, L., Cao, N., Perham, M., Dickinson, R., et al. (2006). Identification and characterization of a small molecule AMPK activator that treats key components of type 2 diabetes and the metabolic syndrome. *Cell Metab* 3, 403–416.

Corton, J.M., Gillespie, J.G., Hawley, S.A., and Hardie, D.G. (1995). 5-aminoimidazole-4-carboxamide ribonucleoside. A specific method for activating AMP-activated protein kinase in intact cells? *Eur J Biochem* 229, 558–565.

Davies, S.P., Helps, N.R., Cohen, P.T., and Hardie, D.G. (1995). 5'-AMP inhibits dephosphorylation, as well as promoting phosphorylation, of the AMP-activated protein kinase. Studies using bacterially expressed human protein phosphatase-2C alpha and native bovine protein phosphatase-2AC. *FEBS Lett* 377, 421–425.

Dennis, P.B., Jaeschke, A., Saitoh, M., Fowler, B., Kozma, S.C., and Thomas, G. (2001). Mammalian TOR: A homeostatic ATP sensor. *Science* 294, 1102–1105.

Di Nardo, A., Wertz, M.H., Kwiatkowski, E., Tsai, P.T., Leech, J.D., Greene-Colozzi, E., Goto, J., et al. (2014). Neuronal Tsc1/2 complex controls autophagy through AMPK-dependent regulation of ULK1. *Hum Mol Genet* 23, 3865–3874.

Egan, D.F., Shackelford, D.B., Mihaylova, M.M., Gelino, S., Kohnz, R.A., Mair, W., Vasquez, D.S., et al. (2011). Phosphorylation of ULK1 (hATG1) by AMP-activated protein kinase connects energy sensing to mitophagy. *Science* 331, 456–461.

Ethier, C., Tardif, M., Arul, L., and Poirier, G.G. (2012). PARP-1 modulation of mTOR signaling in response to a DNA alkylating agent. *PLoS One* 7, e47978.

Feng, Y., Yao, Z., and Klionsky, D.J. (2015). How to control self-digestion: Transcriptional, post-transcriptional, and post-translational regulation of autophagy. *Trends Cell Biol* 25, 354–363.

Fogarty, S., Hawley, S.A., Green, K.A., Saner, N., Mustard, K.J., and Hardie, D.G. (2010). Calmodulin-dependent protein kinase kinase-beta activates AMPK without forming a stable complex: Synergistic effects of Ca2+ and AMP. *Biochem J* 426, 109–118.

Frias, M.A., Thoreen, C.C., Jaffe, J.D., Schroder, W., Sculley, T., Carr, S.A., and Sabatini, D.M. (2006). mSin1 is necessary for Akt/PKB phosphorylation, and its isoforms define three distinct mTORC2s. *Curr Biol* 16, 1865–1870.

Gadalla, A.E., Pearson, T., Currie, A.J., Dale, N., Hawley, S.A., Sheehan, M., Hirst, W., et al. (2004). AICA riboside both activates AMP-activated protein kinase and competes with adenosine for the nucleoside transporter in the CA1 region of the rat hippocampus. *J Neurochem* 88, 1272–1282.

Ganley, I.G., Lam du, H., Wang, J., Ding, X., Chen, S., and Jiang, X. (2009). ULK1.ATG13. FIP200 complex mediates mTOR signaling and is essential for autophagy. *J Biol Chem* 284, 12297–12305.

Gao, X., and Pan, D. (2001). TSC1 and TSC2 tumor suppressors antagonize insulin signaling in cell growth. *Genes Dev* 15, 1383–1392.

Gledhill, J.R., Montgomery, M.G., Leslie, A.G., and Walker, J.E. (2007). Mechanism of inhibition of bovine F1-ATPase by resveratrol and related polyphenols. *Proc Natl Acad Sci U S A* 104, 13632–13637.

Gomez-Galeno, J.E., Dang, Q., Nguyen, T.H., Boyer, S.H., Grote, M.P., Sun, Z., Chen, M., et al. (2010). A potent and selective AMPK activator that inhibits de novo lipogenesis. *ACS Med Chem Lett* 1, 478–482.

Goransson, O., McBride, A., Hawley, S.A., Ross, F.A., Shpiro, N., Foretz, M., Viollet, B., Hardie, D.G., and Sakamoto, K. (2007). Mechanism of action of A-769662, a valuable tool for activation of AMP-activated protein kinase. *J Biol Chem* 282, 32549–32560.

Gowans, G.J., Hawley, S.A., Ross, F.A., and Hardie, D.G. (2013). AMP is a true physiological regulator of AMP-activated protein kinase by both allosteric activation and enhancing net phosphorylation. *Cell Metab* 18, 556–566.

Greer, E.L., Oskoui, P.R., Banko, M.R., Maniar, J.M., Gygi, M.P., Gygi, S.P., and Brunet, A. (2007). The energy sensor AMP-activated protein kinase directly regulates the mammalian FOXO3 transcription factor. *J Biol Chem* 282, 30107–30119.

Gwinn, D.M., Shackelford, D.B., Egan, D.F., Mihaylova, M.M., Mery, A., Vasquez, D.S., Turk, B.E., and Shaw, R.J. (2008). AMPK phosphorylation of raptor mediates a metabolic checkpoint. *Mol Cell* 30, 214–226.

Hara, K., Maruki, Y., Long, X., Yoshino, K., Oshiro, N., Hidayat, S., Tokunaga, C., Avruch, J., and Yonezawa, K. (2002). Raptor, a binding partner of target of rapamycin (TOR), mediates TOR action. *Cell* 110, 177–189.

Hardie, D.G. (2011). AMPK and autophagy get connected. *EMBO J* 30, 634–635.

Hardie, D.G. (2014). AMPK—Sensing energy while talking to other signaling pathways. *Cell Metab* 20, 939–952.

Hardie, D.G. (2015a). AMPK: Positive and negative regulation, and its role in whole-body energy homeostasis. *Curr Opin Cell Biol* 33, 1–7.

Hardie, D.G. (2015b). Molecular pathways: Is AMPK a friend or a foe in cancer? *Clin Cancer Res*, 3836–3840.

Hardie, D.G., Ross, F.A., and Hawley, S.A. (2012). AMPK: A nutrient and energy sensor that maintains energy homeostasis. *Nat Rev Mol Cell Biol* 13, 251–262.

Hardie, D.G., Schaffer, B.E., and Brunet, A. (2015). AMPK: An energy-sensing pathway with multiple inputs and outputs. *Trends Cell Biol* 26(3), 190–201.

Hawley, S.A., Boudeau, J., Reid, J.L., Mustard, K.J., Udd, L., Makela, T.P., Alessi, D.R., and Hardie, D.G. (2003). Complexes between the LKB1 tumor suppressor, STRAD alpha/beta and MO25 alpha/beta are upstream kinases in the AMP-activated protein kinase cascade. *J Biol* 2, 28.

Hawley, S.A., Fullerton, M.D., Ross, F.A., Schertzer, J.D., Chevtzoff, C., Walker, K.J., Peggie, M.W., et al. (2012). The ancient drug salicylate directly activates AMP-activated protein kinase. *Science* 336, 918–922.

Hawley, S.A., Gadalla, A.E., Olsen, G.S., and Hardie, D.G. (2002). The antidiabetic drug metformin activates the AMP-activated protein kinase cascade via an adenine nucleotide-independent mechanism. *Diabetes* 51, 2420–2425.

Hawley, S.A., Pan, D.A., Mustard, K.J., Ross, L., Bain, J., Edelman, A.M., Frenguelli, B.G., and Hardie, D.G. (2005). Calmodulin-dependent protein kinase kinase-beta is an alternative upstream kinase for AMP-activated protein kinase. *Cell Metab* 2, 9–19.

Hawley, S.A., Ross, F.A., Chevtzoff, C., Green, K.A., Evans, A., Fogarty, S., Towler, M.C., et al. (2010). Use of cells expressing gamma subunit variants to identify diverse mechanisms of AMPK activation. *Cell Metab* 11, 554–565.

Hawley, S.A., Ross, F.A., Gowans, G.J., Tibarewal, P., Leslie, N.R., and Hardie, D.G. (2014). Phosphorylation by Akt within the ST loop of AMPK-α1 down-regulates its activation in tumour cells. *Biochem J* 459, 275–287.

Hong-Brown, L.Q., Brown, C.R., Kazi, A.A., Navaratnarajah, M., and Lang, C.H. (2012). Rag GTPases and AMPK/TSC2/Rheb mediate the differential regulation of mTORC1 signaling in response to alcohol and leucine. *Am J Physiol Cell Physiol* 302, C1557–C1565.

Horman, S., Vertommen, D., Heath, R., Neumann, D., Mouton, V., Woods, A., Schlattner, U., et al. (2006). Insulin antagonizes ischemia-induced Thr172 phosphorylation of AMP-activated protein kinase alpha-subunits in heart via hierarchical phosphorylation of Ser485/491. *J Biol Chem* 281, 5335–5340.

Hosokawa, N., Hara, T., Kaizuka, T., Kishi, C., Takamura, A., Miura, Y., Iemura, S., et al. (2009). Nutrient-dependent mTORC1 association with the ULK1-Atg13-FIP200 complex required for autophagy. *Mol Biol Cell* 20, 1981–1991.

Huang, R., Xu, Y., Wan, W., Shou, X., Qian, J., You, Z., Liu, B., et al. (2015). Deacetylation of nuclear LC3 drives autophagy initiation under starvation. *Mol Cell* 57, 456–466.

Hudson, E.R., Pan, D.A., James, J., Lucocq, J.M., Hawley, S.A., Green, K.A., Baba, O., Terashima, T., and Hardie, D.G. (2003). A novel domain in AMP-activated protein kinase causes glycogen storage bodies similar to those seen in hereditary cardiac arrhythmias. *Curr Biol* 13, 861–866.

Hurley, R.L., Anderson, K.A., Franzone, J.M., Kemp, B.E., Means, A.R., and Witters, L.A. (2005). The Ca2+/calmodulin-dependent protein kinase kinases are AMP-activated protein kinase kinases. *J Biol Chem* 280, 29060–29066.

Hurley, R.L., Barre, L.K., Wood, S.D., Anderson, K.A., Kemp, B.E., Means, A.R., and Witters, L.A. (2006). Regulation of AMP-activated protein kinase by multisite phosphorylation in response to agents that elevate cellular cAMP. *J Biol Chem* 281, 36662–36672.

Inoki, K., Li, Y., Xu, T., and Guan, K.L. (2003a). Rheb GTPase is a direct target of TSC2 GAP activity and regulates mTOR signaling. *Genes Dev* 17, 1829–1834.

Inoki, K., Li, Y., Zhu, T., Wu, J., and Guan, K.L. (2002). TSC2 is phosphorylated and inhibited by Akt and suppresses mTOR signalling. *Nat Cell Biol* 4, 648–657.

Inoki, K., Ouyang, H., Zhu, T., Lindvall, C., Wang, Y., Zhang, X., Yang, Q., et al. (2006). TSC2 integrates Wnt and energy signals via a coordinated phosphorylation by AMPK and GSK3 to regulate cell growth. *Cell* 126, 955–968.

Inoki, K., Zhu, T., and Guan, K.L. (2003b). TSC2 mediates cellular energy response to control cell growth and survival. *Cell* 115, 577–590.

Itakura, E., Kishi, C., Inoue, K., and Mizushima, N. (2008). Beclin 1 forms two distinct phosphatidylinositol 3-kinase complexes with mammalian Atg14 and UVRAG. *Mol Biol Cell* 19, 5360–5372.

Itakura, E., and Mizushima, N. (2010). Characterization of autophagosome formation site by a hierarchical analysis of mammalian Atg proteins. *Autophagy* 6, 764–776.

Ito, N., and Rubin, G.M. (1999). gigas, a Drosophila homolog of tuberous sclerosis gene product-2, regulates the cell cycle. *Cell* 96, 529–539.

Jacinto, E., Facchinetti, V., Liu, D., Soto, N., Wei, S., Jung, S.Y., Huang, Q., Qin, J., and Su, B. (2006). SIN1/MIP1 maintains rictor-mTOR complex integrity and regulates Akt phosphorylation and substrate specificity. *Cell* 127, 125–137.

Jacinto, E., Loewith, R., Schmidt, A., Lin, S., Ruegg, M.A., Hall, A., and Hall, M.N. (2004). Mammalian TOR complex 2 controls the actin cytoskeleton and is rapamycin insensitive. *Nat Cell Biol* 6, 1122–1128.

Jung, C.H., Jun, C.B., Ro, S.H., Kim, Y.M., Otto, N.M., Cao, J., Kundu, M., and Kim, D.H. (2009). ULK-Atg13-FIP200 complexes mediate mTOR signaling to the autophagy machinery. *Mol Biol Cell* 20, 1992–2003.

Jung, C.H., Ro, S.-H., Cao, J., Otto, N.M., and Kim, D.-H. (2010). mTOR regulation of autophagy. *FEBS Lett* 584, 1287–1295.

Kaizuka, T., Hara, T., Oshiro, N., Kikkawa, U., Yonezawa, K., Takehana, K., Iemura, S., Natsume, T., and Mizushima, N. (2010). Tti1 and Tel2 are critical factors in mammalian target of rapamycin complex assembly. *J Biol Chem* 285, 20109–20116.

Kemp, B.E., Oakhill, J.S., and Scott, J.W. (2007). AMPK structure and regulation from three angles. *Structure* 15, 1161–1163.

Kim, D.H., Sarbassov, D.D., Ali, S.M., King, J.E., Latek, R.R., Erdjument-Bromage, H., Tempst, P., and Sabatini, D.M. (2002). mTOR interacts with raptor to form a nutrient-sensitive complex that signals to the cell growth machinery. *Cell* 110, 163–175.

Kim, D.H., Sarbassov, D.D., Ali, S.M., Latek, R.R., Guntur, K.V., Erdjument-Bromage, H., Tempst, P., and Sabatini, D.M. (2003). GbetaL, a positive regulator of the rapamycin-sensitive pathway required for the nutrient-sensitive interaction between raptor and mTOR. *Mol Cell* 11, 895–904.

Kim, J., and Guan, K.L. (2011). Regulation of the autophagy initiating kinase ULK1 by nutrients: Roles of mTORC1 and AMPK. *Cell Cycle* 10, 1337–1338.

Kim, J., Kim, Y.C., Fang, C., Russell, R.C., Kim, J.H., Fan, W., Liu, R., Zhong, Q., and Guan, K.L. (2013). Differential regulation of distinct Vps34 complexes by AMPK in nutrient stress and autophagy. *Cell* 152, 290–303.

Kim, J., Kundu, M., Viollet, B., and Guan, K.L. (2011). AMPK and mTOR regulate autophagy through direct phosphorylation of Ulk1. *Nat Cell Biol* 13, 132–141.

Kimura, N., Tokunaga, C., Dalal, S., Richardson, C., Yoshino, K., Hara, K., Kemp, B.E., Witters, L.A., Mimura, O., and Yonezawa, K. (2003). A possible linkage between AMP-activated protein kinase (AMPK) and mammalian target of rapamycin (mTOR) signalling pathway. *Genes Cells* 8, 65–79.

Knebel, A., Morrice, N., and Cohen, P. (2001). A novel method to identify protein kinase substrates: eEF2 kinase is phosphorylated and inhibited by SAPK4/p38delta. *EMBO J* 20, 4360–4369.

Langendorf, C.G., and Kemp, B.E. (2015). Choreography of AMPK activation. *Cell Res* 25, 5–6.

Laplante, M., and Sabatini, D.M. (2012). mTOR signaling in growth control and disease. *Cell* 149, 274–293.

Lee, I.H., Cao, L., Mostoslavsky, R., Lombard, D.B., Liu, J., Bruns, N.E., Tsokos, M., Alt, F.W., and Finkel, T. (2008). A role for the NAD-dependent deacetylase Sirt1 in the regulation of autophagy. *Proc Natl Acad Sci U S A* 105, 3374–3379.

Lee, I.H., and Finkel, T. (2009). Regulation of autophagy by the p300 acetyltransferase. *J Biol Chem* 284, 6322–6328.

Lee, J.W., Park, S., Takahashi, Y., and Wang, H.G. (2010). The association of AMPK with ULK1 regulates autophagy. *PLoS One* 5, e15394.

Lee, Y.S., Kim, W.S., Kim, K.H., Yoon, M.J., Cho, H.J., Shen, Y., Ye, J.M., et al. (2006). Berberine, a natural plant product, activates AMP-activated protein kinase with beneficial metabolic effects in diabetic and insulin-resistant states. *Diabetes* 55, 2256–2264.

Liang, C., Feng, P., Ku, B., Dotan, I., Canaani, D., Oh, B.H., and Jung, J.U. (2006). Autophagic and tumour suppressor activity of a novel Beclin1-binding protein UVRAG. *Nat Cell Biol* 8, 688–699.

Liang, J., Shao, S.H., Xu, Z.X., Hennessy, B., Ding, Z., Larrea, M., Kondo, S., et al. (2007). The energy sensing LKB1-AMPK pathway regulates p27(kip1) phosphorylation mediating the decision to enter autophagy or apoptosis. *Nat Cell Biol* 9, 218–224.

Liang, J., Xu, Z.X., Ding, Z., Lu, Y., Yu, Q., Werle, K.D., Zhou, G., et al. (2015). Myristoylation confers noncanonical AMPK functions in autophagy selectivity and mitochondrial surveillance. *Nat Commun* 6, 7926.

Lin, S.Y., Li, T.Y., Liu, Q., Zhang, C., Li, X., Chen, Y., Zhang, S.M., et al. (2012a). GSK3-TIP60-ULK1 signaling pathway links growth factor deprivation to autophagy. *Science* 336, 477–481.

Lin, S.Y., Li, T.Y., Liu, Q., Zhang, C., Li, X., Chen, Y., Zhang, S.M., et al. (2012b). Protein phosphorylation-acetylation cascade connects growth factor deprivation to autophagy. *Autophagy* 8, 1385–1386.

Loffler, A.S., Alers, S., Dieterle, A.M., Keppeler, H., Franz-Wachtel, M., Kundu, M., Campbell, D.G., Wesselborg, S., Alessi, D.R., and Stork, B. (2011). Ulk1-mediated phosphorylation of AMPK constitutes a negative regulatory feedback loop. *Autophagy* 7, 696–706.

Marino, G., Niso-Santano, M., Baehrecke, E.H., and Kroemer, G. (2014). Self-consumption: The interplay of autophagy and apoptosis. *Nat Rev Mol Cell Biol* 15, 81–94.

Martin, T.D., Chen, X.W., Kaplan, R.E., Saltiel, A.R., Walker, C.L., Reiner, D.J., and Der, C.J. (2014). Ral and Rheb GTPase activating proteins integrate mTOR and GTPase signaling in aging, autophagy, and tumor cell invasion. *Mol Cell* 53, 209–220.

Matsunaga, K., Morita, E., Saitoh, T., Akira, S., Ktistakis, N.T., Izumi, T., Noda, T., and Yoshimori, T. (2010). Autophagy requires endoplasmic reticulum targeting of the PI3-kinase complex via Atg14L. *J Cell Biol* 190, 511–521.

McBride, A., Ghilagaber, S., Nikolaev, A., and Hardie, D.G. (2009). The glycogen-binding domain on the AMPK beta subunit allows the kinase to act as a glycogen sensor. *Cell Metab* 9, 23–34.

Mizushima, N. (2009). Physiological functions of autophagy. *Curr Top Microbiol Immunol* 335, 71–84.

Mizushima, N., and Komatsu, M. (2011). Autophagy: Renovation of cells and tissues. *Cell* 147, 728–741.

Mizushima, N., Levine, B., Cuervo, A.M., and Klionsky, D.J. (2008). Autophagy fights disease through cellular self-digestion. *Nature* 451, 1069–1075.

Mooney, M.H., Fogarty, S., Stevenson, C., Gallagher, A.M., Palit, P., Hawley, S.A., Hardie, D.G., et al. (2008). Mechanisms underlying the metabolic actions of galegine that contribute to weight loss in mice. *Br J Pharmacol* 153, 1669–1677.

Moreira, D., Rodrigues, V., Abengozar, M., Rivas, L., Rial, E., Laforge, M., Li, X., et al. (2015). Leishmania infantum modulates host macrophage mitochondrial metabolism by hijacking the SIRT1-AMPK axis. *PLoS Pathog* 11, e1004684.

Oakhill, J.S., Chen, Z.P., Scott, J.W., Steel, R., Castelli, L.A., Ling, N., Macaulay, S.L., and Kemp, B.E. (2010). β-Subunit myristoylation is the gatekeeper for initiating metabolic stress sensing by AMP-activated protein kinase (AMPK). *Proc Natl Acad Sci U S A* 107, 19237–19241.

Oakhill, J.S., Scott, J.W., and Kemp, B.E. (2012). AMPK functions as an adenylate charge-regulated protein kinase. *Trends Endocrinol Metab* 23, 125–132.

Oakhill, J.S., Steel, R., Chen, Z.P., Scott, J.W., Ling, N., Tam, S., and Kemp, B.E. (2011). AMPK is a direct adenylate charge-regulated protein kinase. *Science* 332, 1433–1435.

Obara, K., and Ohsumi, Y. (2011). PtdIns 3-kinase orchestrates autophagosome formation in yeast. *J Lipids* 2011, 498768.

Pang, T., Xiong, B., Li, J.Y., Qiu, B.Y., Jin, G.Z., Shen, J.K., and Li, J. (2007). Conserved alpha-helix acts as autoinhibitory sequence in AMP-activated protein kinase alpha subunits. *J Biol Chem* 282, 495–506.

Papandreou, I., Lim, A.L., Laderoute, K., and Denko, N.C. (2008). Hypoxia signals autophagy in tumor cells via AMPK activity, independent of HIF-1, BNIP3, and BNIP3L. *Cell Death Differ* 15, 1572–1581.

Park, D.W., Jiang, S., Liu, Y., Siegal, G.P., Inoki, K., Abraham, E., and Zmijewski, J.W. (2014). GSK3beta-dependent inhibition of AMPK potentiates activation of neutrophils and macrophages and enhances severity of acute lung injury. *Am J Physiol Lung Cell Mol Physiol* 307, L735–L745.

Peterson, T.R., Laplante, M., Thoreen, C.C., Sancak, Y., Kang, S.A., Kuehl, W.M., Gray, N.S., and Sabatini, D.M. (2009). DEPTOR is an mTOR inhibitor frequently overexpressed in multiple myeloma cells and required for their survival. *Cell* 137, 873–886.

Polekhina, G., Gupta, A., Michell, B.J., van Denderen, B., Murthy, S., Feil, S.C., Jennings, I.G., et al. (2003). AMPK beta subunit targets metabolic stress sensing to glycogen. *Curr Biol* 13, 867–871.

Polekhina, G., Gupta, A., van Denderen, B.J., Feil, S.C., Kemp, B.E., Stapleton, D., and Parker, M.W. (2005). Structural basis for glycogen recognition by AMP-activated protein kinase. *Structure* 13, 1453–1462.

Potter, C.J., Huang, H., and Xu, T. (2001). Drosophila Tsc1 functions with Tsc2 to antagonize insulin signaling in regulating cell growth, cell proliferation, and organ size. *Cell* 105, 357–368.

Russell, R.C., Tian, Y., Yuan, H., Park, H.W., Chang, Y.Y., Kim, J., Kim, H., Neufeld, T.P., Dillin, A., and Guan, K.L. (2013). ULK1 induces autophagy by phosphorylating Beclin-1 and activating VPS34 lipid kinase. *Nat Cell Biol* 15, 741–750.

Sancak, Y., Thoreen, C.C., Peterson, T.R., Lindquist, R.A., Kang, S.A., Spooner, E., Carr, S.A., and Sabatini, D.M. (2007). PRAS40 is an insulin-regulated inhibitor of the mTORC1 protein kinase. *Mol Cell* 25, 903–915.

Sanchez, A.M., Csibi, A., Raibon, A., Cornille, K., Gay, S., Bernardi, H., and Candau, R. (2012). AMPK promotes skeletal muscle autophagy through activation of forkhead FoxO3a and interaction with Ulk1. *J Cell Biochem* 113, 695–710.

Sanders, M.J., Grondin, P.O., Hegarty, B.D., Snowden, M.A., and Carling, D. (2007). Investigating the mechanism for AMP activation of the AMP-activated protein kinase cascade. *Biochem J* 403, 139–148.

Sarbassov, D.D., Ali, S.M., Kim, D.H., Guertin, D.A., Latek, R.R., Erdjument-Bromage, H., Tempst, P., and Sabatini, D.M. (2004). Rictor, a novel binding partner of mTOR, defines a rapamycin-insensitive and raptor-independent pathway that regulates the cytoskeleton. *Curr Biol* 14, 1296–1302.

Scott, J.W., Hawley, S.A., Green, K.A., Anis, M., Stewart, G., Scullion, G.A., Norman, D.G., and Hardie, D.G. (2004). CBS domains form energy-sensing modules whose binding of adenosine ligands is disrupted by disease mutations. *J Clin Invest* 113, 274–284.

Scott, J.W., van Denderen, B.J., Jorgensen, S.B., Honeyman, J.E., Steinberg, G.R., Oakhill, J.S., Iseli, T.J., et al. (2008). Thienopyridone drugs are selective activators of AMP-activated protein kinase beta1-containing complexes. *Chem Biol* 15, 1220–1230.

Shaw, R.J., Kosmatka, M., Bardeesy, N., Hurley, R.L., Witters, L.A., DePinho, R.A., and Cantley, L.C. (2004). The tumor suppressor LKB1 kinase directly activates AMP-activated kinase and regulates apoptosis in response to energy stress. *Proc Natl Acad Sci U S A* 101, 3329–3335.

Shen, H.M., and Mizushima, N. (2014). At the end of the autophagic road: An emerging understanding of lysosomal functions in autophagy. *Trends Biochem Sci* 39, 61–71.

Suh, Y., Afaq, F., Khan, N., Johnson, J.J., Khusro, F.H., and Mukhtar, H. (2010). Fisetin induces autophagic cell death through suppression of mTOR signaling pathway in prostate cancer cells. *Carcinogenesis* 31, 1424–1433.

Sun, Q., Fan, W., Chen, K., Ding, X., Chen, S., and Zhong, Q. (2008). Identification of Barkor as a mammalian autophagy-specific factor for Beclin 1 and class III phosphatidylinositol 3-kinase. *Proc Natl Acad Sci U S A* 105, 19211–19216.

Suzuki, H., Kaizuka, T., Mizushima, N., and Noda, N.N. (2015). Structure of the Atg101-Atg13 complex reveals essential roles of Atg101 in autophagy initiation. *Nat Struct Mol Biol* 22, 572–580.

Suzuki, T., Bridges, D., Nakada, D., Skiniotis, G., Morrison, S.J., Lin, J.D., Saltiel, A.R., and Inoki, K. (2013). Inhibition of AMPK catabolic action by GSK3. *Mol Cell* 50, 407–419.

Tapon, N., Ito, N., Dickson, B.J., Treisman, J.E., and Hariharan, I.K. (2001). The drosophila tuberous sclerosis complex gene homologs restrict cell growth and cell proliferation. *Cell* 105, 345–355.

Tee, A.R., Manning, B.D., Roux, P.P., Cantley, L.C., and Blenis, J. (2003). Tuberous sclerosis complex gene products, tuberin and hamartin, control mTOR signaling by acting as a GTPase-activating protein complex toward Rheb. *Curr Biol* 13, 1259–1268.

Thedieck, K., Polak, P., Kim, M.L., Molle, K.D., Cohen, A., Jeno, P., Arrieumerlou, C., and Hall, M.N. (2007). PRAS40 and PRR5-like protein are new mTOR interactors that regulate apoptosis. *PLoS One* 2, e1217.

Vander Haar, E., Lee, S.I., Bandhakavi, S., Griffin, T.J., and Kim, D.H. (2007). Insulin signalling to mTOR mediated by the Akt/PKB substrate PRAS40. *Nat Cell Biol* 9, 316–323.

van Slegtenhorst, M., Nellist, M., Nagelkerken, B., Cheadle, J., Snell, R., van den Ouweland, A., Reuser, A., Sampson, J., Halley, D., and van der Sluijs, P. (1998). Interaction between hamartin and tuberin, the TSC1 and TSC2 gene products. *Hum Mol Genet* 7, 1053–1057.

Wang, J., Whiteman, M.W., Lian, H., Wang, G., Singh, A., Huang, D., and Denmark, T. (2009). A non-canonical MEK/ERK signaling pathway regulates autophagy via regulating Beclin 1. *J Biol Chem* 284, 21412–21424.

Wang, L., Harris, T.E., Roth, R.A., and Lawrence, J.C., Jr. (2007). PRAS40 regulates mTORC1 kinase activity by functioning as a direct inhibitor of substrate binding. *J Biol Chem* 282, 20036–20044.

Wang, X., Li, W., Williams, M., Terada, N., Alessi, D.R., and Proud, C.G. (2001). Regulation of elongation factor 2 kinase by p90(RSK1) and p70 S6 kinase. *EMBO J* 20, 4370–4379.

Woods, A., Dickerson, K., Heath, R., Hong, S.P., Momcilovic, M., Johnstone, S.R., Carlson, M., and Carling, D. (2005). Ca2+/calmodulin-dependent protein kinase kinase-beta acts upstream of AMP-activated protein kinase in mammalian cells. *Cell Metab* 2, 21–33.

Woods, A., Johnstone, S.R., Dickerson, K., Leiper, F.C., Fryer, L.G., Neumann, D., Schlattner, U., Wallimann, T., Carlson, M., and Carling, D. (2003). LKB1 is the upstream kinase in the AMP-activated protein kinase cascade. *Curr Biol* 13, 2004–2008.

Wu, H., Yang, J.M., Jin, S., Zhang, H., and Hait, W.N. (2006). Elongation factor-2 kinase regulates autophagy in human glioblastoma cells. *Cancer Res* 66, 3015–3023.

Xiao, B., Heath, R., Saiu, P., Leiper, F.C., Leone, P., Jing, C., Walker, P.A., et al. (2007). Structural basis for AMP binding to mammalian AMP-activated protein kinase. *Nature* 449, 496–500.

Xiao, B., Sanders, M.J., Carmena, D., Bright, N.J., Haire, L.F., Underwood, E., Patel, B.R., et al. (2013). Structural basis of AMPK regulation by small molecule activators. *Nat Commun* 4, 3017.

Xiao, B., Sanders, M.J., Underwood, E., Heath, R., Mayer, F.V., Carmena, D., Jing, C., et al. (2011). Structure of mammalian AMPK and its regulation by ADP. *Nature* 472, 230–233.

Xin, F.J., Wang, J., Zhao, R.Q., Wang, Z.X., and Wu, J.W. (2013). Coordinated regulation of AMPK activity by multiple elements in the alpha-subunit. *Cell Res* 23, 1237–1240.

Yang, Z., and Klionsky, D.J. (2010). Mammalian autophagy: Core molecular machinery and signaling regulation. *Curr Opin Cell Biol* 22, 124–131.

Yi, C., Ma, M., Ran, L., Zheng, J., Tong, J., Zhu, J., Ma, C., et al. (2012). Function and molecular mechanism of acetylation in autophagy regulation. *Science* 336, 474–477.

Young, J., and Povey, S. (1998). The genetic basis of tuberous sclerosis. *Mol Med Today* 4, 313–319.

Yu, H.C., Lin, C.S., Tai, W.T., Liu, C.Y., Shiau, C.W., and Chen, K.F. (2013). Nilotinib induces autophagy in hepatocellular carcinoma through AMPK activation. *J Biol Chem* 288, 18249–18259.

Yu, X., Long, Y.C., and Shen, H.M. (2015). Differential regulatory functions of three classes of phosphatidylinositol and phosphoinositide 3-kinases in autophagy. *Autophagy* 11, 1711–1728.

Zadra, G., Photopoulos, C., Tyekucheva, S., Heidari, P., Weng, Q.P., Fedele, G., Liu, H., et al. (2014). A novel direct activator of AMPK inhibits prostate cancer growth by blocking lipogenesis. *EMBO Mol Med* 6, 519–538.

Zhang, C.S., Jiang, B., Li, M., Zhu, M., Peng, Y., Zhang, Y.L., Wu, Y.Q., et al. (2014). The lysosomal v-ATPase-Ragulator complex is a common activator for AMPK and mTORC1, acting as a switch between catabolism and anabolism. *Cell Metab* 20, 526–540.

Zhang, Y., Gao, X., Saucedo, L.J., Ru, B., Edgar, B.A., and Pan, D. (2003). Rheb is a direct target of the tuberous sclerosis tumour suppressor proteins. *Nat Cell Biol* 5, 578–581.

Zhang, Y.L., Guo, H., Zhang, C.S., Lin, S.Y., Yin, Z., Peng, Y., Luo, H., et al. (2013). AMP as a low-energy charge signal autonomously initiates assembly of AXIN-AMPK-LKB1 complex for AMPK activation. *Cell Metab* 18, 546–555.

Zhong, Y., Wang, Q.J., Li, X., Yan, Y., Backer, J.M., Chait, B.T., Heintz, N., and Yue, Z. (2009). Distinct regulation of autophagic activity by Atg14L and Rubicon associated with Beclin 1-phosphatidylinositol-3-kinase complex. *Nat Cell Biol* 11, 468–476.

Zhou, G., Myers, R., Li, Y., Chen, Y., Shen, X., Fenyk-Melody, J., Wu, M., et al. (2001). Role of AMP-activated protein kinase in mechanism of metformin action. *J Clin Invest* 108, 1167–1174.

Zhu, B., Zhou, Y., Xu, F., Shuai, J., Li, X., and Fang, W. (2012). Porcine circovirus type 2 induces autophagy via the AMPK/ERK/TSC2/mTOR signaling pathway in PK-15 cells. *J Virol* 86, 12003–12012.

Zhu, Z., Yan, J., Jiang, W., Yao, X.G., Chen, J., Chen, L., Li, C., Hu, L., Jiang, H., and Shen, X. (2013). Arctigenin effectively ameliorates memory impairment in Alzheimer's disease model mice targeting both beta-amyloid production and clearance. *J Neurosci* 33, 13138–13149.

2 Signal Regulation of WIPI Protein Function in Macroautophagy

Theresia Zuleger
Eberhard Karls University Tuebingen
Tuebingen, Germany

Tassula Proikas-Cezanne
Eberhard Karls University Tuebingen and
Max Planck Institute for Developmental Biology
Tuebingen, Germany

CONTENTS

2.1 INTRODUCTION

The concept of autophagy (from the Greek meaning self-eating), earlier discussed by Christian de Duve, was based on the discovery of autophagosomes (referred to as round bodies of irregular density), unique intracellular vesicles proposed (1) to derive from organelle membranes such as the endoplasmic reticulum and (2) to sequester cytoplasmic material such as mitochondria and ribosomes for subsequent degradation (Ashford and Porter 1962; Clark 1957; De Duve and Wattiaux 1966; Deter et al. 1967; Hirsimaki and Reunanen 1980; Novikoff and Shin 1978; Reunanen and Hirsimaki 1983; Yang and Klionsky 2010).

The principle physiological function of autophagy is to contribute to the maintenance of cellular homeostasis through low basal level of partial self-consumption in order to constitutively clear and regenerate the cytoplasm. Importantly, autophagy facilitates cell survival upon starvation or other cellular insults, leading to rapid autophagy induction above basal level in order to compensate for nutrient and energy shortage. Critically, autophagy specifically degrades damaged organelles and fights cellular malfunctions and genomic instability (Stolz et al. 2014; Yang and Klionsky 2010).

Autophagy is executed by the autophagy machinery of so-called autophagy-related (ATG) proteins (Klionsky et al. 2003) that initiate stochastic or selective cargo sequestration in newly formed double-membrane autophagosomes. The formation of the autophagosome limits the rate of autophagic degradation, and the half-life of autophagosomes was early determined to be as short as 8 minutes in hepatocytes (Pfeifer 1977, 1978; Schworer and Mortimore 1979). This demonstrates a low steady-state level of autophagosomes under basal, nutrient-rich conditions that also applies to the majority of cell types (Meijer et al. 2015). Autophagosomes are formed upon the elongation and closure of a template membrane, called isolation membrane or phagophore, which derives from endomembranes or semi-autonomous organelles through an as yet unbeknown mechanism (Klionsky 2005; Lamb et al. 2013; Mueller and Proikas-Cezanne 2015). After the fusion of the autophagosome with lysosomes (the chimeric vesicle is now called autolysosome), acidic hydrolases are acquired and the autophagosomal cargo is degraded. Finally, cargo monomers are released to the cytoplasm for recycling or storage purposes (Feng et al. 2014; Shibutani and Yoshimori 2014). As such, autophagy permits the recycling of cytoplasmic material and the rejuvenation of the endomembrane system (Mueller and Proikas-Cezanne 2015). In addition, ATG proteins fulfill an increasingly recognized range of noncanonical functions including phagocytosis, endosome transport, and cytokine secretion (Codogno et al. 2012; Lamb et al. 2013; Martinez et al. 2015).

As the process of autophagy is intrinsic to cellular and organismic health and longevity, it is of intense interest and pressing importance to understand the impact of aberrant autophagy to both the onset and development of human pathologies in order to contribute to the development of rational next-generation therapies (Choi et al. 2013; Jiang and Mizushima 2014; Mizushima et al. 2008). In cancer, autophagy acts as a tumor suppressive pathway that assists to maintain genomic stability and to prevent chronic tissue damage; however, tumor cells heavily engage autophagy in nutrient-poor tumor regions to support metabolism, tumorigenesis, and gaining resistance to therapy (White 2012). In neurodegenerative diseases such as Alzheimer's disease, Parkinson's disease, tauopathies, and polyglutamine expansion diseases, aberrant autophagy is an important aspect of the disease onset and manifestation, leading to insufficient removal of protein aggregates and damaged mitochondria (Cuervo and Zhang 2015; Nixon 2013; Ravikumar et al. 2010b).

2.2   SIGNAL INTEGRATION THROUGH AMPK AND mTORC1 HUBS CONTROL AUTOPHAGY

To form autophagosomes, the tightly regulated and hierarchic action of ATG proteins is required (Itakura and Mizushima 2010; Lamb et al. 2013). The ATG proteins were first identified in yeast and are conserved throughout eukaryotes (Ohsumi 2014) where they are regulated via principle nutrient and energy-dependent signaling cascades, thus exposed to the rigorous regulation by the mechanistic target of rapamycin (mTOR) and the adenosine monophosphate–activated protein kinase (AMPK) (Meijer and Codogno 2011; Meijer and Dubbelhuis 2004; Meijer et al. 2015) (Figure 2.1).

In favor to upregulate anabolic pathways, the insulin/growth factor–mediated activation of the protein kinase B (PKB)-tuberous sclerosis protein 1/2 (TSC1/2)-mTOR pathway inhibits catabolic pathways, including autophagy (Bar-Peled and Sabatini 2014). However, mTOR stimulation via insulin/growth factor–mediated signaling is only sufficiently achieved in the presence of amino acids, with leucine being most effective in inhibiting autophagy (Meijer et al. 2015). This molecular connection was first demonstrated by the finding that administrated amino acids inhibited autophagy in the liver and stimulated S6 phosphorylation, while rapamycin-mediated mTOR inhibition permitted autophagy (Blommaart et al. 1995; Luiken et al. 1994).

The mTOR serine/threonine-specific protein kinase is evolutionarily highly conserved from yeast to human, belongs to the phosphoinositide 3-kinase (PtdIns3K)-related protein kinase family, and is organized in two distinct multi-protein

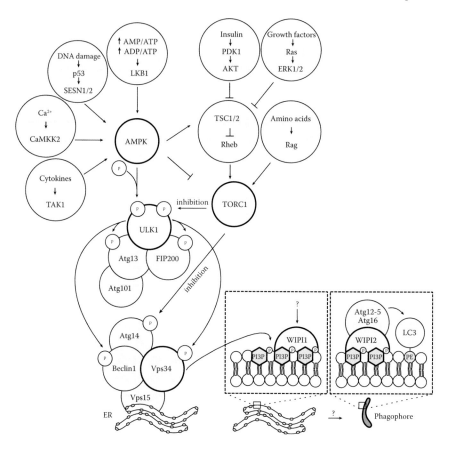

FIGURE 2.1 A model for the regulation of WIPI1 and WIPI2 recruitment to the site of autophagosome formation. See text for further details.

complexes, mTOR complex 1 (mTORC1) and mTOR complex 2 (mTORC2) (Laplante and Sabatini 2009). The mTORC1 complex consists of mTOR, regulatory-associated protein of mTOR (raptor), 40 kDa proline-rich Akt substrate (PRAS40), mammalian lethal with Sec13 protein 8 (mLST8), and dishevelled, egl-10 and pleckstrin (DEP) domain containing mTOR-interacting protein (deptor). The mTORC2 complex consists of mTOR, rapamycin-insensitive companion of mTOR (rictor), protein observed with Rictor-1 (Protor-1), protein kinase–interacting protein (mSIN1), mLST8, and deptor. According to the current understanding, both mTORC1 and mTORC2 regulate metabolism and promote cellular survival. Distinctively, mTORC1 is recognized to master regulate cell metabolism, proliferation, and growth by integrating intracellular and extracellular signaling networks, and mTORC2 is recognized to phosphorylate protein kinase B (PKB, also known as Akt) and to regulate cytoskeleton organization (Laplante and Sabatini 2012, 2013).

Under nutrient-rich conditions, mTORC1 is active at the lysosome and inhibits autophagy (Boya et al. 2013) (Figure 2.1). The activity of mTORC1 at the lysosome is critically regulated by the Ras-related guanosine-5′-triphosphate (GTP)-binding protein (Rag) and Ras homolog enriched in brain (Rheb), GTPases (Figure 2.1). Active, GTP-bound Rheb interacts with mTORC1 upon lysosomal positioning of mTORC1 by the amino acid–sensitive Rag, and hence, mTORC1 activation depends on insulin/growth factor cascades as well as amino acids availability (Laplante and Sabatini 2012) (Figure 2.1).

Critically, mTORC1 activity is controlled through the heterodimer tuberous sclerosis 1 and 2 TSC1/2 (TSC1 is also known as hamartin, TSC2 as tuberin), where multiple upstream signaling cascades converge and manipulate mTORC1 activity (Figure 2.1). TSC1/2 functions as a GTPase-activating protein (GAP) for Rheb, thereby negatively regulating mTORC1 activity by converting Rheb into its inactive guanosine diphosphate (GDP)-bound confirmation (Huang and Manning 2008). Hence, TSC1/2 inactivation allows mTORC1 activity. TSC1/2 inactivation is achieved through direct phosphorylation by different effector protein kinases that function in several upstream signaling cascades, such as PKB/Akt and the extracellular-signal-regulated kinase 1/2 (ERK1/2) (Laplante and Sabatini 2012) (Figure 2.1).

In favor to upregulate catabolic pathways, the inhibitory action of mTORC1 on autophagy is counteracted through AMPK that integrates multiple upstream inputs, including low energy, DNA damage, Ca^{2+}, and cytokine signaling, to activate autophagy (Hardie 2015) (Figure 2.1). In principle, AMPK signaling is considered essential for autophagy and active AMPK mediates direct and indirect mTORC1 inhibition (Laplante and Sabatini 2012; Meijer et al. 2015). When activated, AMPK was shown to (1) phosphorylate TSC2 and activate the TSC1/2 GAP activity on Rheb, thereby inhibiting mTORC1 and (2) phosphorylate raptor, leading to 14-3-3 binding and allosteric inhibition of mTORC1 (Laplante and Sabatini 2012) (Figure 2.1).

Autophagy regulation through both AMPK and mTORC1 converges at the level of controlling the kinase activities of the essential autophagy-related protein unc-51-like kinases 1 and 2 (ULK1, ULK2) (Egan et al. 2011; Kim et al. 2011; Meijer and Codogno 2011) and the ULK1-associated ATG proteins ATG13, focal adhesion kinase family interacting protein of 200 kDa (FIP200), and ATG101 (Alers et al. 2012; Russell et al. 2014) (Figure 2.1). Although the complexity of ULK1/2 regulation is far from being understood in molecular detail, it was demonstrated that AMPK phosphorylates and activates ULK1, whereas mTORC1 phosphorylates and inactivates ULK1 (Wong et al. 2013) (Figure 2.1). Activated ULK1 auto-phosphorylates and phosphorylates its specific downstream targets, including the ULK1-interacting proteins ATG13 and FIP200, to subsequently regulate the initiation of autophagosome formation (Figure 2.1).

Activated ULK1 phosphorylates at least two factors of the PtdIns3KC3 complex, Beclin 1 (ATG6 in yeast) and Vps34 (Egan et al. 2015; Russell et al. 2013) (Figure 2.1). The PtdIns3KC3 complex consists of the lipid kinase Vps34, the serine/threonine-specific protein kinase Vps15 (or p150) and the regulatory ATG proteins Beclin 1 and ATG14L (Baskaran et al. 2014; Russell et al. 2014). The ATG14L protein is considered essential for correct positioning of the PtdIns3KC3 complex at the cytoplasmic site of the endoplasmic reticulum (ER) (Lamb et al. 2013; Matsunaga et al. 2010), where PtdIns(3)P production is initiated for autophagosome formation. ER-positioning of PtdIns3KC3 complex is negatively regulated by mTORC1, which phosphorylates ATG14L and thereby inhibits ER-localized PtdIns(3)P production (Yuan et al. 2013).

PtdIns(3)P production at the onset of autophagy provokes ER dynamics that have been defined as omegasome structures and that contribute to the formation of the phagophore (Axe et al. 2008; Roberts and Ktistakis 2013). Multiple organelles are taken into consideration as membrane origins for the forming phagophore, including the ER, outer membrane of mitochondria, plasma membrane (PM), Golgi apparatus, and recycling endosomes (Geng et al. 2010; Hailey et al. 2010; Hayashi-Nishino et al. 2009; Lamb et al. 2013; Longatti et al. 2012; Ravikumar et al. 2010a; Shibutani and Yoshimori 2014). Moreover, the ER–Golgi intermediate compartment (ERGIC) (Ge et al. 2013) as well as ER–mitochondria contact sites (Hamasaki et al. 2013a) have been shown to serve as platforms for phagophore formation, and ATG9-loaded endosomes should account for phagophore initiation and expansion (Mari et al. 2010).

For the expansion of the phagophore, two autophagy-specific ubiquitin-like conjugation systems are strictly required, the ATG12 and the ATG8/LC3 (microtubule-associated protein 1A/1B light chain 3) ubiquitin-like conjugation systems, both of which finally result in the conjugation of LC3 to phosphatidylethanolamine (PE) (also referred to as LC3 lipidation) (Kabeya et al. 2000). The ATG12 ubiquitin-like conjugation cascade results in the formation of the ATG12-ATG5/ATG16L complex that functions as an E3-like ligase on LC3 to conjugate it to PE (Hamasaki et al. 2013b). Lipidated LC3 is a membrane protein of the phagophore and the autophagosome with two major functions, permitting (1) tethering and hemifusion events during phagophore elongation (Nakatogawa et al. 2007) and (2) specific recognition of autophagy receptors and adaptors that interact with LC3 via LC3-interacting (LIR-) domains to enable cargo selection and recognition (Stolz et al. 2014).

2.3 THE REQUIREMENT FOR PHOSPHATIDYLINOSITOL 3-PHOSPHATE TO FORM AUTOPHAGOSOMES

It is known for long that the formation of autophagosomes requires a large production of PtdIns(3)P at the site of phagophore formation (Meijer et al. 2015). The ER, currently regarded as the main site for phagophore formation, contains very little PtdIns(3)P under normal, nutrient-rich conditions (Gillooly et al. 2000).

Under starvation condition, phosphatidylinositol or phosphoinositides (derivates of phosphorylated phosphatidylinositol) become phosphorylated by the PtdIns3Ks at the 3-position of the inositol ring. In mammals, three classes of PtdIns3Ks have been identified and shown to differentially contribute to the process of autophagy (F et al. 2013). The PtdIns3K class I (PtdIns3KC1) produces PtdIns(3,4,5)P$_3$ in response to growth factor stimulation and acts as an inhibitor of autophagy, as downstream of PtdIns(3,4,5)P$_3$ production mTORC1 is activated via the Akt/PKB pathway (Gingras et al. 2001; Jacinto and Hall 2003). The PtdIns3K class II (PtdIns3KC2) produces PtdIns(3)P; however, it is considered that the bulk of autophagy-relevant PtdIns(3)P is produced by the PtdIns3KC3 complex (Devereaux et al. 2013). Apart from their role in autophagy, PtdIns3KC1, 2, and 3 fulfill diverse functions in membrane trafficking, such as endosome fusion, endosome motility and recycling, endosome-to-Golgi trafficking, degradative sorting, regulation of phagocytosis and macropinocytosis, and regulated exocytosis (Vanhaesebroeck et al. 2010). It is their function in differentially associated complexes that defines their functional specificities, as found for the PtdIns3KC3 complex 1 to be specifically engaged in producing PtdIns(3)P for the forming phagophore (Baskaran et al. 2014; F et al. 2013).

PtdIns(3)P production enables the binding of proteins containing the FYVE domain (named after Fab 1 (yeast orthologue of PIKfyve), YOTB, Vac 1 (vesicle transport protein), and EEA1) directly to PtdIns(3)P (Stenmark et al. 1996), where FYVE domain proteins subsequently function as PtdIns(3)P effectors (Gaullier et al. 1998; Simonsen et al. 1998).

The double FYVE-containing protein 1 (DFCP1) is one of these proteins and under normal conditions localizes to ER and Golgi membranes (Ridley et al. 2001). However, under starvation conditions, DFCP1 translocate to omegasomes in a PtdIns(3)P-dependent manner (Axe et al. 2008). Nevertheless, DFCP1 overexpression inhibits the initiation of phagophore formation probably by occupying PtdIns(3)P, hence preventing the recruitment of the autophagy-specific PtdIns(3) effectors (Axe et al. 2008).

2.4   WIPI PROTEINS AS PHOSPHATIDYLINOSITOL 3-PHOSPHATE EFFECTORS AT THE NASCENT AUTOPHAGOSOME

The PtdIns(3)P-effector proteins carrying out a particular function in autophagy belong to the ancient WD-repeat protein family referred to as β-propellers that bind phosphoinositides (PROPPIN) (Thumm et al. 2013). The PROPPIN family members are divided in two paralogous groups both of which harbor members from unicellular and multicellular eukaryotes (Behrends et al. 2010; Polson et al. 2010; Proikas-Cezanne et al. 2004). The human PROPPIN members are referred to as WIPI for WD-repeat protein interacting with phosphoinositides (WIPI), and four WIPI genes have been identified that give rise to the proteins WIPI1 to WIPI4 along with numerous splice variants (Jeffries et al. 2004; Proikas-Cezanne et al. 2004, 2015) (Figures 2.1 and 2.2). All WIPI genes were found highly expressed in normal human tissues, with elevated levels in skeletal muscle and heart when compared with other tissues (Proikas-Cezanne et al. 2004). Only little is known about the regulation of WIPI gene expression. However, available data showed that the transcription factors

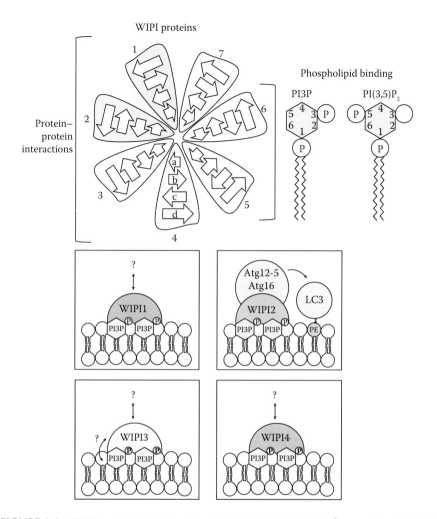

FIGURE 2.2   WIPI proteins fold into phospholipid-binding 7-bladed β-propellers. See text for further details.

TFEB and Spi-1 proto-oncogene (SPI1) regulate WIPI1 gene expression, respectively, in liver cells and neutrophils, and that ZKSCAN3 regulates WIPI2 gene expression (Proikas-Cezanne et al. 2015). Furthermore, the expression for all human WIPI genes are aberrantly in different human tumor entities where WIPI mutations have been identified; however, it is unknown whether or not these represent bystander or driver mutations in a particular context (Proikas-Cezanne et al. 2015).

Human WIPI proteins fold seven-bladed β-propeller proteins, as early demonstrated by homology modeling (Proikas-Cezanne et al. 2004). Via conserved residues that cluster at one site of the propeller (Proikas-Cezanne et al. 2004), WIPI proteins are considered to directly bind to PtdIns(3)P, and also to a lesser extend to PtdIns(3,5)P$_2$ (Gaugel et al. 2012; Jeffries et al. 2004; Lu et al. 2011; Polson et al. 2010; Proikas-Cezanne et al.

2004, 2007). Binding to PtdIns(3)P and to PtdIns(3,5)P_2 is mediated via the identical amino acids, and a crucial feature of two particular arginine residues, R226 and R227 in human WIPI1, as their mutation to alanine residues abolishes phospholipid binding (Dooley et al. 2014; Gaugel et al. 2012; Jeffries et al. 2004; Lu et al. 2011; Polson et al. 2010; Proikas-Cezanne et al. 2004, 2007). This was also demonstrated for the yeast homolog ATG18 (Dove et al. 2004; Krick et al. 2006) and both the propeller structure as well as the specific binding to PtdIns(3)P and PtdIns(3,5)P_2 has been verified by structural analysis of homologous with swollen vacuole phenotype 2 (HSV2), another yeast PROPPIN (Baskaran et al. 2012; Krick et al. 2012).

The autophagic function of the WIPI1 protein was first demonstrated by analyzing the colocalization with *bona fide* autophagic membrane markers, including ATG12, ATG16L, LC3, and p62 under starvation conditions or treatment with the mTOR inhibitor rapamycin (Gaugel et al. 2012; Itakura and Mizushima 2010; Proikas-Cezanne et al. 2004, 2007; van der Vos et al. 2012; Vergne et al. 2009). WIPI1 was found to localize at early phagophores (Proikas-Cezanne et al. 2007) and was also found to decorate the membrane of mature autophagosomes (Proikas-Cezanne and Robenek 2011). Based on the additional finding that WIPI1 translocates to the ER and the PM upon starvation (Proikas-Cezanne and Robenek 2011), both membranes have been considered as membrane origins for autophagy that involves WIPI proteins. A function of WIPI1 as PtdIns(3)P effector at the onset of autophagy can be considered because (1) the inhibition of autophagy by specific PtdIns(3)P inhibitors (e.g., LY294002, wortmannin) nullifies autophagosomal membrane localization of WIPI1, and the WIPI1 protein remains distributed in the cytosol (Gaugel et al. 2012; Proikas-Cezanne et al. 2004, 2007; Vergne et al. 2009) and (2) the inhibition of Jumpy and MTMR3, specific lipid phosphatases that remove the phosphate group on position 3 of the inositol ring in PtdIns(3)P, leads to a significant accumulation of PtdIns(3)P-enriched autophagosomal membranes that are decorated with WIPI1 (Taguchi-Atarashi et al. 2010; Vergne et al. 2009). In addition, siRNA-mediated WIPI1 downregulation counteracted LC3 lipidation, indicating a required function of WIPI1 upstream of LC3 in the formation of autophagosomes (Gaugel et al. 2012; Mauthe et al. 2011).

Several WIPI2 splice variants have been identified (Proikas-Cezanne et al. 2004), however, only two of them, WIPI2B and WIPI2D, were found to function in autophagy (Dooley et al. 2014; Mauthe et al. 2011; Polson et al. 2010; Proikas-Cezanne and Robenek 2011). WIPI2 has been demonstrated to localize at (1) ER-connected omegasomes positive for DFCP1 (Polson et al. 2010), (2) early phagophore structures positive for, for example, ATG16L and LC3 (Dooley et al. 2014; Polson et al. 2010), and (3) mature autophagosomes (Proikas-Cezanne and Robenek 2011). Upon starvation, similar to the translocation of WIPI1, WIPI2 is found not only at the PM but also on vesicles close to Golgi membranes (Proikas-Cezanne and Robenek 2011). Downregulation of WIPI2 by employing siRNAs in human tumor cells indicated earlier that WIPI2 is required for LC3 lipidation (Mauthe et al. 2011; Polson et al. 2010); however, the work of Dooley and coworkers demonstrated that WIPI2B in fact recruits ATG16L for subsequent LC3 lipidation at the nascent autophagosome (Dooley et al. 2014). This work demonstrated the important PtdIns(3)P-effector function of WIPI2B at the nascent autophagosome. Moreover, WIPI2B is required to encode the PtdIns(3)P signal toward LC3 lipidation after *Salmonella* infection (Dooley et al. 2014). After infection with *Shigella*, WIPI2 was also shown to function together with tectonin β-propeller repeat containing 1 (TECPR1), which binds to the ATG12-ATG5 conjugate and is needed for autophagosomal fusion (Chen et al. 2012; Ogawa et al. 2011).

Human WIPI3 is as yet uncharacterized, and it is not known if WIPI3 specifically binds to PtdIns(3)P and PtdIns(3,5)P$_2$ or if WIPI3 functions in the process of autophagy (Proikas-Cezanne et al. 2015). In contrast, this has been demonstrated for human WIPI4. Interestingly, identified human WIPI4 mutations represent the first example where a mutant autophagy-related protein is causative for a human disease, in particular a form of neurodegeneration (Proikas-Cezanne et al. 2015). WIPI4 was found to be mutated *de novo* in SENDA (static encephalopathy of childhood with neurodegeneration in adulthood), a sporadic NBIA (neurodegeneration with brain iron accumulation) subtype (Hayflick et al. 2013; Saitsu et al. 2013). In lymphoblastoid cell lines derived from SENDA patients, it has been shown that mutation in the *WIPI4* gene, located on the X chromosome, causes a strong decrease in autophagic function and an accumulation of LC3-positive autophagosomal membranes, leading to a malfunction in the degradation via autophagosomes (Saitsu et al. 2013). Another subtype of NBIA is the β-propeller protein–associated neurodegeneration (BPAN), where a knockdown of WIPI4 in mice brain showed the symptoms of BPAN patients and also a defect in the autophagic flux, as indicated by the accumulation of p62 and protein aggregates in neurons, providing a new model for further investigation of BPAN (Zhao et al. 2015).

2.5 REGULATION OF WIPI FUNCTION

The regulation of the PtdIns(3)P-effector function of the WIPI proteins is largely unknown, however, WIPI1 and WIPI2 were found to function downstream of mTORC1 and ULK1 but upstream of LC3, and WIPI4 may function downstream of LC3 (Codogno et al. 2012; Lamb et al. 2013; Proikas-Cezanne et al. 2015) (Figure 2.1).

WIPI1 was demonstrated to function downstream of mTORC1 because forced decrease of mTORC1 activity by siRNA-mediated mTOR downregulation resulted in an accumulation of WIPI1-positive autophagosomal membranes (Gaugel et al. 2012), as well as upon the employment of mTORC1 inhibitors (Proikas-Cezanne et al. 2007) or through the activation of FOXO-mediated glutamine production (van der Vos et al. 2012). As the inhibition of mTORC1 releases the negative control of mTORC1 on ULK1, ULK1 initiates PtdIns3K3 complex activation and PtdIns(3)P production (Figure 2.1). Hence, autophagosomal membrane localization of WIPI1 is abolished upon siRNA-mediated ULK1 or Vps34 downregulation (Itakura and Mizushima 2010; McAlpine et al. 2013). However, WIPI1 was also considered to function upstream of mTORC1 in the context of melanosome formation (Ho et al. 2011) but it is unknown how these results match the considered place of WIPI1 in the hierarchy of canonical autophagy (Itakura and Mizushima 2010). Another interesting finding that is not understood is the notion that both WIPI1 and WIPI2 do not respond to glucose starvation with an increase in their autophagosomal membrane localization, despite the fact that glucose starvation stimulates autophagy (Figure 2.1).

AMPK is activated by many upstream pathways, but prominently through calmodulin-dependent kinase α/β (CAMKKα and -β or CAMKK1 and -2) signaling (Means 2008). Based on this notion, it was demonstrated that an increase in cytoplasmic Ca^{2+} levels that leads to AMPK activation in fact induces autophagy (Hoyer-Hansen et al. 2007). However, it was demonstrated that cytoplasmic Ca^{2+} also permits autophagy induction independent of AMPK (Grotemeier et al. 2010). In this context, another downstream module of cytoplasmic Ca^{2+} signaling, independent of, or parallel to, AMPK has been suggested to involve CAMKI in the regulation of WIPI1 (Pfisterer et al. 2011). The above findings show, with many others, that the regulation of AMPK-mediated autophagy is complex and occurs via canonical and noncanonical routes (Codogno et al. 2012).

Regulation of WIPI1 can also occur at an epigenetic level, as it was shown that under normal conditions in naïve T cells, the methyltransferase G9a represses epigenetically the promoter region of both WIPI1 and LC3 (Artal-Martinez de Narvajas et al. 2013). This suppressive property is removed during starvation as well as receptor-stimulated activation of naïve T cells, indicating a regulatory mechanism important for autophagy initiation (Artal-Martinez de Narvajas et al. 2013).

Regulation of WIPI2 can also occur by the leucine-rich repeat kinase 2 (LRRK2), which plays an important role in the onset of Parkinson's disease. Mutations in the LRRK2 gene cause an alteration in the response to starvation, impairing an increase in the abundance of WIPI2 (Manzoni et al. 2013). In another setting, WIPI1 and WIPI2 were found influenced after introduction of recombinant capsid protein VP1 (rVP1) of the foot-and-mouth disease virus, probably by a Beclin-1-independent, noncanonical manner (Liao et al. 2013).

2.6   METHODOLOGY TO ASSESS WIPI-MEDIATED AUTOPHAGY

Fluorescence microscopy is routinely employed to assess autophagy. For example, the proportion of formed phagophores and autophagosomes upon different treatment conditions or genetic manipulations is usually measured by quantitative LC3 (LC3 puncta) imaging (Klionsky et al. 2016). In addition, fluorescence-based WIPI imaging (WIPI puncta) was established to assess PtdIns(3)P-dependent autophagy, and is used in combination with LC3 and other autophagy markers (Klionsky et al. 2016; Proikas-Cezanne et al. 2007). In this setting, the number of both WIPI puncta per cell as well as WIPI puncta-positive cells reflects the abundance of phagophores and autophagosomes (Thost et al. 2015). The manual counting of WIPI puncta in mammalian cells was extended to protocols for automated high-throughput/high-content image acquisition and analysis using monoclonal cell lines expressing tagged WIPI versions in low levels (Mauthe et al. 2011, 2012; Pfisterer et al. 2011, 2014; Thost et al. 2015). As for LC3 assessments, an increase of WIPI puncta either reflects the induction or inhibition of autophagy, or both, and hence, it is advised to employ lysosomal inhibitors, such as bafilomycin A1 (Thost et al. 2015).

2.7 SUMMARY

Unanswered question with regard to the role of WIPI proteins in autophagy includes whether or not the four human WIPI proteins (1) share overlapping and/or distinct functions in autophagy, (2) fulfill additional roles in other vesicular trafficking processes functions in autophagy, and (3) are co-regulated or regulated by distinct signaling cascades. It is considered that answering these questions will contribute to the further mechanistic insight to the regulation of autophagosome formation.

ACKNOWLEDGMENTS

Theresia Zuleger received a predoctoral fellowship by the Landesgraduiertenförderung Baden-Württemberg, Germany.

REFERENCES

Alers, S., A.S. Loffler, S. Wesselborg, and B. Stork. 2012. The incredible ULKs. *Cell Communication and Signaling*. 10:7.

Artal-Martinez de Narvajas, A., T.S. Gomez, J.S. Zhang, A.O. Mann, Y. Taoda, J.A. Gorman, M. Herreros-Villanueva, et al. 2013. Epigenetic regulation of autophagy by the methyltransferase G9a. *Molecular and Cellular Biology*. 33:3983–3993.

Ashford, T.P., and K.R. Porter. 1962. Cytoplasmic components in hepatic cell lysosomes. *The Journal of Cell Biology*. 12:198–202.

Axe, E.L., S.A. Walker, M. Manifava, P. Chandra, H.L. Roderick, A. Habermann, G. Griffiths, and N.T. Ktistakis. 2008. Autophagosome formation from membrane compartments enriched in phosphatidylinositol 3-phosphate and dynamically connected to the endoplasmic reticulum. *The Journal of Cell Biology*. 182:685–701.

Bar-Peled, L., and D.M. Sabatini. 2014. Regulation of mTORC1 by amino acids. *Trends in Cell Biology*. 24:400–406.

Baskaran, S., L.A. Carlson, G. Stjepanovic, L.N. Young, J. Kim do, P. Grob, R.E. Stanley, E. Nogales, and J.H. Hurley. 2014. Architecture and dynamics of the autophagic phosphatidylinositol 3-kinase complex. *Elife*. 3.

Baskaran, S., M.J. Ragusa, E. Boura, and J.H. Hurley. 2012. Two-site recognition of phosphatidylinositol 3-phosphate by PROPPINs in autophagy. *Molecular Cell*. 47:339–348.

Behrends, C., M.E. Sowa, S.P. Gygi, and J.W. Harper. 2010. Network organization of the human autophagy system. *Nature*. 466:68–76.

Blommaart, E.F., J.J. Luiken, P.J. Blommaart, G.M. van Woerkom, and A.J. Meijer. 1995. Phosphorylation of ribosomal protein S6 is inhibitory for autophagy in isolated rat hepatocytes. *The Journal of Biological Chemistry*. 270:2320–2326.

Boya, P., F. Reggiori, and P. Codogno. 2013. Emerging regulation and functions of autophagy. *Nature Cell Biology*. 15:713–720.

Chen, D., W. Fan, Y. Lu, X. Ding, S. Chen, and Q. Zhong. 2012. A mammalian autophagosome maturation mechanism mediated by TECPR1 and the Atg12-Atg5 conjugate. *Molecular Cell*. 45:629–641.

Choi, A.M., S.W. Ryter, and B. Levine. 2013. Autophagy in human health and disease. *The New England Journal of Medicine*. 368:1845–1846.

Clark, S.L., Jr. 1957. Cellular differentiation in the kidneys of newborn mice studies with the electron microscope. *The Journal of Biophysical and Biochemical Cytology*. 3:349–362.

Codogno, P., M. Mehrpour, and T. Proikas-Cezanne. 2012. Canonical and non-canonical autophagy: Variations on a common theme of self-eating? *Nature Reviews. Molecular Cell Biology*. 13:7–12.

Cuervo, A.M., and S. Zhang. 2015. Selective autophagy and Huntingtin: Learning from disease. *Cell Cycle*. 14:1617–1618.

De Duve, C., and R. Wattiaux. 1966. Functions of lysosomes. *Annual Review of Physiology*. 28:435–492.

Deter, R.L., P. Baudhuin, and C. De Duve. 1967. Participation of lysosomes in cellular autophagy induced in rat liver by glucagon. *The Journal of Cell Biology*. 35:C11–16.

Devereaux, K., C. Dall'Armi, A. Alcazar-Roman, Y. Ogasawara, X. Zhou, F. Wang, A. Yamamoto, P. De Camilli, and G. Di Paolo. 2013. Regulation of mammalian autophagy by class II and III PI 3-kinases through PI3P synthesis. *PLoS One*. 8:e76405.

Dooley, H.C., M. Razi, H.E. Polson, S.E. Girardin, M.I. Wilson, and S.A. Tooze. 2014. WIPI2 links LC3 conjugation with PI3P, autophagosome formation, and pathogen clearance by recruiting Atg12-5-16L1. *Molecular Cell*. 55:238–252.

Dove, S.K., R.C. Piper, R.K. McEwen, J.W. Yu, M.C. King, D.C. Hughes, J. Thuring, et al. 2004. Svp1p defines a family of phosphatidylinositol 3,5-bisphosphate effectors. *The EMBO Journal*. 23:1922–1933.

Egan, D.F., M.G. Chun, M. Vamos, H. Zou, J. Rong, C.J. Miller, H.J. Lou, et al. 2015. Small molecule inhibition of the autophagy kinase ULK1 and identification of ULK1 substrates. *Molecular Cell*. 59:285–297.

Egan, D.F., D.B. Shackelford, M.M. Mihaylova, S. Gelino, R.A. Kohnz, W. Mair, D.S. Vasquez, et al. 2011. Phosphorylation of ULK1 (hATG1) by AMP-activated protein kinase connects energy sensing to mitophagy. *Science*. 331:456–461.

F, O.F., T.E. Rusten, and H. Stenmark. 2013. Phosphoinositide 3-kinases as accelerators and brakes of autophagy. *The FEBS Journal*. 280:6322–6337.

Feng, Y., D. He, Z. Yao, and D.J. Klionsky. 2014. The machinery of macroautophagy. *Cell Research*. 24:24–41.

Gaugel, A., D. Bakula, A. Hoffmann, and T. Proikas-Cezanne. 2012. Defining regulatory and phosphoinositide-binding sites in the human WIPI-1 beta-propeller responsible for autophagosomal membrane localization downstream of mTORC1 inhibition. *Journal of Molecular Signaling*. 7:16.

Gaullier, J.M., A. Simonsen, A. D'Arrigo, B. Bremnes, H. Stenmark, and R. Aasland. 1998. FYVE fingers bind PtdIns(3)P. *Nature*. 394:432–433.

Ge, L., D. Melville, M. Zhang, and R. Schekman. 2013. The ER-Golgi intermediate compartment is a key membrane source for the LC3 lipidation step of autophagosome biogenesis. *Elife*. 2:e00947.

Geng, J., U. Nair, K. Yasumura-Yorimitsu, and D.J. Klionsky. 2010. Post-Golgi Sec proteins are required for autophagy in *Saccharomyces cerevisiae*. *Molecular Biology of the Cell*. 21:2257–2269.

Gillooly, D.J., I.C. Morrow, M. Lindsay, R. Gould, N.J. Bryant, J.M. Gaullier, R.G. Parton, and H. Stenmark. 2000. Localization of phosphatidylinositol 3-phosphate in yeast and mammalian cells. *The EMBO Journal*. 19:4577–4588.

Gingras, A.C., B. Raught, and N. Sonenberg. 2001. Regulation of translation initiation by FRAP/mTOR. *Genes & Development*. 15:807–826.

Grotemeier, A., S. Alers, S.G. Pfisterer, F. Paasch, M. Daubrawa, A. Dieterle, B. Viollet, S. Wesselborg, T. Proikas-Cezanne, and B. Stork. 2010. AMPK-independent induction of autophagy by cytosolic Ca2+ increase. *Cellular Signalling*. 22:914–925.

Hailey, D.W., A.S. Rambold, P. Satpute-Krishnan, K. Mitra, R. Sougrat, P.K. Kim, and J. Lippincott-Schwartz. 2010. Mitochondria supply membranes for autophagosome biogenesis during starvation. *Cell*. 141:656–667.

Hamasaki, M., N. Furuta, A. Matsuda, A. Nezu, A. Yamamoto, N. Fujita, H. Oomori, et al. 2013a. Autophagosomes form at ER-mitochondria contact sites. *Nature*. 495:389–393.

Hamasaki, M., S.T. Shibutani, and T. Yoshimori. 2013b. Up-to-date membrane biogenesis in the autophagosome formation. *Current Opinion in Cell Biology*. 25:455–460.

Hardie, D.G. 2015. AMPK: Positive and negative regulation, and its role in whole-body energy homeostasis. *Current Opinion in Cell Biology*. 33:1–7.

Hayashi-Nishino, M., N. Fujita, T. Noda, A. Yamaguchi, T. Yoshimori, and A. Yamamoto. 2009. A subdomain of the endoplasmic reticulum forms a cradle for autophagosome formation. *Nature Cell Biology*. 11:1433–1437.

Hayflick, S.J., M.C. Kruer, A. Gregory, T.B. Haack, M.A. Kurian, H.H. Houlden, J. Anderson, et al. 2013. beta-Propeller protein-associated neurodegeneration: a new X-linked dominant disorder with brain iron accumulation. *Brain*. 136:1708–1717.

Hirsimaki, P., and H. Reunanen. 1980. Studies on vinblastine-induced autophagocytosis in mouse liver. II. Origin of membranes and acquisition of acid phosphatase. *Histochemistry*. 67:139–153.

Ho, H., R. Kapadia, S. Al-Tahan, S. Ahmad, and A.K. Ganesan. 2011. WIPI1 coordinates melanogenic gene transcription and melanosome formation via TORC1 inhibition. *The Journal of Biological Chemistry*. 286:12509–12523.

Hoyer-Hansen, M., L. Bastholm, P. Szyniarowski, M. Campanella, G. Szabadkai, T. Farkas, K. Bianchi, et al. 2007. Control of macroautophagy by calcium, calmodulin-dependent kinase kinase-beta, and Bcl-2. *Molecular Cell*. 25:193–205.

Huang, J., and B.D. Manning. 2008. The TSC1-TSC2 complex: A molecular switchboard controlling cell growth. *The Biochemical Journal*. 412:179–190.

Itakura, E., and N. Mizushima. 2010. Characterization of autophagosome formation site by a hierarchical analysis of mammalian Atg proteins. *Autophagy*. 6:764–776.

Jacinto, E., and M.N. Hall. 2003. Tor signalling in bugs, brain and brawn. *Nature Reviews. Molecular Cell Biology*. 4:117–126.

Jeffries, T.R., S.K. Dove, R.H. Michell, and P.J. Parker. 2004. PtdIns-specific MPR pathway association of a novel WD40 repeat protein, WIPI49. *Molecular Biology of the Cell*. 15:2652–2663.

Jiang, P., and N. Mizushima. 2014. Autophagy and human diseases. *Cell Research*. 24:69–79.

Kabeya, Y., N. Mizushima, T. Ueno, A. Yamamoto, T. Kirisako, T. Noda, E. Kominami, Y. Ohsumi, and T. Yoshimori. 2000. LC3, a mammalian homologue of yeast Apg8p, is localized in autophagosome membranes after processing. *The EMBO Journal*. 19:5720–5728.

Kim, J., M. Kundu, B. Viollet, and K.L. Guan. 2011. AMPK and mTOR regulate autophagy through direct phosphorylation of Ulk1. *Nature Cell Biology*. 13:132–141.

Klionsky, D.J. 2005. The molecular machinery of autophagy: Unanswered questions. *Journal of Cell Science*. 118:7–18.

Klionsky, D.J., K. Abdelmohsen, A. Abe, M.J. Abedin, H. Abeliovich, A. Acevedo Arozena, H. Adachi, et al. 2016. Guidelines for the use and interpretation of assays for monitoring autophagy (3rd edition). *Autophagy*. 12:1–222.

Klionsky, D.J., J.M. Cregg, W.A. Dunn, Jr., S.D. Emr, Y. Sakai, I.V. Sandoval, A. Sibirny, et al. 2003. A unified nomenclature for yeast autophagy-related genes. *Developmental Cell*. 5:539–545.

Krick, R., R.A. Busse, A. Scacioc, M. Stephan, A. Janshoff, M. Thumm, and K. Kuhnel. 2012. Structural and functional characterization of the two phosphoinositide binding sites of PROPPINs, a beta-propeller protein family. *Proceedings of the National Academy of Sciences of the United States of America*. 109:E2042–2049.

Krick, R., J. Tolstrup, A. Appelles, S. Henke, and M. Thumm. 2006. The relevance of the phosphatidylinositolphosphat-binding motif FRRGT of Atg18 and Atg21 for the Cvt pathway and autophagy. *FEBS Letters*. 580:4632–4638.

Lamb, C.A., T. Yoshimori, and S.A. Tooze. 2013. The autophagosome: Origins unknown, biogenesis complex. *Nature Reviews. Molecular Cell Biology.* 14:759–774.

Laplante, M., and D.M. Sabatini. 2009. mTOR signaling at a glance. *Journal of Cell Science.* 122:3589–3594.

Laplante, M., and D.M. Sabatini. 2012. mTOR signaling in growth control and disease. *Cell.* 149:274–293.

Laplante, M., and D.M. Sabatini. 2013. Regulation of mTORC1 and its impact on gene expression at a glance. *Journal of Cell Science.* 126:1713–1719.

Liao, C.C., M.Y. Ho, S.M. Liang, and C.M. Liang. 2013. Recombinant protein rVP1 upregulates BECN1-independent autophagy, MAPK1/3 phosphorylation and MMP9 activity via WIPI1/WIPI2 to promote macrophage migration. *Autophagy.* 9:5–19.

Longatti, A., C.A. Lamb, M. Razi, S. Yoshimura, F.A. Barr, and S.A. Tooze. 2012. TBC1D14 regulates autophagosome formation via Rab11- and ULK1-positive recycling endosomes. *The Journal of Cell Biology.* 197:659–675.

Lu, Q., P. Yang, X. Huang, W. Hu, B. Guo, F. Wu, L. Lin, A.L. Kovacs, L. Yu, and H. Zhang. 2011. The WD40 repeat PtdIns(3)P-binding protein EPG-6 regulates progression of omegasomes to autophagosomes. *Developmental Cell.* 21:343–357.

Luiken, J.J., E.F. Blommaart, L. Boon, G.M. van Woerkom, and A.J. Meijer. 1994. Cell swelling and the control of autophagic proteolysis in hepatocytes: Involvement of phosphorylation of ribosomal protein S6? *Biochemical Society Transactions.* 22:508–511.

Manzoni, C., A. Mamais, S. Dihanich, R. Abeti, M.P. Soutar, H. Plun-Favreau, P. Giunti, S.A. Tooze, R. Bandopadhyay, and P.A. Lewis. 2013. Inhibition of LRRK2 kinase activity stimulates macroautophagy. *Biochimica et Biophysica Acta.* 1833:2900–2910.

Mari, M., J. Griffith, E. Rieter, L. Krishnappa, D.J. Klionsky, and F. Reggiori. 2010. An Atg9-containing compartment that functions in the early steps of autophagosome biogenesis. *The Journal of Cell Biology.* 190:1005–1022.

Martinez, J., R.K. Malireddi, Q. Lu, L.D. Cunha, S. Pelletier, S. Gingras, R. Orchard, et al. 2015. Molecular characterization of LC3-associated phagocytosis reveals distinct roles for Rubicon, NOX2 and autophagy proteins. *Nature Cell Biology.* 17:893–906.

Matsunaga, K., E. Morita, T. Saitoh, S. Akira, N.T. Ktistakis, T. Izumi, T. Noda, and T. Yoshimori. 2010. Autophagy requires endoplasmic reticulum targeting of the PI3-kinase complex via Atg14L. *The Journal of Cell Biology.* 190:511–521.

Mauthe, M., A. Jacob, S. Freiberger, K. Hentschel, Y.D. Stierhof, P. Codogno, and T. Proikas-Cezanne. 2011. Resveratrol-mediated autophagy requires WIPI-1-regulated LC3 lipidation in the absence of induced phagophore formation. *Autophagy.* 7:1448–1461.

Mauthe, M., W. Yu, O. Krut, M. Kronke, F. Gotz, H. Robenek, and T. Proikas-Cezanne. 2012. WIPI-1 Positive autophagosome-like vesicles entrap pathogenic *Staphylococcus aureus* for Lysosomal Degradation. *International Journal of Cell Biology.* 2012:179207.

McAlpine, F., L.E. Williamson, S.A. Tooze, and E.Y. Chan. 2013. Regulation of nutrient-sensitive autophagy by uncoordinated 51-like kinases 1 and 2. *Autophagy.* 9:361–373.

Means, A.R. 2008. The year in basic science: Calmodulin kinase cascades. *Molecular Endocrinology.* 22:2759–2765.

Meijer, A.J., and P. Codogno. 2011. Autophagy: Regulation by energy sensing. *Current Biology: CB.* 21:R227–229.

Meijer, A.J., and P.F. Dubbelhuis. 2004. Amino acid signalling and the integration of metabolism. *Biochemical and Biophysical Research Communications.* 313:397–403.

Meijer, A.J., S. Lorin, E.F. Blommaart, and P. Codogno. 2015. Regulation of autophagy by amino acids and MTOR-dependent signal transduction. *Amino Acids.* 47:2037–2063.

Mizushima, N., B. Levine, A.M. Cuervo, and D.J. Klionsky. 2008. Autophagy fights disease through cellular self-digestion. *Nature.* 451:1069–1075.

Mueller, A.J., and T. Proikas-Cezanne. 2015. Function of human WIPI proteins in autophagosomal rejuvenation of endomembranes? *FEBS Letters.* 589:1546–1551.

Nakatogawa, H., Y. Ichimura, and Y. Ohsumi. 2007. Atg8, a ubiquitin-like protein required for autophagosome formation, mediates membrane tethering and hemifusion. *Cell.* 130:165–178.

Nixon, R.A. 2013. The role of autophagy in neurodegenerative disease. *Nature Medicine.* 19:983–997.

Novikoff, A.B., and W.Y. Shin. 1978. Endoplasmic reticulum and autophagy in rat hepatocytes. *Proceedings of the National Academy of Sciences of the United States of America.* 75:5039–5042.

Ogawa, M., Y. Yoshikawa, T. Kobayashi, H. Mimuro, M. Fukumatsu, K. Kiga, Z. Piao, et al. 2011. A Tecpr1-dependent selective autophagy pathway targets bacterial pathogens. *Cell Host & Microbe.* 9:376–389.

Ohsumi, Y. 2014. Historical landmarks of autophagy research. *Cell Research.* 24:9–23.

Pfeifer, U. 1977. Inhibition by insulin of the physiological autophagic breakdown of cell organelles. *Acta Biologica et Medica Germanica.* 36:1691–1694.

Pfeifer, U. 1978. Inhibition by insulin of the formation of autophagic vacuoles in rat liver. A morphometric approach to the kinetics of intracellular degradation by autophagy. *The Journal of Cell Biology.* 78:152–167.

Pfisterer, S.G., D. Bakula, T. Frickey, A. Cezanne, D. Brigger, M.P. Tschan, H. Robenek, and T. Proikas-Cezanne. 2014. Lipid droplet and early autophagosomal membrane targeting of Atg2A and Atg14L in human tumor cells. *Journal of Lipid Research.* 55:1267–1278.

Pfisterer, S.G., M. Mauthe, P. Codogno, and T. Proikas-Cezanne. 2011. Ca^{2+}/calmodulin-dependent kinase (CaMK) signaling via CaMKI and AMP-activated protein kinase contributes to the regulation of WIPI-1 at the onset of autophagy. *Molecular Pharmacology.* 80:1066–1075.

Polson, H.E., J. de Lartigue, D.J. Rigden, M. Reedijk, S. Urbe, M.J. Clague, and S.A. Tooze. 2010. Mammalian Atg18 (WIPI2) localizes to omegasome-anchored phagophores and positively regulates LC3 lipidation. *Autophagy.* 6:506–522.

Proikas-Cezanne, T., and H. Robenek. 2011. Freeze-fracture replica immunolabelling reveals human WIPI-1 and WIPI-2 as membrane proteins of autophagosomes. *Journal of Cellular and Molecular Medicine.* 15:2007–2010.

Proikas-Cezanne, T., S. Ruckerbauer, Y.D. Stierhof, C. Berg, and A. Nordheim. 2007. Human WIPI-1 puncta-formation: a novel assay to assess mammalian autophagy. *FEBS Letters.* 581:3396–3404.

Proikas-Cezanne, T., Z. Takacs, P. Donnes, and O. Kohlbacher. 2015. WIPI proteins: Essential PtdIns3P effectors at the nascent autophagosome. *Journal of Cell Science.* 128:207–217.

Proikas-Cezanne, T., S. Waddell, A. Gaugel, T. Frickey, A. Lupas, and A. Nordheim. 2004. WIPI-1alpha (WIPI49), a member of the novel 7-bladed WIPI protein family, is aberrantly expressed in human cancer and is linked to starvation-induced autophagy. *Oncogene.* 23:9314–9325.

Ravikumar, B., K. Moreau, L. Jahreiss, C. Puri, and D.C. Rubinsztein. 2010a. Plasma membrane contributes to the formation of pre-autophagosomal structures. *Nature Cell Biology.* 12:747–757.

Ravikumar, B., S. Sarkar, J.E. Davies, M. Futter, M. Garcia-Arencibia, Z.W. Green-Thompson, M. Jimenez-Sanchez, et al. 2010b. Regulation of mammalian autophagy in physiology and pathophysiology. *Physiological Reviews.* 90:1383–1435.

Reunanen, H., and P. Hirsimaki. 1983. Studies on vinblastine-induced autophagocytosis in mouse liver. IV. Origin of membranes. *Histochemistry.* 79:59–67.

Ridley, S.H., N. Ktistakis, K. Davidson, K.E. Anderson, M. Manifava, C.D. Ellson, P. Lipp, et al. 2001. FENS-1 and DFCP1 are FYVE domain-containing proteins with distinct functions in the endosomal and golgi compartments. *Journal of Cell Science.* 114:3991–4000.

Roberts, R., and N.T. Ktistakis. 2013. Omegasomes: PI3P platforms that manufacture autophagosomes. *Essays in Biochemistry.* 55:17–27.

Russell, R.C., Y. Tian, H. Yuan, H.W. Park, Y.Y. Chang, J. Kim, H. Kim, T.P. Neufeld, A. Dillin, and K.L. Guan. 2013. ULK1 induces autophagy by phosphorylating Beclin-1 and activating VPS34 lipid kinase. *Nature Cell Biology.* 15:741–750.

Russell, R.C., H.X. Yuan, and K.L. Guan. 2014. Autophagy regulation by nutrient signaling. *Cell Research.* 24:42–57.

Saitsu, H., T. Nishimura, K. Muramatsu, H. Kodera, S. Kumada, K. Sugai, E. Kasai-Yoshida, et al. 2013. De novo mutations in the autophagy gene WDR45 cause static encephalopathy of childhood with neurodegeneration in adulthood. *Nature Genetics.* 45:445–449, 449e441.

Schworer, C.M., and G.E. Mortimore. 1979. Glucagon-induced autophagy and proteolysis in rat liver: Mediation by selective deprivation of intracellular amino acids. *Proceedings of the National Academy of Sciences of the United States of America.* 76:3169–3173.

Shibutani, S.T., and T. Yoshimori. 2014. A current perspective of autophagosome biogenesis. *Cell Research.* 24:58–68.

Simonsen, A., R. Lippe, S. Christoforidis, J.M. Gaullier, A. Brech, J. Callaghan, B.H. Toh, C. Murphy, M. Zerial, and H. Stenmark. 1998. EEA1 links PI(3)K function to Rab5 regulation of endosome fusion. *Nature.* 394:494–498.

Stenmark, H., R. Aasland, B.H. Toh, and A. D'Arrigo. 1996. Endosomal localization of the autoantigen EEA1 is mediated by a zinc-binding FYVE finger. *The Journal of Biological Chemistry.* 271:24048–24054.

Stolz, A., A. Ernst, and I. Dikic. 2014. Cargo recognition and trafficking in selective autophagy. *Nature Cell Biology.* 16:495–501.

Taguchi-Atarashi, N., M. Hamasaki, K. Matsunaga, H. Omori, N.T. Ktistakis, T. Yoshimori, and T. Noda. 2010. Modulation of local PtdIns3P levels by the PI phosphatase MTMR3 regulates constitutive autophagy. *Traffic.* 11:468–478.

Thost, A.K., P. Donnes, O. Kohlbacher, and T. Proikas-Cezanne. 2015. Fluorescence-based imaging of autophagy progression by human WIPI protein detection. *Methods.* 75:69–78.

Thumm, M., R.A. Busse, A. Scacioc, M. Stephan, A. Janshoff, K. Kuhnel, and R. Krick. 2013. It takes two to tango: PROPPINs use two phosphoinositide-binding sites. *Autophagy.* 9:106–107.

van der Vos, K.E., P. Eliasson, T. Proikas-Cezanne, S.J. Vervoort, R. van Boxtel, M. Putker, I.J. van Zutphen, et al. 2012. Modulation of glutamine metabolism by the PI(3)K-PKB-FOXO network regulates autophagy. *Nature Cell Biology.* 14:829–837.

Vanhaesebroeck, B., J. Guillermet-Guibert, M. Graupera, and B. Bilanges. 2010. The emerging mechanisms of isoform-specific PI3K signalling. *Nature Reviews. Molecular Cell Biology.* 11:329–341.

Vergne, I., E. Roberts, R.A. Elmaoued, V. Tosch, M.A. Delgado, T. Proikas-Cezanne, J. Laporte, and V. Deretic. 2009. Control of autophagy initiation by phosphoinositide 3-phosphatase Jumpy. *The EMBO Journal.* 28:2244–2258.

White, E. 2012. Deconvoluting the context-dependent role for autophagy in cancer. *Nature Reviews. Cancer.* 12:401–410.

Wong, P.M., C. Puente, I.G. Ganley, and X. Jiang. 2013. The ULK1 complex: Sensing nutrient signals for autophagy activation. *Autophagy.* 9:124–137.

Yang, Z., and D.J. Klionsky. 2010. Eaten alive: A history of macroautophagy. *Nature Cell Biology.* 12:814–822.

Yuan, H.X., R.C. Russell, and K.L. Guan. 2013. Regulation of PIK3C3/VPS34 complexes by MTOR in nutrient stress-induced autophagy. *Autophagy.* 9:1983–1995.

Zhao, Y.G., L. Sun, G. Miao, C. Ji, H. Zhao, H. Sun, L. Miao, S.R. Yoshii, N. Mizushima, X. Wang, and H. Zhang. 2015. The autophagy gene Wdr45/Wipi4 regulates learning and memory function and axonal homeostasis. *Autophagy.* 11:881–890.

Section II

Autophagy and Cell Fate

3 Cross Talk between Autophagy and Cell Death Pathways
Toward Understanding a System

Chrisna Swart, Andre Du Toit, and Ben Loos
University of Stellenbosch
Stellenbosch, South Africa

CONTENTS

3.1 INTRODUCTION

Although we have witnessed remarkable changes in understanding cell death from a phenomenon strongly guided by morphology to the molecular origins of that morphology, we still have much to learn. The success of controlling cell death onset in a clinical environment has been very limited, albeit better in a controlled laboratory environment. As a reader of this book, you are cognizant that the complexity of autophagy and cellular decision-making in the mammalian organism is greater than originally anticipated; that signaling pathways are not linear; that causation between protein and gene activity is not unidirectional; and that the microenvironment, cell history, and the current molecular makeup of the cell strongly determines the molecular land-scape that impacts on the outcome of a cellular stress response. "Context-dependent" is the key word often used to indicate this great degree of complexity and variability. "Autophagy may, context dependently, favor cell death onset or cell survival," is a major phrase often used in the recent literature that suggests its central position in cur-rent times of cell death research. In order to enable cell death control for therapeutic purposes, be it through delaying cell death onset in diseases of degeneration or injury or by profoundly inducing cell death in cancers, it is required to untangle this context, to map the molecular players that not only regulate but also control particular cell death pathways. If we fail to do so, the increasingly revealed molecular and functional complexity that is brought about by the omics era will be of little benefit, rather, it will lead deeper into a state of "drowning in a sea of data" (Brenner 2012).

In this chapter, we wish to review some of the current literature that may enable us to better discern the principle molecular pathways that regulate the cellular stress response. There are many ways in which the autophagy machinery is regulated and controlled, and, likewise, there are many ways in which a cell can die. Here, we hope to highlight the main nexus of cell death decision-making, by focusing on the dynamic interplay between ATP, mitochondria, and macroautophagy (hereafter referred to as autophagy), with a "spin" on systems integration. Why? Because it is cellular systems and networks that are collapsing and failing, for reasons that may be controlled or accidental, when the cell transits from life to death.

3.2   FROM CLEAR MORPHOLOGY TO SYSTEMS COMPLEXITY

The early classification attempts of cell death were based on morphological obser-vation, and hence received their name initially based on what was observed mostly under the light microscope. Access to published material was limited; hence, infor-mation did not always reach the global scientific community. Before a "formal" classification for the differentiation between cell death was made, cell death was discussed in a documented manner in Rudolph Virchow's lectures in the 1850s and 1860s, where the term necrosis was subcategorized as necrobiosis, to indicate "slow death," somewhat different compared to what was known as necrosis (Virchow1859). The drive and passion initiated by the German pathologist to "think microscopi-cally," led to the careful description of cell death in various tissues. In 1885, Walther Flemming, who also coined the terms "chromatin" and "mitosis," described apop-tosis in the ovarian follicle as "death by chromatolysis" (Majno and Joris 1995), and his student Franz Nissen followed this terminology and described apoptosis in the mammary gland as chromatolysis. In 1914, German anatomist Ludwig Gräper indicated chromatolysis, and concluded that the progressive shrinking of the yolk sac must be brought about by a physiological process of cell elimination (Majno and Joris 1995). In the 1950s, Glücksmann described apoptosis in the developing embryo as "chromatopyknosis," further revealing the morphological changes during apoptosis. In 1960, the first data that indicated the dynamic changes during the cell death process was observed, and the term "point of no return" (PONR) was coined as part of the cell death process (Majno et al. 1960). What follows is an era that substantially contributed to the shaping of cell death nomenclature. The American biologist Richard Lockshin described the programmed demise in intersegmental insect muscle and coined the term "programmed cell death" (PCD) in 1964. Eight years later, Australian pathologist John F. Kerr with Scottish colleagues Andrew Wyllie and Alastair Currie coined the term "shrinkage necrosis" (1971) and shortly thereafter "apoptosis" (1972). Subsequently, with growing advancements in molecular biology and microscopy techniques, a differentiation between Type I cell death (apoptosis), Type II cell death (autophagy), and Type III cell death (necrosis) was recommended (Clarke 1990; Schweichel and Merker 1973), and served as a powerful visual catalogue for the science community. Now a quarter of a century later, the autophagy machinery has been identified, its proteome mapped (Behrends et al. 2010), and it is becoming increasingly clear that the various components of the cell death machinery differ in their impact on the systems performance, that is, whether and how cells may stay alive (Kimchi 2007). It is being voiced that the cell death concept may extend into nonlinear dynamical pathways, demanding the construction of more suitable network model systems (DeGracia et al. 2012; Kimchi 2007). Hence, the categorisation based on morphological characteristics of cell death appears increasingly less fit to serve as a model system that is compatible with the amount and type of data output that has informed our understanding of cell death in the recent years. Hence, a gradual transition toward an inclusive map that informs on cell death subroutines, their essential and less essential aspects, discerning between their accidental or regulated and programmed nature, is being put forward (Galluzzi et al. 2014; Kroemer et al. 2005, 2009). Here, the dynamic nature of a stress response that may lead to cell death is being placed centrally and includes the PONR that distinguishes between the initiation and execution of regulated cell death dynamically, in context with the severity of the insult and homeostatic perturbations. We can look forward to an exciting future in the field of cell death research, as we are adapting our approach of understanding and definition constantly, closely led by the data that unfold. Here, the quote by the French physiologist Claude Bernard is fitting: "In the experimental method, observable reality is our only authority" (Noble 2007). This will with no doubt allow us to better unravel the complex networks that govern the control of cell death.

3.3 CELL DEATH AND A SYSTEMS PERSPECTIVE

Cell death is strongly metabolically controlled. The response of cells to metabolic perturbations that may lead to cell death or cell survival (Loos and Engelbrecht 2009) is governed by a complex network of highly interconnected checkpoints (Green et al. 2014). It is now becoming increasingly clear that cellular metabolism and signaling systems cannot be dissociated from one another but require an integrative approach that is inclusive of the intimate cross talk between metabolic circuits and signal transduction systems (Green at al. 2014). Several proteins serve multiple functions in regulating cell death as part of a signaling cascade on the one hand and cell survival as a function of a metabolic integration on the other. Metabolites may act epigenetically affecting cell death, mitochondrial function, or autophagy. Autophagy is a major protein degradation system that operates at basal levels in eukaryotic systems, mediating the sequestration of intracellular cargo, such as long-lived proteins or intracellular entities. This cargo is delivered to lysosomes in the form of double-membraned vesicles, the so-called autophagosomes. This facilitates not only bulk cytoplasm degradation but may also include the sequestration of organelles, such as the endoplasmic reticulum or mitochondria (Klionsky 2005). The degradation rate of cargo under basal conditions is indicated by the basal flux or basal autophagy. The change of autophagic flux, that is, an increase or a decrease in the rate of protein degradation, can be assessed by probing for key proteins such as LC3 or p62, and is indicative of the status of autophagic flux. Figure 3.1a indicates that the autophagic flux status is best revealed when using saturated concentrations of

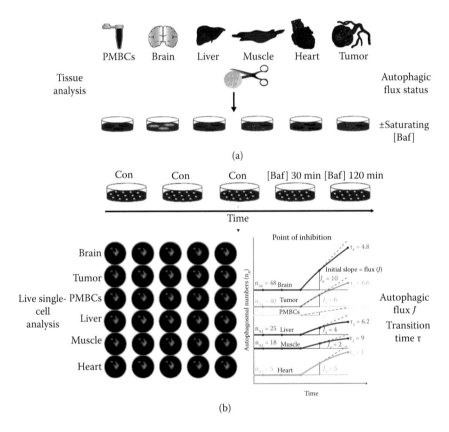

(a)

(b)

FIGURE 3.1 Progress curves can be generated to measure autophagic flux, by quantifying the accumulation of autophagosomes over time in the presence of saturating concentrations of bafilomycin A1. Counting of autophagosomes (nA) over time under control conditions, here for three time points, is required to show that the autophagic system is at steady state. The quantitative measurement of the basal autophagosome flux, J, at steady state is expressed as autophagosomes produced/cell/time. The autophagosome pool size nA differs between the six (hypothetical) sample tissues, which have been prepared for single-cell analysis. Here, brain tissue is indicating the highest autophagic flux ($J = 10$) and the smallest transition time ($\tau = 4.8$), followed by the tumor tissue. Note that, although the tumor tissue has a higher autophagic flux ($J = 6$) compared with the heart ($J = 5$), however, due to its small autophagogomal pool size, the cardiac transition time is smaller ($\tau = 1$) than that of the tumor tissue ($\tau = 6.6$).

autophagosomal/lysosomal fusion inhibitors, such as bafilomycin A1, which inhibits the lysosomal H^+-ATP-ase, thereby rendering fusion between autophagosomes and lysosomes dysfunctional. Cells may, however, change their proficiency in autophagy operation (Galluzzi et al. 2014), leaving the cell with an impaired or enhanced autophagy proficiency. This is often observed in disease onset or tumor progression (Galluzzi et al. 2014). A heightened autophagy proficiency would then indicate an increased degradation rate through autophagy; hence, an increased autophagic flux, and likewise, a decreased autophagy proficiency would refer to a decreased degradation rate through autophagy, that is, a decreased autophagic flux. It is important to note that a change in autophagic flux brings about a change in cell death susceptibility, being often, but not always, protective when flux is heightened prior to cellular injury (Hamacher-Brady et al. 2006; Loos et al. 2011a). It also becomes clear that flux, with the unit/time, requires an assessment over time (Loos et al. 2014).

Until recently, these levels of complexities have been largely disregarded, but increasing impact through a systems biology viewpoint on information gained on network organization of the autophagy system (Behrends et al. 2010; Caron et al. 2010; Huett et al. 2010; Ng 2010), as well as the mitochondrial fission and fusion behavior (Chauhan et al. 2014) and cellular injury (DeGracia et al. 2012) has been achieved. The integration of metabolic circuits and signaling systems is demanded in order to functionally unravel the network complexity of cell death and cell survival.

3.4 MITOCHONDRIA—THE NEXUS OF CELL DEATH

Central to the network that governs cellular fate are the mitochondria, functioning as a major switchboard, integrating metabolic sensors with metabolites and signal transduction pathways. It is here where the cellular demand for ATP is regulated in an integrative manner, and fine-tuned in the context of ATP consuming processes (Buttgereit and Brand 1995). Autophagy in this context takes a central position in integrating this stress response (Kroemer et al. 2010), as it is intrinsically linked to metabolite- and ATP sensing, to the provision of additional substrates such as amino acids or fatty acids, that may be fed into the tricarboxylic acid cycle TCA cycle for mitochondrial oxidative phosphorylation, as well as for quality control of damaged mitochondria and primarily long-lived proteins (Blackstone and Chang 2011; Loos et al. 2013). The extent of integration becomes especially clear when observed from an ATP-centered perspective. The kinases mammalian target of rapamycin (mTOR) and activated adenosine monophosphate kinase (AMPK) are key sensors that link the metabolite and energetic environment to a change in autophagic flux and protein synthesis. ATP availability is impacted through a number of avenues, some associated with switching off ATP consuming processes, thereby preserving ATP through metabolic reprogramming, some brought about by additionally providing metabolite substrates, and some by affecting the efficiency of ATP production at the mitochondrial cristae level (Loos et al. 2013). Here, the mTOR pathway, among the AMPK and PKA-sensing systems, plays a major role, since its regulation of protein degradation versus protein synthesis impacts ATP availability multifolds (Loos et al. 2013).

It is now becoming increasingly clear that a change in protein degradation rate through autophagy, that is, a change in autophagic flux, is associated with the onset of many human pathologies. Not only does a heightened autophagic flux provide metabolite substrates and affect proteostasis but also by inhibiting protein synthesis it decreases one of the primary ATP consuming processes, thereby affecting ATP homeostasis significantly (Buttgereit and Brand 1995). It becomes increasingly apparent that the successful cellular stress response is associated and compatible with an energetically favorable intracellular and extracellular microenvironment that matches the current cellular energy demands brought about by metabolic perturbations. Hence, not only intracellular signals, sensors, and transducers but also the cellular matrix that is being set through the cell's history (Lockshin et al. 2000) as well as the extracellular microenvironment contribute to the signaling and metabolic boundary conditions. The study of their molecular parts in isolation may not be sufficient to understand and predict cellular fate. In order to discern the likelihood for a given cellular outcome in the control and regulation of cell death, adaptive or suicidal in nature, an integration of these systems, based on most quantifiable parameters, is required. It was the French physiologist Claude Bernard, who, despite coining the term "milieu intérieur" and suggesting it as the most integral part of preserving cellular viability, indicated that "this application of mathematics to natural phenomena is the aim of all science, because the expression of the laws of phenomena should always be mathematical." (Noble 2007). Today, the power and impact of such a systems approach has already in part been demonstrated and forms an integral part of predictive, preventative, and precision medicine (Hood et al. 2004).

3.5   GAIN THROUGH INTEGRATION—AUTOPHAGIC FLUX AND CELL DEATH CONTROL

First, cell physiological processes are dynamic; likewise, when reporting on the kinetic properties of a cellular stress response or cell death, the true cell physiology is revealed most powerfully. Cells *are* not alive but rather may *stay* alive and survive; neither *are* cells dead with immediate action but rather may *become* nonviable and die. Staying alive or dying is a dynamic process and requires a methodological assessment that enables to capture this process most accurately. A dynamic assessment of cell death onset does not only allow to characterize and quantify the PONR of a given cell death modality (Loos et al. 2011), enabling the control and manipulation of its position, but also sometimes reveals causality of cell death (Arrasate et al. 2004). Here, it was demonstrated that overexpression of the mutated huntingtin protein in neurons leads to aggregates and inclusion bodies; however, the cells with inclusion bodies die last and not first, suggesting that the disease-modifying toxic entity is the proteinaceous species prior to the aggregate state. Moreover, the dynamic assessment of cell death, when applying statistically relevant tests (Jager et al. 2008), allows the indication of risk of cell death onset, as opposed to reporting merely on a significant change in viability (Arrasate et al. 2004). Calculating the likelihood for a particular event to take place has more meaning and brings about more understanding about the cellular system than verifying a change of a parameter that may be a single part of the event. Hence, the integration of data points and parameters that inform this cellular decision-making, such as mitochondrial function or autophagic flux (Hamacher-Brady et al. 2006; Loos et al. 2011), may enable us to better comprehend cell loss. Moreover, the kinetic properties of both autophagy and mitochondrial function are at the very core of their functions. Autophagic flux by its nature of definition is characterized by a given protein degradation rate and can hence only be quantified when assessing the change of cargo or autophagosomes in time (Klionsky et al. 2012). Here, live cell imaging and the use of powerful fluorochromes (Koga et al. 2011; Loos et al. 2014; Mizushima and Klionsky 2007; Tsvetkov et al. 2013) enable the sensitive reporting of autophagic flux (Figure 3.1B). Likewise, the functioning of mitochondria, which operate as a switchboard controlling cellular decision-making at the nexus between life and death, can only be best characterized when assessing kinetically the oxygen flux, consumption rate, or mitochondrial fission and fusion rate (Karbowski et al. 2004). The steady-state parameters themselves, such as intracellular ATP concentration, do not give any indication about ATP synthesis rate or ATP consumption (Meijer 2009). This brings about the next important principle of a systems approach, which suggests that there is no privileged level of causality (Noble 2007) but rather multiple levels of systems exist with feedback control mechanisms. This becomes clear when assessing the vast interplay between metabolite networks and signaling networks that govern cellular fate (Green et al. 2014). Of course, there are a number of subsystems, or checkpoints, such as the mitochondrial checkpoints, metabolic checkpoints, the autophagy checkpoint, the acetyl-CoA/CoA checkpoint, or the AMPK–mTOR checkpoint to name a few. These checkpoints are now comprehensively mapped, such as the mTOR-signaling network (Caron et al. 2010), revealing 380 proteins that are connected by 777 reactions, or the human autophagy system, with 409 candidate interacting proteins and 751 interactions (Behrends et al. 2010), respectively. Extensive connectivity of subnetworks increases this complexity. The relationship between these checkpoints is so profound and intricate that the isolation of a privileged level of causality is unlikely successful. This is confirmed by the immense functional overlap between molecules that play a role in metabolism and signaling alike, as well as the immense integration of stress pathways, that lead, for example, to a change in autophagic flux (Kroemer et al. 2010). Higher level insight is required to understand higher level biological functions, such as the maintenance of proteostasis or mitochondrial membrane potential, and how their dysregulation impacts on cell viability. In this chapter, we begin this process by highlighting the extent of molecular overlap between the cell death modalities and by referring, where suitable, to the systems' nature.

A key step in this approach is not only to identify and characterize the molecular players that take part in the autophagy and cell death machinery but also to assess how their interaction with each other and with their wider environment contributes to the maintenance of the machinery. This will, for example, allow to ask when the autophagic machinery transits from being functional to a merely operational or collapsing state? It assumes that its functionality is multilevel, with a bidirectional transmission of information. This becomes clear when assessing how the autophagic machinery requires, for example, both functional ATG5 or ATG7 expression but in addition also the fine-tuning of autophagic flux through, for example, acetylation processes (Morselli et al. 2011). It becomes increasingly clear that both "bottom up" causal chain reactions as well as downward causation drive the extent of autophagic flux. In addition, it is required to assess to what extent a change in one parameter, for example, autophagic flux ΔJ leads to a change in a given endpoint measure, for example, ATP availability $\Delta[ATP]$? Although it is known that ATP availability *per se* is central in the manifestation of cell death (Leist et al. 1997), and that autophagy may contribute to ATP generation systems by providing metabolite substrates, the exact and mathematically described relationship between autophagic flux, mitochondrial respiration, and ATP synthesis remains less clear. Here, well and precisely measurable data sets that can inform the physical laws of nature, such as forces, fluxes, or voltage changes as well as geometrical constraints such as diameters, form factors, and shapes of key structures or organelles are of importance to enable sound statistical analysis as well as mathematical modeling or programming (Mortiboys et al. 2008; Noble 2007).

3.6 THE CENTRALITY OF ATP AND AUTOPHAGY AS PRIMARY STRESS RESPONSE

Life requires ATP. It is not surprising that the primary ATP generating system, the mitochondria, is operating as a major nexus between life and death. By burning calories derived from our diet, with the oxygen we breathe, ATP is produced for allowing work to be performed and temperature to be maintained. However, in so doing, endogenous reactive oxygen species (ROS) are being formed, which may damage mitochondria, DNA, and cellular functions (Wallace 2005). Quality control is hence an intricate part of mitochondrial function, enabling an equilibrium that is compatible with viability. Autophagy, like mitochondria, also operates as a nexus between a living and a dying cell, as it not only enables the production of substrates for ATP generation (Harris et al. 1989; Loos et al. 2011; Onodera and Ohsumi 2005), in particular when the cell becomes nutrient deprived, but also participates in the quality control of damaged mitochondria and oxidized protein species. It thereby plays a major part in the cell's energy homeostasis, anchored within an energetic feedback loop (Loos et al. 2013), which is supported by its large network with metabolic sensors and transducers (Green et al. 2014). Autophagy also requires ATP for the proteolytic activities to take place (Schellens et al. 1988) and is, in particular due to protein synthesis control, connected to the principal ATP consuming process of the cell (Buttgereit and Brandt 1995). The cell's ability to rapidly enhance autophagy degradation rate positions this pathway not only as the primary stress response mechanism but also as the most central component of stress response integration (Kroemer et al. 2010; Loos and Engelbrecht 2009). This is important for two main reasons; first, it stresses that the accurate quantification of autophagic flux is an absolute necessity to understand the cellular stress response (Klionsky et al. 2012; Koga et al. 2011; Loos et al. 2014; Tsvetkov et al. 2013) and second, it underpins the dynamic and overlapping nature of cell death modalities, as ATP connects the transitions within and between the essential aspects of cell death (Leist et al. 1997, 1999; Loos and Engelbrecht 2009; Samara et al. 2008). The complexity between these two nexus becomes clear when observing dysfunction in either system in cells that have both high ATP demand and high autophagic clearance, such as described in diseases of neurodegeneration or cancer (Boland et al. 2008; White 2013). It is now becoming

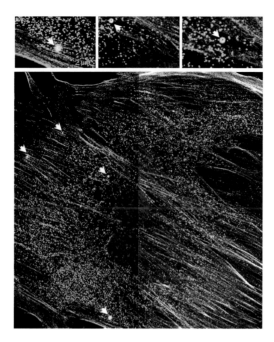

FIGURE 3.2 A combination of targeting methods and fluorescent probes, revealing the interaction between damaged mitochondria and autophagosomes through mitophagy (arrow heads). Both systems operate as nexus between ATP generation, life, and death. Mouse embryonic fibroblasts transfected with GFP-LC3 (green). After fixation cells were immuno-labeled for mitochondrial DNA with a mouse monolonal antibody (red) and counterstained using Alexa fluor 647 phalloidin (magenta) and Hoechst 33342 (blue). Images were acquired from z-planes and projected in maximum intensity. Scale bars: 20 μm.

clear that various levels of cross talk between mitochondrial networks and autophagy exist at various systems levels such as the BCL-2/Beclin signaling- (Mauiuri et al. 2007b), metabolism-, and the organellar mitophagy level (Blackstone and Chang 2011). Figure 3.2 illustrates the morphological manifestation of mitophagy, when labeling for mitochondrial DNA and autophagosomes simultaneously. Most intriguing is the multilevel nature of cross talk between these systems, such as the recently revealed interaction between ROS, the acetylation status of the cell (Morselli et al. 2011), and histone acetylation control (Wallace 2005). This reinforces a systems approach, in order to accommodate the downward causation that operates to respond primarily to mitochondrial derived changes in the oxidation/reduction profile of the cell, to maintain cell viability.

3.7 MOLECULAR OVERLAP WITHIN AND BETWEEN CELL DEATH MODALITIES

It is crucial to note that although each of the proposed cell death modalities consists of a number of distinct processes, major molecular overlap exist (reviewed in Nikoletopoulou et al. 2013)—overlap that often obscures precise characterization of the observed cell death that manifests. In fact, often cell fatality is triggered through a dual action encompassing more than one modality, suggesting that these modalities should be grouped and studied comprehensively, as a process where the incorporated pathways and molecules depend on the mode of toxicity incurred. A closer look at what circumstances trigger or utilize the molecular tools provided by apoptosis, autophagy, and necrosis is required in order to more accurately determine the original stressor that leads to the onset of cell death in a particular instance.

In this chapter, we will discuss current modalities and the molecular overlap that exist. Furthermore, we will discuss the impact of autophagic flux variability on cell death (Loos et al. 2013) and cancer, as well as the occurrence of autosis, a uniquely identified form of cell death, and its relationship with autophagic cell death (ACD). For many decades, the modes of cell death have been studied in isolation, primarily due to the morphology-based guidance in the cell death identification, assuming that they represent mutually exclusive cellular states. However, accumulative evidence suggests that they are regulated by similar pathways, share initiator and effector molecules, and engage in common subcellular sites (Nikoletopoulou et al. 2013). The two major modes of cell death, apoptosis and necrosis, often cooperate in a balanced interplay that involves autophagy, which serves either a pro-survival or pro-death purpose, depending on the molecular framework presented.

3.8   AUTOPHAGY AND APOPTOSIS

Autophagy has been involved in both types of cell death described in the previous section. During autophagy, cells exhibit extensive internal membrane remodeling by engulfment of portions of the cytoplasm into large double-membrane vacuoles which dock and fuse with lysosomes (Kuma et al. 2004). Autophagy was first characterized molecularly in autophagy-defective yeast (Nakatogawa et al. 2009) and is now thought to be driven by over 30 well-conserved autophagic proteins/genes (ATGs), from yeast to mammals (Mizushima et al. 2011). During the initial stages of autophagy (nucleation), the ULK/ATG1 complex activates the class III phosphoinositide 3-kinases (PI3K; a component of a multi-protein complex that includes ATG6/Beclin 1), which in turn recruits Atg proteins to the isolation membrane (also called the phagophore) (Simonsen and Tooze 2009). This is followed by vesicle elongation and completion through two ubiquitin-like conjugation systems (ATG12- and microtubule-associated protein 1A/1B light chain 3 [LC3] conjugation), which govern the covalent conjugation of ATG5 to ATG12, as well as the conversion of LC3-I to its phosphatidylethanolamine-conjugated form, LC3-II (Ohsumi 2001). The outer membrane of mature autophagosomes fuses with lysosomes (forming autolysosomes) to allow cargo degradation. PI3K also regulates cell growth through Akt kinase/TOR kinase signaling which inhibits autophagy and promotes cell growth. In turn, the Ras pathway regulates PI3K (Blommaart et al. 1997). Alternative pathways for autophagy have also been suggested, such as an ATG5-independent (Nishida et al. 2009) pathway, as well as a Beclin 1-independent type of autophagy (Chu et al. 2007).

Autophagy and apoptosis cooperate to influence cellular fate. Both are essential in development and homeostasis, and interactions among pathway components reveal a complex cross talk often under identical stimulus induction. Conditions such as growth factor deprivation, nutrient starvation, and energy metabolism activate the LKB1-5′ adenosine monophosphate–activated protein kinase (LKB1-AMPK) pathway, which leads to an increase in stability of cyclin-dependent kinase inhibitor p27kip1 and promotion of cell survival via autophagy induction. On the contrary, knockdown of p27kip1 under the same conditions also activates apoptosis (Liang et al. 2007), indicating a dual role for this pathway in both modalities. Furthermore, perturbation of ER calcium homeostasis or ER function was shown to increase both autophagy and apoptotic cell death (Ding et al. 2007). Death-associated protein kinase DAP-kinase (DAPK), a calcium/calmodulin-regulated ser/thr kinase, mediates cell death induced by diverse death signals and has been found to act as a tumor suppressor. DAPK becomes activated during ER stress and in turn triggers a mixture of apoptotic and autophagic death responses (Gozuacik et al. 2008).

The mTOR kinase integrates signals to regulate growth and aging. A number of mTOR activators have been implicated in the regulation of apoptosis (Thedieck et al. 2007). Starvation inhibits mTOR to activate autophagy (Kenyon 2010). Interestingly, studies suggest that mTOR is inhibited during autophagy initiation, but reactivated under prolonged starvation (Yu et al. 2010). This reactivation leads to the generation of proto-lysosomal tubules and vesicles that restore lysosomal homeostasis, a negative feedback that ensures reversion of autophagy upon nutrient replenishment, thus preventing excess cytoplasmic vacuolization which could lead to ACD (Yu et al. 2004a, 2004b, 2008a). Protein 53 (p53) represents another mTOR-mediated link between apoptosis and autophagy. p53 and its role in apoptosis have been well established; however, recently, the functional link between p53 and autophagy has been highlighted. p53 is able to stimulate autophagy by inhibiting mTOR via the activation of AMPK (Feng et al. 2005). Cross talk between autophagy and apoptosis has also been found to be mediated, in part, by the interaction between Beclin 1 (bec-1; the mammalian ortholog of yeast ATG6 and component of the P13K complex) and anti-apoptotic proteins Bcl-2 and Bcl-XL (Pattingre et al. 2005; Takacs-Vellai et al. 2005). Inactivation of bec-1 triggers apoptosis, suggesting that bec-1 acts as a crucial regulator of both pathways.

Caspases, the primary drivers of apoptosis, have also been found to serve as modulators of autophagy (reviewed in Wu et al. 2014), as indicated in Figure 3.3. Pro-apoptotic signals unite to activate caspases to execute apoptosis. At the same time, activated caspases can also shut down the autophagic response by the degradation of autophagy proteins (e.g., Beclin-1, ATG5, and ATG7). Caspases are also able to convert pro-autophagic proteins into pro-apoptotic proteins to initiate apoptosis instead (Bergmann 2007). A novel function for the anti-apoptotic protein, Flice-inhibitory protein (FLIP), was recently revealed as a negative regulator of autophagy (Lee et al. 2009). It was reported that FLIPs compete with LC3 for binding with ATG3, preventing the ATG3-mediated elongation of autophagosomes and thus decreasing the levels of autophagy. However, under stress conditions, the ATG3-LC3 interaction is allowed to induce autophagy instead.

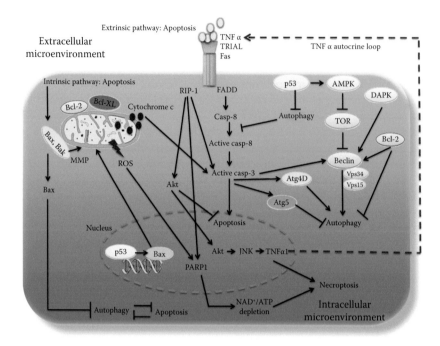

FIGURE 3.3 A large degree of molecular overlap exists between cell death modalities: autophagy, apoptosis, and necrosis involving TNFα and mitochondria signaling, ROS, and energy sensing. Although many arrows indicate (at present) a single direction only, however, it is becoming clear that many proteins may serve multiple functions in regulating cell death as part of a signaling cascade on the one hand and cell survival as a function of a metabolic integration on the other. Cells *are* not alive, but rather may *stay* alive and survive; neither *are* cells dead with immediate action nor they may *become* nonviable and die.

3.9 AUTOPHAGY AND NECROSIS

The most important direct pro-survival function of autophagy is likely its ability to block apoptosis or to suppress necrotic cell death. The overlap between autophagy and necrosis, however, is rather complex and the exact underlying mechanisms remain largely unknown and somewhat controversial. It has been shown that autophagy is able to either promote (Bonapace et al. 2010) or suppress (Bell et al. 2008) necrosis or necroptosis, a programmed form of necrosis, or not be molecularly associated with it at all (Osborn and Jaffer 2008). Studies show that autophagy is activated to block necroptosis in several cell lines in response to tumor necrosis factor-α (TNFα) antigen stimulation and starvation (Bell et al. 2008; Farkas et al. 2011). The short peptide zVAD, which acts as a general caspase inhibitor, is able to prevent apoptosis and trigger necroptosis in response to TNFα treatment (Wu et al. 2011). zVAD can also prevent autophagy through its inhibitory effect on lysosomal cathepsins, highlighting the pro-survival function of autophagy in the context of necroptosis. At the same time, inhibition of autophagy via mTOR signaling enhances necroptosis, while starvation protects against zVAD-induced necroptosis (Wu et al. 2008, 2009, 2011).

Sirtuins, which are NAD+-dependent protein deacetylases, involved in the regulation of apoptosis, stress resistance, and aging, have been implicated in both autophagy and necroptosis. SIRT1 forms a complex with several autophagy components to induce autophagy (Lee et al. 2008), and recent findings indicate SIRT2 in the regulation of necroptosis (Narayan et al. 2012). The nuclear enzyme, PARP1, underlies another link between autophagy and necrosis. Its overactivation leads to ATP depletion, necrotic cell death induction, and prevention of energy-dependent apoptosis. PARP1 activation also interfaces with pathways that promote autophagy. It was found that PARP1 activation in response to DNA damage can activate AMPK, inhibit mTOR, and thus induces autophagy (Alexander et al. 2010; Huang and Shen 2009; Muñoz-Gámez et al. 2009). It would thus seem that DNA-induced autophagy may act as a pro-survival mechanism to protect against necrosis resulting from PARP1 activation, indicating that PARP1 plays opposing roles upon activation, inducing necrosis due to ATP depletion, and at the same time facilitating protection against cell death through autophagy induction. A fine balance between autophagy and necrosis determines the final life-or-death outcome (Nikoletopoulou et al. 2013), with ATP being a central factor in this context.

3.10 APOPTOSIS AND NECROSIS

Apoptosis, defined as a dependence on activation via a family of proteases known as caspases, has been the most thoroughly studied mode of cell death, and its underlying processes receive consensus among researchers. Apoptosis is employed upon cell damage, stress, during physiological development, and morphogenesis (Bibel and Barde 2000). This pathway may be triggered by extrinsic stimuli through cell death receptors, such as TNFα, Fas (CD95/APO1), and TNF-related apoptosis inducing ligand (TRIAL) receptors, or by intrinsic stimuli incorporating mitochondrial signaling (Adams 2003; Kroemer et al. 2007) (Figure 3.3). Activation of caspases results in mitochondrial membrane permeabilization (MMP), chromatin condensation, and DNA fragmentation, giving rise to the distinct apoptotic morphological features (Green 2005; Salvesen and Dixit 1999). Increased MMP leads to the release of apoptotic proteins such as cytochrome c, Smac, and Omi. Cytochrome c release enables its association with apoptotic protease activation factor-1, which in turn activates the caspase cascade to execute cell death by apoptosis. Smac and Omi enhance caspase activation via inhibition of apoptosis family members. The B cell lymphoma 2 (Bcl-2) family controls apoptosis-mediated MMP and includes anti-apoptotic proteins (Bcl-2, Bcl-xL, and Mcl-1), multi-domain pro-apoptotic proteins (Bak and Bax), and proteins with a BH3 domain (Bid, Bim, Bik, Noxa, and Puma). Bak and Bax play essential roles in apoptogenic protein release.

Necrosis was initially thought to represent a mode of cell death that is in major contrast to apoptosis, due to its unordered and passive nature and the cellular "explosion" observed when the membrane integrity is lost in response to trauma. The morphology of necrotic cells includes organellar swelling, rupture of the plasma membrane, and complete lysis of the cell (Leist and Jäättelä 2001; Schweichel and Merker 1973). Unlike apoptosis, in necrosis, the nucleus becomes distended and remains largely intact. Necrotic cells release factors that are sensed by NLRP3, a core inflammasome protein, resulting in activation of inflammatory responses and the release of proinflammatory cytokine interleukin-1 beta (IL1β) (Los et al. 2002; Zong and Thompson 2006). Inflammasome activation is triggered mainly through mitochondrial ATP released from damaged cells (Iyer et al. 2009), indicating the first link between apoptosis and necrosis, via the mitochondria. Typical necrosis is not associated with caspase activation or with physiological development. However, a programmed form of necrosis, necroptosis, has been identified. Necroptosis occurs often in physical traumas and has been found to be associated with diverse forms of neurodegeneration and cell death inflicted by ischemia or infection (reviewed in Nikoletopoulou et al. 2013). Unlike the more disordered form of necrosis, programmed necroptosis, especially, shares several key processes with apoptosis.

First, TNFα treatment was found to induce a classical form of apoptosis in F17 cells but a necrotic form of cell death in L-M cells (Laster et al. 1988). Alternative death receptors that typically induce apoptosis have also been shown to induce necroptosis in different cell types (reviewed in Nikoletopoulou et al. 2013). This occurs especially when ATP levels are low or when apoptosis is blocked (Los et al. 2002). Thus, intracellular ATP levels play a determining role in the cross talk between apoptosis and necrosis. Where high levels of ATP typically induce energy-requiring apoptosis, low ATP levels have been shown to favor necrosis instead (Eguchi et al. 1997; Los et al. 2002). It is important to note that necrosis still requires some levels of ATP, and complete ATP depletion was found to trigger a cell death pathway distinct from both apoptosis and necrosis (Skulachev 2006). Energy sensing plays a central role in determining the mode of cell death activated, indicating mitochondria as key organelles in this context. TNFα is directly involved in mitochondrial ATP production, as well as in the generation of ROS (Skulachev 2006), and has been found to induce the activation of poly [ADP-ribose] polymerase1 (PARP1) via mitochondrial ROS, leading to ATP depletion and subsequent necrosis (Los et al. 2002). PARP1, activated by DNA damage, is a nuclear enzyme involved in DNA repair, stability, and transcriptional regulation. Its overactivation consumes large amounts of NAD^+ that results in massive ATP depletion (Sims et al. 1983). PARP1, therefore, appears to function as a molecular switch between apoptosis and necroptosis through the regulation of intracellular ATP levels. Taken together, it becomes clear that multiple signaling avenues connect mitochondrial function and ROS to ATP homeostasis and a cellular injury response.

Beyond the above, additional evidence for the cross talk between necrosis and apoptosis involves the tumor suppressor p53 and members of the Bcl-2 family of proteins. p53 plays a crucial role in stress sensing by responding to a wide range of signals, such as DNA damage, oxidative stress, or ischemia. It controls programs for apoptosis, typically by inducing components of the death receptor and mitochondrial pathways which cooperate to promote subsequently cellular demise (Brady et al. 2011; Riley et al. 2008). p53 can also directly promote MMP to trigger apoptosis via the Bcl-2 family (Green and Kroemer 2009). Recently, an additional molecular link between p53 and necrosis became evident, as Bax/Bak double-knockout mouse embryonic fibroblasts were found to die via necrotic mechanisms controlled by p53-mediated transcription of cathepsin Q together with DNA damage–induced ROS (Tu et al. 2009). The Bcl-2 family has well-established roles in apoptosis and only recently a link with necroptosis was established. In insulin-dependent diabetes mellitus, β-cells die via apoptosis (Delaney et al. 1997), a demise prevented by the overexpression of Bcl-2 (Iwahashi et al. 1996). Necrosis is thought to occur in the early phases of the disease, and it was shown that both apoptosis and necrosis in these cells can be prevented by overexpression of Bcl-2 (Saldeen 2000). Bcl-2 overexpression also prevented necrosis onset in both neuronal (Kane et al. 1995) and nonneuronal cells (Shimizu et al. 1995).

Above sections describe only a few examples of the molecular overlap and cross talk that exist between the cell death pathways; however, many more exist and are expanding the multilevel system of the cellular stress response. It is hence required to not only describe the network components in their separate entities but also to unravel how they connect to one another and to the microenvironment, thereby contributing toward the maintenance of a robust cellular homeostasis.

3.11 AUTOPHAGIC CELL DEATH

Much uncertainty exists in the field of ACD, and to what extent the cell is dying with or through autophagy. A full review of this debate would be beyond the scope of this chapter. However, it becomes clear that the measurement of protein degradation rate and its impacted parameters have to form part of the analysis as to when autophagic flux deviation becomes pathological or when it sensitizes cell death onset (Loos et al. 2013). It is generally accepted that the autophagic machinery regulates cell survival during nutrient deprivation (Kuma et al. 2004; Scott et al. 2004), ridding the cell of internal pathogens (Gutierrez et al. 2004; Pareja and Colombo 2013) or protein aggregate species, or to aid in tumor suppression (Qu et al. 2003; Yue et al. 2003), antigen presentation (Paludan et al. 2005), providing metabolites, or regulating the organism lifespan (Meléndez et al. 2003). Complexity sets in when taking into account the observation of robust autophagy in dying cells during development, infectious processes, and neurodegenerative diseases, where it is either failing to protect the cell, or where it has been initiated with a pathway that leads to cellular demise. The term ACD thus indicates a pathological condition, where an increase in autophagic vacuoles eventually seals the cell fate. Here, it will become crucial to identify the pathway flux that is present at that time, and to assess how it compares to basal levels, as the pool size of autophagosomes *per se* is by no means an indication of the autophagic flux (Loos et al. 2014). However, a decrease or lack of autophagy can also be harmful and a clear discrepancy is necessary in order to elucidate under which circumstances autophagy affects cellular well-being. Two crucial criteria have been put forth with regard to recognizing ACD: first, that an inhibition of the autophagic pathway would infer true cytoprotection; and second, that features of apoptosis or necrosis would be absent (Galluzzi et al. 2012). In essence, inhibitors against any form of PCD should be cytoprotective. It was further suggested that, for ACD to be regarded as a true form of PCD, complete dismantling of the cell needs to be mediated directly by autophagy, leading to cell death (Shen et al. 2012). This last criterion has been met with concern (Clarke and Puyal 2012), and it was suggested that autophagy would play a role either in the induction of cell death or in the final dismantlement of cells, but that it would be a remarkable coincidence for it to play both roles in a single cell. Furthermore, it was argued that rapidly dividing mammalian cells, such as cancer cells (Shen et al. 2012) may not represent the most likely physiological model system for identifying a "pure" ACD. From the above literature, it becomes clear that the network of a cellular stress cannot be restricted to an induction phase and dismantlement phase only, but that a multilevel systems approach is required to indicate systems' operation or systems failure. Autophagy and its degradation rate are certainly an integral part of that system and require special attention.

3.12   TAKE FLUX CONTROL

When investigating inhibitors or enhancers for autophagy in the context of controlling cell death onset, a few parameters have to be assessed. First, the degree of change in autophagic flux status in the diseased versus the normal cell tissue will be of importance. Here, it needs to be established to what extent the tissue is characterized by a dependency on autophagy. Some groups speak here of an "autophagy addiction," when autophagic flux appears to be particularly heightened. A combinational approach for autophagic flux assessment is recommended, which includes western blot analysis for key autophagy proteins and fluorescence microscopy as well as electron microscopy (Klionsky et al. 2012). The dynamic component of live-cell imaging and the power of fluorescent probes are most fitting to address the requirements for a real-time quantification. The use of saturated concentration of autophagosomal/lysosomal fusion inhibitors, such as bafilomycin A1, is now generally accepted as a requirement to assess the status of autophagic flux (Klionsky et al. 2012; Loos et al. 2014). A dynamic single-cell analysis–based approach may reveal not only flux status (Figure 3.1a) but also autophagic flux J (Figure 3.1b), indicated as autophagosomes/cell/hour, as well as the transition time τ in hours, reporting the time required to clear a complete intracellular autophagosomal pool (Loos et al. 2014). Here, complete pool size analyses, such as autophagosomal pool nA, based on whole cell x, y, and z dimensional acquisitions in time allow a robust comparison and discernment of fluxes and pool sizes upon treatment interventions (ΔnA and ΔJ) and may, particularly in combination with other techniques (Klionsky et al. 2012) lay a foundation toward a standardized approach to assess autophagic flux in mammalian systems.

3.13   TOWARD CELL DEATH PREDICTION AND FUTURE OUTLOOK

The connection between autophagy and cell death is well observed in neurodegenerative diseases, where protein misfolding and proteostasis decline as well as autophagy pathology are common (Boland et al. 2008; Skibinski and Finkbeiner 2013). Recent advances involve utilization of long time lapse single-cell analysis systems to follow neurons for weeks to predict cellular fate (Skibinski and Finkbeiner 2013). A key feature of such analysis is the application of statistical survival tools, such as the Kaplan–Meier method or Cox proportional hazard analysis (Jager et al. 2008; Scrucca et al. 2010) to a single-cell analysis system, allowing prediction of relationships between particular molecular events and cell death onset. Such an approach is important, as it allows the construction of comprehensive predictive models (Hughes-Alford and Lauffenburger 2012; Rehm and Prehn 2013) of cell death. These will become particularly powerful when informed by multiple systems, such as mitochondrial systems analysis (Chauhan et al. 2014), autophagic flux (Tsvetkov et al. 2013) as well as autophagy networks (Behrends et al. 2010) combined with data indicative of bioenergetics (Huber et al. 2011), embedded in theories of nonlinearity (DeGracia et al. 2012) and robustness (Rizk et al. 2009). Such an approach, combined with novel data derived from multiple systems levels, scaling from molecular density maps based on super-resolution microscopy techniques (Hell 2009) to whole organism *in vivo* imaging through light-sheet microscopy (Pampaloni et al. 2013) will undoubtedly contribute toward predicting viability (Tsukamoto et al. 2014) and enable informed predictive as well as precision medicine. The future looks bright indeed.

REFERENCES

Adams, J.M. 2003. Ways of dying: Multiple pathways to apoptosis. *Genes Dev.* 17, 2481–2495.

Alexander, A., Cai, S.-L., Kim, J., Nanez, A., Sahin, M., MacLean, K.H., Inoki, K., et al. 2010. ATM signals to TSC2 in the cytoplasm to regulate mTORC1 in response to ROS. *Proc. Natl. Acad. Sci. U. S. A.* 107, 4153–4158.

Arrasate, M., Mitra, S., Schweitzer, E.S., Segal, M.R., and Finkbeiner, S. 2004. Inclusion body formation reduces levels of mutant huntingtin and the risk of neuronal death. *Nature.* 431, 805.

Behrends, C., Sowa, M.E., Gygi, S.P., and Harper, J.W. 2010. Network organization of the human autophagy system. *Nature.* 466, 68–77.

Bell, B.D., Leverrier, S., Weist, B.M., Newton, R.H., Arechiga, A.F., Luhrs, K.A., Morrissette, N.S., and Walsh, C.M. 2008. FADD and caspase-8 control the outcome of autophagic signaling in proliferating T cells. *Proc. Natl. Acad. Sci. U. S. A.* 105, 16677–16682.

Bergmann, A. 2007. Autophagy and cell death: No longer at odds. *Cell* 131, 1032–1034.

Bibel, M., and Barde, Y.A. 2000. Neurotrophins: Key regulators of cell fate and cell shape in the vertebrate nervous system. *Genes Dev.* 14, 2919–2937.

Blackstone, C., and Chang, C.R. 2011. Mitochondria unite to survive. *Nat Cell Biol.* 13, 521–522.

Blommaart, E.F., Krause, U., Schellens, J.P., Vreeling-Sindelárová, H., and Meijer, A.J. 1997. The phosphatidylinositol 3-kinase inhibitors wortmannin and LY294002 inhibit autophagy in isolated rat hepatocytes. *Eur. J. Biochem. FEBS* 243, 240–246.

Boland, B., Kumar, A., Lee, S., Platt, F.M., Wegiel, J., Yu, W.H., and Nixon, R.A. 2008. Autophagy induction and autophagosome clearance in neurons: Relationship to autophagic pathology in Alzheimer's disease. *J. Neurosci.* 28, 6926–6937.

Bonapace, L., Bornhauser, B.C., Schmitz, M., Cario, G., Ziegler, U., Niggli, F.K., Schäfer, B.W., Schrappe, M., Stanulla, M., and Bourquin, J.-P. 2010. Induction of autophagy-dependent necroptosis is required for childhood acute lymphoblastic leukemia cells to overcome glucocorticoid resistance. *J. Clin. Invest.* 120, 1310–1323.

Brady, C.A., Jiang, D., Mello, S.S., Johnson, T.M., Jarvis, L.A., Kozak, M.M., Kenzelmann Broz, D., et al. 2011. Distinct p53 transcriptional programs dictate acute DNA-damage responses and tumor suppression. *Cell* 145, 571–583.

Brenner, S. 2012. Turing centenary: Life's code script. *Nature* 482(7386), 461.

Buttgereit, F., and Brand, M.D. 1995. A hierarchy of ATP consuming processes in mammalian cells. *Biochem. J.* 312, 163–167.

Caron, E., Ghosh, S., Matsuoka, Y., Ashton-Braucage, D., Therrien, M., Lemieux, S., Perreault, C., Roux, P.P., and Kitano, H. 2010. A comprehensive map of the mTOR signaling network. *Mol. Syst. Biol.* 6, 453.

Chauhan, A., Vera, J., and Wolkenhauer, O. 2014. The systems biology of mitochondrial fission and fusion and implications for disease and aging. *Biogerontology* 15, 1–12.

Chu, C.T., Zhu, J., and Dagda, R. 2007. Beclin 1-independent pathway of damage-induced mitophagy and autophagic stress: Implications for neurodegeneration and cell death. *Autophagy* 3, 663–666.

Clarke, P.G. 1990. Developmental cell death: Morphological diversity and multiple mechanisms. *Anat. Embryol. (Berl);* 181, 195–213.

Clarke, P.G.H., and Puyal, J. 2012. Autophagic cell death exists. *Autophagy* 8, 867–869.

DeGracia, D.J., Huang, Z.F., and Huang, S. 2012. A nonlinear dynamical theory of cell injury. *J. Cereb. Blood Flow Metab.* 32, 1000–1013.

Delaney, C.A., Cunningham, J.M., Green, M.H., and Green, I.C. 1997. Nitric oxide rather than superoxide or peroxynitrite inhibits insulin secretion and causes DNA damage in HIT-T15 cells. *Adv. Exp. Med. Biol.* 426, 335–339.

Ding, W.-X., Ni, H.-M., Gao, W., Hou, Y.-F., Melan, M.A., Chen, X., Stolz, D.B., Shao, Z.-M., and Yin, X.-M. 2007. Differential effects of endoplasmic reticulum stress-induced autophagy on cell survival. *J. Biol. Chem.* 282, 4702–4710.

Eguchi, Y., Shimizu, S., and Tsujimoto, Y. 1997. Intracellular ATP levels determine cell death fate by apoptosis or necrosis. *Cancer Res.* 57, 1835–1840.

Farkas, T., Daugaard, M., and Jäättelä, M. 2011. Identification of small molecule inhibitors of phosphatidylinositol 3-kinase and autophagy. *J. Biol. Chem.* 286, 38904–38912.

Feng, Z., Zhang, H., Levine, A.J., and Jin, S. 2005. The coordinate regulation of the p53 and mTOR pathways in cells. *Proc. Natl. Acad. Sci. U. S. A.* 102, 8204–8209.

Galluzzi, L., Bravo-San Pedro, J.M, Vitale, I., Aaronson, S.A., Abrams, J.M., Adam, D., Alnemri, E.S., et al. 2014. Essential versus accessory aspects of cell death: Recommendations of the NCCD 2015. *Cell Death Differ.* 22, 1–16.

Galluzzi, L., Vitale, I., Abrams, J.M., Alnemri, E.S., Baehrecke, E.H., Blagosklonny, M.V., Dawson, T.M., et al. 2012. Molecular definitions of cell death subroutines: Recommendations of the Nomenclature Committee on Cell Death 2012. *Cell Death Differ.* 19, 107–120.

Gozuacik, D., Bialik, S., Raveh, T., Mitou, G., Shohat, G., Sabanay, H., Mizushima, N., Yoshimori, T., and Kimchi, A. 2008. DAP-kinase is a mediator of endoplasmic reticulum stress-induced caspase activation and autophagic cell death. *Cell Death Differ.* 15, 1875–1886.

Green, D., Galluzzi, L., and Kroemer G. 2014. Metabolic control of cell death. *Science* 345, 1250256.

Green, D.R. 2005. Apoptotic pathways: Ten minutes to dead. *Cell* 121, 671–674.

Green, D.R., and Kroemer, G. 2009. Cytoplasmic functions of the tumour suppressor p53. *Nature* 458, 1127–1130.

Gutierrez, M.G., Master, S.S., Singh, S.B., Taylor, G.A., Colombo, M.I., and Deretic, V. 2004. Autophagy is a defense mechanism inhibiting BCG and *Mycobacterium tuberculosis* survival in infected macrophages. *Cell* 119, 753–766.

Hamacher-Brady, A., Brady, N.R., and Gottlieb, R.A. 2006. Enhancing macroautophagy protects against ischemia/reperfusion injury in cardiac myocytes. *J. Biol. Chem.* 281, 29776–29787.

Harris, R.A., Goodwin, G.W., Paxton, R., Dexter, P., Powell, S.M., Zhang, B., Han, A., Shimomura, Y., and Gibson, R. 1989. Nutritional and hormonal regulation of the activity state of hepatic branched-chain a-keto acid dehydrogenase complex. *Ann. N. Y. Acad. Sci.* 573, 306–313.

Hell, S.W. 2009. Microscopy and its focal switch. *Nat. Methods.* 6, 24–32.

Hood, L., Heath, J.R., Phelps, M.E., and Lin, B. 2004. Systems biology and new technologies enable predictive and preventative medicine. *Science* 306, 640–643.

Huang, Q., and Shen, H.-M. 2009. To die or to live: The dual role of poly(ADP-ribose) polymerase-1 in autophagy and necrosis under oxidative stress and DNA damage. *Autophagy* 5, 273–276.

Huber, H.J., Dussmann, H., Kilbride, S.M., Rehm, M., and Prehn, J.H. 2011. Glucose metabolism determines resistance of cancer cells to bioenergetic crisis after cytochorme-c release. *Mol. Syst. Biol.* 7, 470.

Huett, A., Goel, G., and Xavier, R.J. 2010. A systems biology viewpoint on autophagy in health and disease. *Curr. Opin. Gastroenterol.* 26, 302–309.

Hughes-Alford, S.K., and Lauffenburger, D.A. 2012. Quantitative analysis of gradient sensing: Towards building predictive models of chemotaxis in cancer. *Curr. Opin. Cell Biol.* 24, 284–291.

Iwahashi, H., Hanafusa, T., Eguchi, Y., Nakajima, H., Miyagawa, J., Itoh, N., Tomita, K., et al. 1996. Cytokine-induced apoptotic cell death in a mouse pancreatic beta-cell line: Inhibition by Bcl-2. *Diabetologia* 39, 530–536.

Iyer, S.S., Pulskens, W.P., Sadler, J.J., Butter, L.M., Teske, G.J., Ulland, T.K., Eisenbarth, S.C., et al. 2009. Necrotic cells trigger a sterile inflammatory response through the Nlrp3 inflammasome. *Proc. Natl. Acad. Sci. U. S. A.* 106, 20388–20393.

Jager, K.J., van Dijk, P.C., Zoccali, C., and Dekker, F.W. 2008. The analysis of survival data: The Kaplan–Meier method. *Kidney Int.* 74, 560–565.

Kane, D.J., Örd, T., Anton, R., and Bredesen, D.E. 1995. Expression of bcl-2 inhibits necrotic neural cell death. *J. Neurosci. Res.* 40, 269–275.

Karbowski, M., Arnoult, D., Chen, H., Chan, D.C., Smith, C.L. and Youle, R. 2004. Quantitation of mitochondrial dynamics by photolabeling of individual organelles shows that mitochondrial fusion is blocked during the Bax activation phase of apoptosis. *J. Cell Biol.* 164, 493–499.

Kenyon, C.J. 2010. The genetics of ageing. *Nature* 464, 504–512.

Kerr, J.F.R., Wyllie, A.H., and Currie, A.R. 1972. Apoptosis: A basic biological phenomenon with wide-ranging implications in tissue kinetics. *Br. J. Cancer* 26, 239–257.

Kimchi, A. 2007. Programmed cell death: From novel gene discovery to studies on network connectivity and emerging biomedical implications. *Cytokine Growth Factor Rev.* 18, 435–440.

Klionsky, D.J. 2005. The molecular machinery of autophagy: Unanswered questions. *J. Cell Sci.* 118, 7–18.

Klionsky, D.J., Abdalla, F.C., Abeliovich, H., Abraham, R.T., Acevedo-Arozena, A., Adeli, K., Agholme, L., et al. 2012. Guidelines for the use and interpretation of assays for monitoring autophagy. *Autophagy* 8, 445–544.

Koga, H., Martinez-Vicente, M., Macian, F., Verkusha, V.V., and Cuervo, A.M. 2011. A photoconvertible fluorescent reporter to track chaperone-mediated autophagy. *Nat. Commun.* 2, 386.

Kroemer, G., El-Deiry, W.S., Golstein, P., Peter, M.E., Vaux, D., Vandenabeele, P., Zhivotovsky, B., et al. 2005. Classification of cell death: Recommendations of the Nomenclature Committee on Cell Death. *Cell Death Differ.* 12 (Suppl 2), 1463–1467.

Kroemer, G., Galluzzi, L., and Brenner, C. 2007. Mitochondrial membrane permeabilization in cell death. *Physiol. Rev.* 87, 99–163.

Kroemer, G., Galluzzi, L., Vandenabeele, P., Abrams, J., Alnemri, E., Baehrecke, E., Blagosklonny, M., et al. 2009. Classification of cell death. *Cell Death Differ.* 16, 3–11.

Kroemer, G., Marino, G., and Levine, B. 2010. Autophagy and the integrated stress response. *Mol. Cell.* 40, 280–293.

Kuma, A., Hatano, M., Matsui, M., Yamamoto, A., Nakaya, H., Yoshimori, T., Ohsumi, Y., Tokuhisa, T., and Mizushima, N. 2004. The role of autophagy during the early neonatal starvation period. *Nature* 432, 1032–1036.

Laster, S.M., Wood, J.G., and Gooding, L.R. 1988. Tumor necrosis factor can induce both apoptic and necrotic forms of cell lysis. *J. Immunol.* 141, 2629–2634.

Lee, I.H., Cao, L., Mostoslavsky, R., Lombard, D.B., Liu, J., Bruns, N.E., Tsokos, M., Alt, F.W., and Finkel, T. 2008. A role for the NAD-dependent deacetylase Sirt1 in the regulation of autophagy. *Proc. Natl. Acad. Sci. U. S. A.* 105, 3374–3379.

Lee, J.-S., Li, Q., Lee, J.-Y., Lee, S.-H., Jeong, J.H., Lee, H.-R., Chang, H., et al. 2009. FLIP-mediated autophagy regulation in cell death control. *Nat. Cell Biol.* 11, 1355–1362.

Leist, M., and Jäättelä, M. 2001. Four deaths and a funeral: From caspases to alternative mechanisms. *Nat. Rev. Mol. Cell Biol.* 2, 589–598.

Leist, M., Single, B., Castoldi, A.F., Kuhnle, S., and Nicotera, P. 1997. Intracellular adenosine triphosphase (ATP) concentration: A switch in the decision between apoptosis and necrosis. *J. Exp. Med.* 185, 1481–1486.

Leist, M., Single, B., Naumann, H., Fava, E., Simon, B., Kühnle, S., and Nicotera, P. 1999. Inhibition of mitochondrial ATP generation by nitric oxide switches apoptosis to necrosis. *Exp. Cell Res.* 249, 396–403.

Liang, J., Shao, S.H., Xu, Z.-X., Hennessy, B., Ding, Z., Larrea, M., Kondo, S., et al. 2007. The energy sensing LKB1-AMPK pathway regulates p27(kip1) phosphorylation mediating the decision to enter autophagy or apoptosis. *Nat. Cell Biol.* 9, 218–224.

Lockshin, R.A., Osborne, B., and Zakeri, Z. 2000. Cell death in the third millennium. *Cell Death Differ.* 7, 2–7.

Lockshin, R.A., and Williams, C.M. 1964. Programmed cell death—II. Endocrine potentiation of the breakdown of the intersegmental muscles of silkmoths. *J. Insect Physiol.* 10, 643–649.

Loos, B., Du Toit, A., and Hofmeyr, J.H. 2014. Defining and measuring autophagosome flux—Concept and reality. *Autophagy* 10(11), 2087–2096.

Loos, B., and Engelbrecht, A.M. 2009. Cell death: A dynamic response concept. *Autophagy* 5, 590–603.

Loos, B., Genade, S., Ellis, B., Lochner, A., and Engelbrecht, A.M. 2011a. At the core of survival: Autophagy delays the onset of both apoptotic and necrotic cell death in a model of ischemic cell injury. *Exp. Cell Res.* 317, 1437–1453.

Loos, B., Lochner, A., and Engelbrecht, A.M. 2011b. Autophagy in heart disease: A strong hypothesis for an untouched metabolic reserve. *Med. Hypotheses.* 77, 52–57.

Loos, B., Lockshin, R., Klionsky, D.J., Engelbrecht, A.M., and Zakheri, Z. 2013. On the variability of autophagy and cell death susceptibility. *Autophagy* 9(9), 1270–1285.

Los, M., Mozoluk, M., Ferrari, D., Stepczynska, A., Stroh, C., Renz, A., Herceg, Z., Wang, Z.-Q., and Schulze-Osthoff, K. 2002. Activation and caspase-mediated inhibition of PARP: A molecular switch between fibroblast necrosis and apoptosis in death receptor signaling. *Mol. Biol. Cell* 13, 978–988.

Maiuri, M.C., Criollo, A., Tasdemir, E., Vicencio, J.M., Tajeddine, N., Hickman, J.A., Geneste, O., and Kroemer, G. 2007a. BH3-only proteins and BH3 mimetics induce autophagy by competetively disrupting the interaction between Beclin-1 and Bcl-2/Bcl-XL. *Autophagy* 3, 374–376.

Maiuri, M.C., Zalckvar, E., Kimchi, A., and Kroemer, G. 2007b. Self-eating and self-killing: Crosstalk between autophagy and apoptosis. *Nat. Rev. Mol. Cell Biol.* 8, 741–752.

Majno, G., and Joris, I. 1995. Apoptosis, oncosis, and necrosis. An overview of cell death. *Am. J. Pathol.* 146, 3–15.

Majno, G., La Gattuta, M., and Thompson, T.E. 1960. Cellular death and necrosis: Chemical, physical and morphologic changes in rat liver. *Virchows Arch. Pathol. Anat. Physiol. Klin. Med.* 333, 421–465.

Meijer, A.J. 2009. Autophagy research: Lessons from metabolism. *Autophagy* 5(1), 3–5.

Meléndez, A., Tallóczy, Z., Seaman, M., Eskelinen, E.-L., Hall, D.H., and Levine, B. 2003. Autophagy genes are essential for dauer development and life-span extension in *C. elegans. Science* 301, 1387–1391.

Mizushima, N., and Klionsky, D.J. 2007. Protein turnover via autophagy: Implications for metabolism. *Annu. Rev. Nutr.* 27, 19–40.

Mizushima, N., Yoshimori, T., and Ohsumi, Y. 2011. The role of Atg proteins in autophagosome formation. *Annu. Rev. Cell Dev. Biol.* 27, 107–132.

Morselli, E., Galluzzi, L., Kepp, O., Mariño, G., Michaud, M., Vitale, I., Maiuri, M.C., and Kroemer, G. 2011a. Oncosuppressive functions of autophagy. *Antioxid. Redox Signal.* 14, 2251–2269.

Morselli, E., Mariño, G., Bennetzen, M.V., Eisenberg, T., Megalou, E., Schroeder, S., Cabrera, S., et al. 2011b. Spermidine and reseveratrol induce autophagy by distinct pathways converging on the acetylproteome. *J. Cell. Biol.* 192(4), 615–629.

Mortiboys, H., Thomas, K.J., Koopman, W.J.H., Klaffke, S., Abou-Sleiman, P., Olpin, S., Wood, N.W., et al. 2008. Mitochondrial function and morphology are impaired in parkin-mutant fibroblasts. *Ann. Neurol.* 64, 555–565.

Muñoz-Gámez, J.A., Rodríguez-Vargas, J.M., Quiles-Pérez, R., Aguilar-Quesada, R., Martín-Oliva, D., de Murcia, G., Menissier de Murcia, J., Almendros, A., Ruiz de Almodóvar, M., and Oliver, F.J. 2009. PARP-1 is involved in autophagy induced by DNA damage. *Autophagy* 5, 61–74.

Nakatogawa, H., Suzuki, K., Kamada, Y., and Ohsumi, Y. 2009. Dynamics and diversity in autophagy mechanisms: Lessons from yeast. *Nat. Rev. Mol. Cell Biol.* 10, 458–467.

Narayan, N., Lee, I.H., Borenstein, R., Sun, J., Wong, R., Tong, G., Fergusson, M.M., et al. 2012. The NAD-dependent deacetylase SIRT2 is required for programmed necrosis. *Nature* 492, 199–204.

Ng, A.C.Y. 2010. Integrative systems biology and networks in autophagy. *Semin. Immunopathol.* 32, 355–361.

Nikoletopoulou, V., Markaki, M., Palikaras, K., and Tavernarakis, N. 2013. Crosstalk between apoptosis, necrosis and autophagy. *Biochim. Biophys. Acta* 1833, 3448–3459.

Nishida, Y., Arakawa, S., Fujitani, K., Yamaguchi, H., Mizuta, T., Kaneseki, T., Komatsu, M., Otsu, K., Tsujimoto, Y., and Shimizu, S. 2009. Discovery of Atg5/Atg7-independent alternative macroautophagy. *Nature* 461, 654–658.

Noble, D. 2007. Claude Bernard, the first systems biologist, and the future of physiology. *Exp. Physiol.* 93, 16–26.

Ohsumi, Y. 2001. Molecular dissection of autophagy: Two ubiquitin-like systems. *Nat. Rev. Mol. Cell Biol.* 2, 211–216.

Onodera, J., and Ohsumi, Y. 2005. Autophagy is required for maintenance of amino acid levels and protein synthesis under nitrogen starvation. *J. Biol. Chem.* 280, 31582–31586.

Osborn, E.A., and Jaffer, F.A. 2008. Advances in molecular imaging of atherosclerotic vascular disease. *Curr. Opin. Cardiol.* 23, 620–628.

Paludan, C., Schmid, D., Landthaler, M., Vockerodt, M., Kube, D., Tuschl, T., and Münz, C. 2005. Endogenous MHC class II processing of a viral nuclear antigen after autophagy. *Science* 307, 593–596.

Pampaloni, F., Ansari, N., and Stelzer, E.H.K. 2013. High-resolution deep imaging of live cellular spheroids with light-sheet-based fluorescence microscopy. *Cell Tissue Res.* 352, 161–177.

Pareja, M.E., and Colombo, M.I. 2013. Autophagic clearance of bacterial pathogens: Molecular recognition of intracellular microorganisms. *Front. Cell. Infect. Microbiol.* 3, 54.

Pattingre, S., Tassa, A., Qu, X., Garuti, R., Liang, X.H., Mizushima, N., Packer, M., Schneider, M.D., and Levine, B. 2005. Bcl-2 antiapoptotic proteins inhibit Beclin 1-dependent autophagy. *Cell* 122, 927–939.

Qu, X., Yu, J., Bhagat, G., Furuya, N., Hibshoosh, H., Troxel, A., Rosen, J., et al. 2003. Promotion of tumorigenesis by heterozygous disruption of the beclin 1 autophagy gene. *J. Clin. Invest.* 112, 1809–1820.

Rehm, M., and Prehn, J.H.M. 2013. Systems modelling methodology for the analysis of apoptosis signal transduction and cell death decisions. *Methods* 61, 165–173.

Riley, T., Sontag, E., Chen, P., and Levine, A. 2008. Transcriptional control of human p53-regulated genes. *Nat. Rev. Mol. Cell Biol.* 9, 402–412.

Rizk, A., Batt, G., Fages F., and Soliman S. 2009. A general computational method for robustness analysis with applications to synthetic gene networks. *Bioinformatics* 25, 169–178.

Saldeen, J. 2000. Cytokines induce both necrosis and apoptosis via a common Bcl-2-inhibitable pathway in rat insulin-producing cells. *Endocrinology* 141, 2003–2010.

Salvesen, G.S., and Dixit, V.M. 1999. Caspase activation: The induced-proximity model. *Proc. Natl. Acad. Sci. U. S. A.* 96, 10964–10967.

Samara, C., Syntichaki, P., and Tavernarakis, N. 2008. Autophagy is required for necrotic cell death in *Caenorhabditis elegans*. *Cell Death Differ.* 15, 105–112.

Schellens, J.P.M., Vreeling-Sindelarova, H., Plomp, P.J.A.M., and Meijer, A.J. 1988. *Exp. Cell Res.* 177, 103–108.

Schweichel, J.U., and Merker, H.J. 1973. The morphology of various types of cell death in prenatal tissues. *Teratology* 7, 253–266.

Scott, R.C., Schuldiner, O., and Neufeld, T.P. 2004. Role and regulation of starvation-induced autophagy in the Drosophila fat body. *Dev. Cell* 7, 167–178.

Scrucca, L., Santucci, A., and Aversa, F. 2010. Regression modeling of competing risk using R: An in depth guide for clinicians. *Bone Marrow Transplant* 45, 1388–1395.

Shen, S., Kepp, O., and Kroemer, G. 2012. The end of autophagic cell death? *Autophagy* 8, 1–3.

Shimizu, S., Eguchi, Y., Kosaka, H., Kamiike, W., Matsuda, H., and Tsujimoto, Y. 1995. Prevention of hypoxia-induced cell death by Bcl-2 and Bcl-xL. *Nature* 374, 811–813.

Simonsen, A., and Tooze, S.A. 2009. Coordination of membrane events during autophagy by multiple class III PI3-kinase complexes. *J. Cell Biol.* 186, 773–782.

Sims, J.L., Berger, S.J., and Berger, N.A. 1983. Poly(ADP-ribose) Polymerase inhibitors preserve nicotinamide adenine dinucleotide and adenosine 5'-triphosphate pools in DNA-damaged cells: Mechanism of stimulation of unscheduled DNA synthesis. *Biochemistry (Mosc.)* 22, 5188–5194.

Skibinski G., and Finkbeiner, S. 2013. Longitudinal measures of proteostasis in live neurons: Features that determine fate in models of neurodegenerative disease. *FEBS Lett.* 587, 1139–1146.

Skulachev, V.P. 2006. Bioenergetic aspects of apoptosis, necrosis and mitoptosis. *Apoptosis Int. J. Program. Cell Death* 11, 473–485.

Strohecker, A.M., Guo, J.Y., Karsli-Uzunbas, G., Price, S.M., Chen, G.J., Mathew, R., McMahon, M., and White, E. 2013. Autophagy sustains mitochondrial glutamine metabolism and growth of BrafV600E-driven lung tumors. *Cancer Discov.* 3, 1272–1285.

Takacs-Vellai, K., Vellai, T., Puoti, A., Passannante, M., Wicky, C., Streit, A., Kovacs, A.L., and Müller, F. 2005. Inactivation of the autophagy gene bec-1 triggers apoptotic cell death in *C. elegans*. *Curr. Biol.* 15, 1513–1517.

Thedieck, K., Polak, P., Kim, M.L., Molle, K.D., Cohen, A., Jenö, P., Arrieumerlou, C., and Hall, M.N. 2007. PRAS40 and PRR5-like protein are new mTOR interactors that regulate apoptosis. *PLoS One* 2, e1217.

Tsukamoto, S., Hara, T., Yamamoto, A., Kito, S., Minami, N., Kubota, T., Sato, K., and Kokubo, T. 2014. Fluorescence-based visualization of autophagic activity predicts mouse embryo viability. *Sci. Rep.* 4, 4533.

Tsvetkov, A.S., Arrasate, M., Barmada, S., Ando, D.M., Sharma, P., Shaby, B.A., and Finkbeiner, S. 2013. Proteostasis of polyglutamine varies among neurons and predicts neurodegeneration. *Nat. Chem. Biol.* 9, 586–592.

Tu, H.-C., Ren, D., Wang, G.X., Chen, D.Y., Westergard, T.D., Kim, H., Sasagawa, S., Hsieh, J.J.-D., and Cheng, E.H.-Y. 2009. The p53-cathepsin axis cooperates with ROS to activate programmed necrotic death upon DNA damage. *Proc. Natl. Acad. Sci. U. S. A.* 106, 1093–1098.

Virchow, R. 1859. *Cellular Pathology as Based upon Physiological and Pathological Histology*, ed. 2. Translated from German by B Chance. Reproduced by Dover Publications, New York, 1971, pp. 356–382.

Wallace, D.C. 2005. A mitochondrial paradigm of metabolic and degenerative diseases, aging and cancer: A dawn for evolutionary medicine. *Annu. Rev. Genet.* 39, 359–407.

White, E. 2013. Exploiting the bad eating habits of Ras-driven cancers. *Genes Dev.* 27, 2065–2071.

Wu, H., Che, X., Zheng, Q., Wu, A., Pan, K., Shao, A., Wu, Q., Zhang, J., and Hong, Y. 2014. Caspases: A molecular switch node in the crosstalk between autophagy and apoptosis. *Int. J. Biol. Sci.* 10, 1072–1083.

Wu, Y.-T., Tan, H.-L., Huang, Q., Kim, Y.-S., Pan, N., Ong, W.-Y., Liu, Z.-G., Ong, C.-N., and Shen, H.-M. 2008. Autophagy plays a protective role during zVAD-induced necrotic cell death. *Autophagy* 4, 457–466.

Wu, Y.-T., Tan, H.-L., Huang, Q., Ong, C.-N., and Shen, H.-M. 2009. Activation of the PI3K-Akt-mTOR signaling pathway promotes necrotic cell death via suppression of autophagy. *Autophagy* 5, 824–834.

Wu, Y.-T., Tan, H.-L., Huang, Q., Sun, X.-J., Zhu, X., and Shen, H.-M. 2011. zVAD-induced necroptosis in L929 cells depends on autocrine production of TNFα mediated by the PKC-MAPKs-AP-1 pathway. *Cell Death Differ.* 18, 26–37.

Yu, L., Alva, A., Su, H., Dutt, P., Freundt, E., Welsh, S., Baehrecke, E.H., and Lenardo, M.J. 2004a. Regulation of an ATG7-beclin 1 program of autophagic cell death by caspase-8. *Science* 304, 1500–1502.

Yu, L., Lenardo, M.J., and Baehrecke, E.H. 2004b. Autophagy and caspases: A new cell death program. *Cell Cycle.* 3, 1124–1126.

Yu, L., McPhee, C.K., Zheng, L., Mardones, G.A., Rong, Y., Peng, J., Mi, N., et al. 2010. Termination of autophagy and reformation of lysosomes regulated by mTOR. *Nature* 465, 942–946.

Yu, L., Strandberg, L., and Lenardo, M.J. 2008a. The selectivity of autophagy and its role in cell death and survival. *Autophagy* 4, 567–573.

Yu, S.-W., Baek, S.-H., Brennan, R.T., Bradley, C.J., Park, S.K., Lee, Y.S., Jun, E.J., et al. 2008b. Autophagic death of adult hippocampal neural stem cells following insulin withdrawal. *Stem Cells.* 26, 2602–2610.

Yue, Z., Jin, S., Yang, C., Levine, A.J., and Heintz, N. 2003. Beclin 1, an autophagy gene essential for early embryonic development, is a haploinsufficient tumor suppressor. *Proc. Natl. Acad. Sci. U. S. A.* 100, 15077–15082.

Zong, W.-X., and Thompson, C.B. 2006. Necrotic death as a cell fate. *Genes Dev.* 20, 1–15.

4 Autophagy, Cellular Senescence, and Cancer

Cheng Bing
Nanyang Technological University
Singapore

Karen Crasta
Nanyang Technological University and
Agency for Science, Technology and Research
Singapore
Imperial College London
London, UK

CONTENTS

4.1   INTRODUCTION

There are three major types of autophagy (Mizushima et al. 2008): (1) Macroautophagy, characterized by the formation of double-membrane vesicles called autophagosomes; (2) Microautophagy, where lysosomes can directly engulf cytoplasmic materials by inward invagination of the lysosomal membrane; and (3) Chaperone-mediated autophagy (CMA), which functions with the help of co-chaperones. This chapter will focus on the role of Macroautophagy (hereafter referred to as autophagy) in tumor biology and cancer therapy (Mizushima and Komatsu 2011; Kaushik and Cuervo 2012).

Autophagy is a fundamental, evolutionarily conserved catabolic process in which cytoplasmic constituents are targeted for removal or turnover in autophagosomes that fuse with the lysosome (Mizushima et al. 2008). Autophagy is tightly regulated through multiple steps: Initial steps include (1) formation of a phagoro-phore (a double-membraned structure), (2) expansion of these nascent phagophores to form autophagosomes, (3) fusion with lysosomes to form autolysosomes, and (4) breakdown of engulfed cargo by lysosomal hydrolases-mediated degradation (Galluzzi et al. 2015). Autophagy plays a critical role in energy homeostasis and cel-lular fitness maintenance, in both normal and stressed conditions. Stress-responsive autophagy confers upon itself both pro- and anti-tumorigenic roles, depending on the cellular and environmental context (Liu and Ryan 2012).

There has been an explosion in the molecular study of autophagy in the past decade, since it was first genetically defined in yeast. The identification of 31 genes, referred to as autophagy-related (ATG) genes, led to the identification of orthologues in higher eukaryotes (He and Klionsky 2009; Yang and Klionsky 2009). A major subset of these genes essential for autophagosome formation, referred to as the "core" molecular machinery, can be divided into four subgroups: (1) The Atg1-Atg13-Atg17 complex: there are two mammalian homologs of Atg1 that function in autophagy—the Unc-51-like kinase 1 and 2 (ULK1 and ULK2), and one homolog of Atg17 known as FIP200 (FAK, family kinase-interacting protein of 200 kDa). (2) Two ubiquitin-like conjugation systems: the Atg12 and Atg8, the mammalian homolog of which is LC3. (3) The class III phosphatidylinositol 3-kinase (PtdIns3K)/Vps34 complex I. (4) Two transmembrane proteins, Atg9/mAtg9 (and associated proteins involved in its movement such as Atg18/WIPI-1) and VMP1 (Xie and Klionsky 2007; Yang and Klionsky 2010).

Here, we discuss the cross talk between autophagy and senescence in cancer. Senescence is defined as a permanent and irreversible cell cycle arrest shown to occur upon endogenous and exogenous stresses, such as oncogene induction, or other cellular stress, such as chemotherapies or radiotherapies (Munoz-Espin and Serrano 2014). Enhanced understanding of this cross talk in the context of various cell types and stimuli has important implications for the idea of targeting autophagy for improved patient outcomes.

4.2 REGULATION OF AUTOPHAGY

In this section, we will discuss the core molecular players involved in autophagy regulation, particularly the AMPK-mTOR-ULK1/2 axis as well as some other key factors involved in transcriptional, post-transcriptional, and post-translational regu-lation of autophagy. Figure 4.1 depicts the multiple signaling pathways involved in regulation of autophagy.

4.2.1 AMPK-mTOR-ULK1/2 and Autophagy

AMP-activated protein kinase (AMPK), a heterotrimeric protein complex, was originally identified as a serine/threonine kinase that negatively regulates several key enzymes involved in lipid metabolism (Hardie 2007). AMPK is regarded as the major cellular energy–sensing kinase and can be activated by metabolic stress or ATP consumption (Alers et al. 2012). It promotes a variety of catabolic processes such as glucose uptake, while simultaneously inhibiting several anabolic processes such as

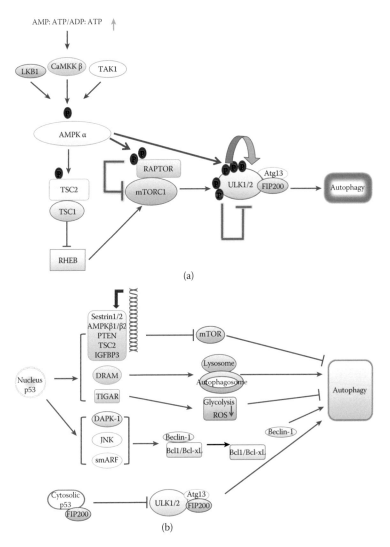

FIGURE 4.1 Multiple signaling networks involved in autophagic regulation. (a) The role of AMPK/mTORC/ULK axis in autophagy regulation. AMPK is an important energy sensor, which is phosphorylated by several upstream kinases such as LKB1, CaMKKβ, and TAK1 upon response to AMP:ATP/ADP:ATP reduction. Activated AMPK inhibits mTORC activity by phosphorylating substrates, leading to activation of ULK1/2 activity and autophagy. (b) Dual roles of p53 in autophagic regulation are associated with its subcellular localization. In the nucleus, p53 mainly functions as an autophagic inducer via transcriptional upregulation of pro-autophagic proteins such as DRAM, Sestrin1/2, AMPKβ1/β2, PTEN, TSC2, and IGFBP3. The p53-inducible protein TIGAR, which attenuates glycolysis and decreases ROS level, contributes to negative regulation of p53 in autophagy. p53 transactivates proteins such as JNK, DAPK-1, and smARF, which release Beclin-1 from Bcl-2/Bcl-xL sequestration, thereby promoting autophagy. In the cytoplasm, p53 interacts with FIP200, impairing formation of the positive autophagy regulator-ULK1/2/Atg13/FIP200 complex, thereby inhibiting autophagy. *(Continued)*

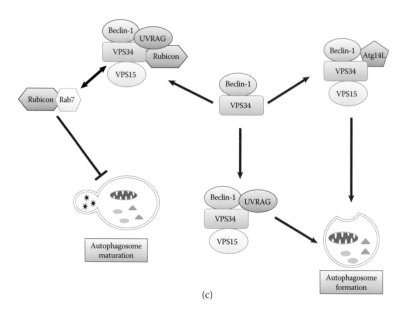

(c)

FIGURE 4.1 (Continued) Multiple signaling networks involved in autophagic regulation. (c) Distinct roles of multiple protein complexes containing the core Beclin-1/Vps34 complex. The core Beclin-1/Vps34 complex form different multiprotein associations with binding partners such as UVRAG, Atg14L, and Rubicon. Beclin-1/vps34/vps15/UVRAG and Beclin-1/vps34/vps15/ Atg14L complexes act as positive regulators of autophagy via autophagosome formation. Rubicon either forms the Beclin-1/Vps34/Vps15/UVRAG/Rubicon complex or interacts with Rab7 independent of Beclin-1, inhibiting autophagosome maturation.

lipid biosynthesis. The activity of AMPK is precisely regulated in different ways. Firstly, phosphorylation of a conserved threonine residue (Thr172) in the activation loop of the catalytic α-subunit by upstream kinases is a prerequisite for activation of AMPK. Several upstream kinases that phosphorylate AMPK have been identified, such as the ubiquitously expressed liver kinase B1 (LKB1) (Hawley et al. 2003; Woods et al. 2003). In addition to this constitutively active kinase, Ca²⁺/calmodulin-dependent kinase kinaseβ (CaMKKβ) and transforming growth factor β-activated kinase-1(TAK1) are also known as upstream activators of AMPK (Hawley et al. 2005; Herrero-Martin et al. 2009). Secondly, AMPK activity can be regulated by the ratio of AMP:ATP/ADP:ATP, which is modulated by allosteric binding to the other two regulatory subunits, β and γ subunits. As cellular energy status decreases, AMP or ADP can directly bind to the γ subunit, leading to a conformational change that promotes AMPK phosphorylation and prevents de-phosphorylation at Thr172. In addition, recent studies have shown that ADP binds to the nucleotide-binding pockets in the γ subunit and enhances AMPK phosphorylation at Thr172, although this does not allosterically activate AMPK (Xiao et al. 2011; Alers et al. 2012). Many types of cellular stresses can lead to AMPK activation such as low nutrients or pharmacological agents that can alter the cellular AMP:ATP/ADP:ATP ratio, which reciprocally promotes global catabolic processes to satisfy energy demand (Mihaylova and Shaw 2011).

Mammalian target of rapamycin (mTOR), a downstream target of AMPK, is an evolutionarily conserved serine/threonine protein kinase and major central regulator of cellular metabolism. It is the major negative regulator axis of autophagy. It forms two distinct signaling complexes, mTOR complex1 (mTORC1) and mTOR complex2 (mTORC2), by binding with multiple companion proteins. mTORC1 integrates various stimuli and signaling networks and functions as an intracellular nutrient sensor to control protein synthesis, metabolism, and cell growth. It stimulates synthesis of protein, lipid, and nucleotides and blocks catabolic processes such as autophagy at different levels (Wullschleger et al. 2006; Kim and Guan 2015). The tuberous sclerosis tumor suppressor complex (TSC1/TSC2) is the most important upstream negative regulator of mTORC1. A well-established upstream regulator of mTORC1 is the growth factor PI3K/AKT signaling pathway. Upon growth factor binding, AKT is recruited to the plasma membrane and activated through phosphorylation by PDK1. Activated AKT in turn phosphorylates TSC2, which prevents the formation of the inhibitory TSC1/TSC2 heterodimer. Since TSC2 is a GTPase-activating protein (GAP) and can inactivate the small GTPase Rheb, its inhibition will subsequently allow Rheb to activate mTORC1 and inhibit autophagy (Alers et al. 2012; Zhang et al. 2013; Feng et al. 2015). Hence, growth factor withdrawal can culminate in mTORC1 inactivation and induction of autophagy. Furthermore, when energy status decreases (higher ratio of AMP:ATP/ADP:ATP), AMPK is activated and phosphorylates TSC2 at Ser1387 and RAPTOR (regulatory-associated protein of mTOR) at Ser792, which also leads to mTORC1 inhibition. After inhibition of mTORC1 activity, phosphorylation of a range of substrates including most characterized S6K (ribosomal S6 kinase), 4EBP (eIF4E-binding protein 1), and ULK1 (UNC-51-like kinase 1) will decrease (Mihaylova and Shaw 2011; Kim and Guan 2015).

ULK1 and ULK2 are two mammalian homologs of Atg1. ULK1 was initially thought to have a primary role in autophagic induction, since knockdown of ULK1 was able to inhibit autophagy in several cell lines (Chan et al. 2007). However, different from the knockout mouse model of other essential autophagy-related genes, ULK1-/- and ULK2-/- mice were born viable and tolerant to brief starvation after birth (Kundu et al. 2008). Mice with ULK1-/- deficiency exhibit a compromised clearance of mitochondria during reticulocyte maturation, referred to as mitophagy, suggesting a specific role for ULK1 in mitophagy (Kundu et al. 2008). There is also evidence showing that ULK1-/- mouse embryonic fibroblasts (MEFs) display normal LC3 lipidation in response to glucose starvation, although they do not respond to other autophagic inducers such as rapamycin treatment (Lee and Tournier 2011). Collectively, these results indicate that ULK1 and ULK2 may have partially redundant functions in autophagic induction. ULK1/2 form a large complex with Atg13 and FIP200 in mammalian cells, but the binding affinity between these components is rarely affected by the nutrient status (Chang and Neufeld 2009; Ganley et al. 2009). The phosphorylation status within ULK1/2-Atg13-FIP200 complex can dramatically change according to nutrient status. Under normal conditions, mTORC1 associates with ULK1/2-Atg13-FIP200 complex and phosphorylates Atg13 and ULK1/2, thereby suppressing ULK1/2 activity. On the contrary, in the absence of sufficient energy, mTORC1 activity is inhibited and these phosphorylated sites are rapidly dephosphorylated, thereby promoting ULK1/2 activation and translocation of the entire complex to the pre-autophagosomal membrane, inducing autophagy (Hosokawa et al. 2009; Jung et al. 2009).

In addition to regulation of ULK1/2 complex activity by mTOR, AMPK contributes to autophagy induction in a two-pronged manner, that is, directly activating ULK1 as well as inhibiting the suppressive effect of mTORC1 on ULK1. Upon nutrient starvation, AMPK directly phosphorylates ULK1 at Ser317 and Ser777, leading to ULK1 activation. AMPK also triggers autophagy by suppressing mTOR activity. First, AMPK phosphorylates and activates TSC2, which in turn inhibits RHEB. AMPK also directly phosphorylates the mTOR binding partner RAPTOR at Ser722 and Ser792, reducing mTOR kinase activity. Phosphorylation and inactivation of RAPTOR involves binding to a YWHA/14-3-3 (tyrosine 3-monooxygenase/tryptophan 5-monooxygenase activation protein) protein (He and Klionsky 2009; Mihaylova and Shaw 2011; Alers et al. 2012). Collectively, the AMPK–mTOR system regulates ULK1 kinase activity to fine-tune the autophagy process.

4.2.2 P53 AND AUTOPHAGY

p53 is an extensively studied tumor-suppressive protein having various roles in the cell cycle, cell proliferation, and apoptosis. About 50% of human cancers harbor p53 deficiency or mutation (Soussi and Lozano 2005). Similar to the relationship between cancer and autophagy, p53 also has contrasting roles in the regulation of autophagy. Here, we focus on the role of p53 in autophagic regulation, which depends on its subcellular localization and transcription/transcription-independent mechanism. Under normal conditions, the intracellular levels of p53 are tightly regulated by MDM2-involved ubiquitination and proteasome-mediated degradation. The p53 expression level or activity is rapidly upregulated and acts as an anticancer barrier in response to diverse stimuli, including genotoxic damage, oncogene activation, and hypoxia (Soussi 2007; Zilfou and Lowe 2009).

In the nucleus, stress-activated p53 mainly plays a pro-autophagic role in a transcriptional-dependent manner (Maiuri et al. 2010; Tang et al. 2015). Several mechanisms have been proposed that explain the role of p53 in the modulation of mTOR signaling cascades and autophagic stimulation. The products of two p53 target genes, Sestrin1 and Sestrin2, have been shown to activate AMPK for phosphorylation of TSC2, thereby stimulating its GAP activity and shutting down mTOR activity (Budanov and Karin 2008). p53 also activates the β1 and β2 subunits of AMPK, PTEN, TSC2, and IGFBP3, which functionally antagonize the autophagy-suppressive role of mTOR (Buckbinder et al. 1995; Feng et al. 2007; Feng 2010).

p53 also regulates other genes that regulate autophagy. (DRAM1 damage-regulated autophagy modulator 1), a p53 target gene encoding a lysosomal protein, was shown to contribute to autophagosome accumulation by modulating autophagosome–lysosome fusion in response to adverse stress (Crighton et al. 2006).

(DAPK1 death-associated protein kinase 1), which has tumor-suppressive functions and is upregulated by activated p53, is a positive regulator of autophagy. DAPK1 binds to and suppresses the LC3-interacting microtubule-associated protein MAP1B. In a addition, DAPK1 also phosphorylates the BH3 domain of Beclin 1, an essential modulator of autophagy, resulting in the release of Beclin 1 from Bcl-2/Bcl-xL-mediated sequestration, thereby facilitating autophagy (Gozuacik and Kimchi 2006; Harrison et al. 2008; Zalckvar et al. 2009).

Finally, p53 can contribute to autophagy induction partially via c-Jun N-terminal kinase (JNK) activation, which promotes autophagy via various mechanisms. JNK phosphorylates Bcl-2 in its N-terminal loop, thus releasing Beclin 1 and inducing autophagy (Wei et al. 2008). JNK also facilitates DRAM upregulation and functions as an autophagic inducer (Lorin et al. 2009, 2010). Additionally, JNK can mediate expression of several autophagy-related genes such as Beclin 1, Atg5, Atg7, and p62 (Park et al. 2009; Lorin et al. 2010).

As mentioned above, not all p53-targeted genes have pro-autophagic roles. Accumulating evidence has shown that knockout, knockdown, or pharmacological inhibition of p53 can induce autophagy in human, mouse, and nematode cells under conditions of hypoxia and nutrient depletion, leading to survival of p53-deficient cancer cells (Rosenfeldt and Ryan 2009). p53 is usually present in the cytosol and translocates to the nucleus under stressed conditions following its phosphorylation by a number of distinct stress-activated kinases. This leads to a decrease in the cytosolic p53 pool (Tasdemir et al. 2008). Cytosolic p53 represses autophagy by interacting with the autophagy protein FIP200, thus blocking the activation of the ULK1–FIP200–ATG13 complex and inhibiting autophagosome formation (Morselli et al. 2011). Under cellular stress, a portion of cytosolic p53 can translocate to the mitochondrial matrix, where it interacts with cyclophin D and promotes opening of the permeability transition pore (PTP; a multiprotein pore that permeabilizes the inner mitochondrial membrane), thereby stimulating the autophagic elimination of dysfunctional mitochondria (Vaseva et al. 2012).

Cytosolic p53 also displayed anti-autophagic function via TP53-induced glycolysis and apoptosis regulator (TIGAR). This lowers fructose-2,6-bisphosphate levels in cells, leading to glycolysis inhibition and an overall decline in intracellular reactive oxygen species (ROS) levels (Bensaad et al. 2006; Green and Chipuk 2006; Kimata et al. 2010; Hoshino et al. 2012).

Recent literature points to an intriguing correlation between p53 status and autophagic suppression in a mouse model of pancreatic tumors driven by Kras mutations (Rosenfeldt et al. 2013). When p53 is intact, deletion of key autophagy genes inhibits progression of low-grade tumors to aggressive carcinoma. However, in the absence of p53, autophagy inhibition accelerates tumorigenesis with concomitant metabolism pathway reprogramming. This finding has important implications for targeting autophagy in cancers.

Taken together, it is clear that similar to the complex function of autophagy in cancer, the regulatory role of p53 in autophagy is context dependent. It is noteworthy that deciphering whether p53 has a pro- or anti-autophagic role in a given setting should be taken into consideration during treatment.

4.2.3 BECLIN 1 AND AUTOPHAGY

Much of the protein–protein interaction and multiprotein complexes that function as key regulators of autophagy were first identified in yeast and subsequent work expanded to mammalian cells. One primary multiprotein complex identified in mammals is the Beclin 1–VPS34 (the class III phosphatidylinositol 3-kinase) core autophagy complex which resembles the yeast Atg6–Vps34 complex that plays an important role in vesicle nucleation in autophagy (Yang and Klionsky 2009; Funderburk et al. 2010). Beclin 1 ("Bcl-2 interacting protein"), the first mammalian autophagy protein to be described, was originally studied due to its 24.4% amino acid sequence identity to the yeast autophagy-related protein Atg6. Indeed, it was found to restore autophagic activity in Atg6-disrupted yeast (Liang et al. 1999). It is a 60 kD coiled-coil protein consisting of a Bcl-2-homology-3 (BH3) domain, a central coiled-coil domain (CCD), and an evolutionarily conserved domain (ECD) (Funderburk et al. 2010; Wesselborg and Stork 2015). Notably, Beclin 1 has been shown to be a haplo-insufficient tumor suppressor (Funderburk et al. 2010).

Several new Beclin 1-binding partners and multiprotein complex containing Beclin 1–VPS34 that showed distinct functions in autophagy regulation were subsequently identified. Affinity-purification of Beclin 1 coupled with sequence-homology searches revealed three important binding partners of Beclin 1: (1) UVRAG (UV irradiation resistance-associated gene), close homolog of yeast Vps38; (2) Rubicon (RUN domain, a cysteine-rich domain containing Beclin 1-interacting protein); and (3) Atg14L, a putative mammalian homolog for yeast Atg14 (Itakura et al. 2008; Liang et al. 2008). These proteins appear to form a stable complex with the core Beclin 1-Vps34 complex, in contrast to previously identified proteins that formed unstable and transient association with the core complex.

The UVRAG-containing complex is more stable and abundant compared with Atg14L or Rubicon under normal conditions. UVRAG, similar to Beclin 1, is a tumor suppressor and is often monoallelically deleted in human colon cancer cells and tissues. It forms an α-helical heterodimer with the CCD of Beclin 1 via its own central CCD. Under normal conditions, UVRAG present in Beclin 1–UVRAG–VPS34–VPS15 complex induces autophagosome formation. Despite its role in autophagy initiation, UVRAG has also been described to act in a separate complex independent of Beclin 1 to promote autophagosome maturation and endocytic trafficking. UVRAG was shown to associate and recruit class C Vps (C-Vps) complex to the autophagosome, stimulating Rab GTPase activity required for autophagosome maturation.

Atg14L, with its CCD needed for binding to the CCD regions of Beclin 1 and VPS34, generally exists in a Beclin 1–Atg14L–VPS34–VPS15 complex essential for autophagosome formation. Its intracellular expression level is largely dependent on Beclin 1 (Matsunaga et al. 2009; Ma et al. 2014a).

Rubicon is shown to exist in a Beclin 1 complex excluding Atg14L but is always found in the same complex with UVRAG. In contrast to the role of Atg14L, Rubicon is known as a negative autophagic regulator. For example, overexpression of Rubicon caused an obvious accumulation of immature autophagosomes, indicating that Rubicon could hinder autophagosome maturation. Furthermore, Rubicon could decrease VPS34 kinase activity in a Beclin 1-independent manner (Matsunaga et al. 2009; Sun et al. 2010).

4.3 AUTOPHAGY IN TUMORIGENESIS

Autophagy has been shown to be pro-tumorigenic or tumor suppressive depending on the stimulus and the cell type. The effects of autophagy vary according to the stage of tumor development and exhibits intratumoral heterogeneity. Figure 4.2 depicts the dual roles of autophagy in tumorigenesis.

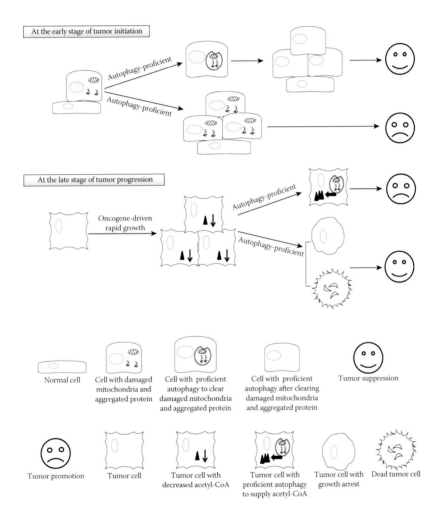

FIGURE 4.2 Dual role of autophagy in tumor progression. At the early stage of tumor initiation, autophagy functions in cellular homeostasis by degrading damaged organelles such as mitochondria or preventing aggregated protein accumulation, thereby suppressing tumor progression. Once tumors are formed, cells grow rapidly and soon undergo insufficient energy supply. Shown is an example of decrease in acetyl-CoA levels, the key intermediate for the TCA cycle. Autophagy is able to replenish these metabolically stressed tumor cells with amino acids and fatty acid substrates by recycling cellular constituents, thus compensating for lack of energy and further supporting proliferation.

On the basis of current understanding, it is generally thought that autophagy tends to limit tumor initiation, but once tumorigenesis is initiated, autophagy promotes establishment and progression of the tumor (Liu and Ryan 2012). During initial stages of tumor progression, autophagy generally acts as a "safe guardian" to prevent cells from evolving to malignant cells by clearing damaged organelles or aggregated proteins. Mice tissues with defective autophagy show large accumulation of damaged and dysfunctional mitochondria, and aggregated toxic proteins, leading to increased ROS production and ER stress (Kouroku et al. 2007; Dewaele et al. 2010). In addition, the mammalian gene encoding Beclin 1 has been shown to be monoallelically deleted in 40–75% sporadic breast cancers and ovarian cancers (Qu et al. 2003). However, once the tumor has already formed, autophagy plays a protumorigenic role. Here, tumor cells rely on autophagy for cell survival and metabolic reprogramming for rapid cell growth and proliferation. Recycling of cellular macromolecules provides cells with and energy and biosynthetic precursors including sugars, amino acids, fatty acids, and nucleotides. Indeed, cancer cells can induce autophagy as a means of adaptation to the harsh tumor microenvironment characterized by nutrient deprivation, hypoxia, lack of growth factors, and metabolic stress (White 2015).

Evidence from solid tumors shows that defects in autophagy can lead to decrease in cell growth as well as tumor regression. One such example is the constitutive activation of autophagy in Kras-driven pancreatic tumors. Inhibition of autophagy by pharmacological and genetic means was found to result in robust tumor regression, indicating that oncogene-induced autophagy is an indispensable step in tumor development of pancreatic cancers (Rosenfeldt et al. 2013). This was accompanied by alteration in metabolism, decrease in OCR and ATP production, as well as increase in glucose uptake (Yang et al. 2011). The link between autophagy and metabolism regulation points to the utility of autophagy-recycled substrates being able to support cell mitochondrial metabolism. RAS is one of the most powerful oncogenes that can drive tumors as well as simultaneously diminish the cellular pool of acetyl-CoA and retard mitochondrial metabolism through three major ways: (1) RAS activates lactate dehydrogenase (LDH) that converts pyruvate to lactate, which is subsequently excreted instead of being converted to acetyl-CoA. (2) RAS activates hypoxia-inducible factor (HIF), thus inhibiting pyruvate dehydrogenase (PDH) and the conversion of pyruvate to acetyl-CoA. (3) RAS inhibits LKB1 activity, followed by blocking of AMPK activity and β-oxidation, thus shutting down another source of acetyl-CoA. Autophagy is involved in metabolic reprogramming induced by RAS, as the recycled cellular constituents can provide cells with amino acid and fatty acid substrates to produce acetyl-CoA, essential for the tricarboxylic acid (TCA) cycle which fuels rapid proliferation of tumor cells (White 2012).

Besides its crucial role in supporting tumor survival under basal condition, autophagy has also shown important outcomes in response to different anticancer therapies. Autophagy has also been associated with chemoresistance via its role in the adaptive response, resulting in increased cell survival (Turcotte and Giaccia 2010). The combination of autophagy inhibitors with chemotherapeutic agents have been shown to inhibit stress adaptation, leading to cell death. Indeed, several clinical trials involving autophagy inhibitors such as hydroxychloroquine (HCQ) or CQ are currently being conducted (Turcotte and Giaccia 2010; Levy and Thorburn 2011) (http://clinicaltrials.gov). Autophagy has also been shown to increase the effectiveness of anticancer radiotherapies (Hou et al. 2013; Sui et al. 2013; Parkhitko et al. 2014; Galluzzi et al. 2015).

For example, in melanomas harboring BRAF mutation, particularly the BRAF[V600E], the inhibitors vemurafenib (also named PLX4032) and dabrafenib targeting BRAF[V600E] mutation often show encouraging outcomes at the beginning of clinical use. However, nearly all patients eventually show relapse. Intensive efforts to investigate mechanisms leading to BRAF inhibitor resistance revealed involvement of reactivation of the mitogen-activated protein kinase (MAPK) kinase pathways or concurrent activation of alternative growth factor signaling pathways. Recent studies have shown that targeting ER-stress-induced autophagy helps to overcome BRAF inhibitor resistance in melanoma, which provides a potential combinatorial use of BRAF inhibitor and autophagy inhibitor as treatments in patients that show relapse (Qin et al. 2010; Ma et al. 2014b).

Antimitotic drugs such as paclitaxel and vincristine have been used as frontline therapies in various types of cancers for several decades (Manchado et al. 2012). However, the role of autophagy in response to antimitotic drugs has garnered interest only recently. Thus far, experimental conclusions are controversial due to different cell contexts and experimental schemes. The Goping group recently demonstrated that paclitaxel inhibits autophagy through two distinct mechanisms in mitotic and nonmitotic cells, leading to apoptosis induced by drug treatment (Veldhoen et al. 2013). In mitotic cells, paclitaxel blocked activation of the class III phosphatidylinositol 3-kinase (Vps34), inhibiting autophagosome formation. In nonmitotic cells, autophagosomes were formed but their movement and maturation was blocked by paclitaxel treatment. Genetic or chemical inhibition of autophagy diminished paclitaxel-induced cell death. In line with their *in vitro* observation, primary breast tumors with lower expression of autophagy-initiating genes were resistant to taxane therapy (Veldhoen et al. 2013). On the contrary, the Liao group showed that the role of autophagy shifts to a cytoprotective response to paclitaxel in cervical cancer cells (Peng et al. 2014). Their established Hela-derived paclitaxel-resistant cell lines (Hela-R) showed increased autophagy levels in Hela-R compared to parental Hela. Genetic and chemical inhibition of autophagy could restore sensitivity of Hela-R to paclitaxel (Peng et al. 2014). The following could account for the opposite roles of autophagy in response to paclitaxel observed by the two groups: First, both groups used different cell lines from one another, breast cancer cell line MCF-7 by the Goping group and cervical cell line Hela by the Liao group. Second, their experimental schemes differ. For the first group, paclitaxel was used to transiently treat cells and autophagy-related markers detected at indicated time points. However, the second group used resistant cell lines that acquired resistance to paclitaxel following continuous exposure to low-dose drugs that were used.

At the other end of the spectrum, several lines of evidence point to a link between a defective autophagic process and cancer development. Autophagy genes are frequently mutated in human cancers (Liu and Ryan 2012). For example, mutations in ATG2B, ATG5, ATG9B, and ATG12 have been reported in gastric and colorectal cancers (Kang et al. 2009). Additionally, ample evidence obtained from murine models and patient tumors show that defects in Beclin 1 can accelerate tumorigenesis (Green et al. 2011; Guo et al. 2013). For instance, mice bearing heterogeneous depletion of Beclin 1 spontaneously develop various malignancies. The Beclin 1 gene has been found to be deleted in about 40% of human prostate cancers, 50% of breast cancers, and 75% of ovarian cancers. Additionally, reduced Beclin 1 expression has been found in other cancers such as human brain, liver, cervical, and colon cancers (Qu et al. 2003; Gozuacik and Kimchi 2004).

Taken together, autophagy has a complex and paradoxical role in tumorigenesis and cancer therapeutics. Understanding its role in different contexts will aid patient treatment outcomes.

4.4 CELLULAR SENESCENCE, MOLECULAR SIGNALING, AND CANCER

Cellular senescence has emerged as one of the most attractive fields of study for biologists because of its diverse role in multiple physiological processes such as aging, tumor progression, and tissue repair. Here, we focus on the "double-edged sword" role of senescence in the "war" against cancer.

Cellular senescence, a stable cell cycle arrest, reflects a program that limits the proliferative capacity of the cell exposed to endogenous or exogenous stress signals. Senescence was originally identified in the context of "replicative senescence," first described by Hayflick and Moorhead, and is triggered by two main factors: (1) telomere shortening and uncapping, followed by activation of the DNA damage response and (2) de-repression of the INK4/ARF (inhibitor of CDK4A/alternative reading frame) locus, which is associated with activation of key tumor suppressor networks (Kim and Sharpless 2006; Campisi 2013).

Oncogene-induced senescence (OIS), first observed in normal cells by expression of oncogenic Ras^{G12V}, has emerged as a barrier for tumor progression. Several groups have independently identified the phenomenon of OIS in premalignant mouse and human lesions, along with their conspicuous absence in advanced-stage tumors. In addition to RAS, other oncogenes are also involved in senescent cells. Tissue-specific expression of BRAF^{V600E} mutant results in lung adenomas or melanocytic nevi with increased senescent cells. C-myc, usually linked to apoptosis or proliferation, can also trigger senescence in murine lymphomas via macrophage-mediated TGF-b secretion. Furthermore, disabling of this senescence response is needed for tumor progression. For example, melanocytic nevi induced by oncogenic BRAF mutant expression may take months or even years to spontaneously evolve to melanomas. This process is markedly accelerated upon inactivation of p53 or Cdkn2a, which can impair senescence (Acosta and Gil 2012; Cisowski et al. 2015).

Anticancer therapies mainly exert their effects via extensive DNA damage, resulting in gross aneuploidies, mitotic catastrophe, and apoptosis. Interestingly, senescence has also been identified in cancer cells treated by chemotherapies or radiotherapies, termed as therapy-induced senescence (TIS). TIS is able to play a tumor-suppressive role by inducing stable growth arrest and can be used as an alternative therapy for cancer treatment. Concomitant with this arrest, senescent cells also secrete a complex mixture of factors referred to as senescence-associated secretory phenotype (SASP) (Acosta and Gil 2012; Perez-Mancera et al. 2014). A plethora of studies have shown that senescent cells can exert a pro-tumorigenic role by affecting the microenvironment and neighboring cells through the SASP. This can lead to increased migration, invasiveness, or metastasis.

A large number of senescence hallmarks have been characterized including enlarged and flattened morphology, positive senescence-associated β-galactosidase (SA-β-gal) staining, lack of proliferation markers such as BrdU and Ki67, activation of cell cycle arrest machinery such as p53/p21/p16, activation of DNA damage response (DDR), as well as SASP in a cell context–dependent fashion (Rodier and Campisi 2011).

Senescence is considered "irreversible" because no known physiological stimuli can promote senescent cells to re-enter the cell cycle, except for molecular biological manipulations such as inactivation of tumor suppressors, which can cause re-proliferation of senescent cells. The onset and maintenance of senescence is stringently regulated by at least two major signaling pathways, the p53/p21 and p16^{INK4a}/pRB pathways (Campisi 2013). There is a myriad of stimuli that can lead to senescence. Here, we highlight five major causes: telomere erosion, genomic damage, mitogens and proliferation-associated signals, epigenomic damage, and activation of tumor suppressors.

Human telomeres are specialized structures at the ends of linear chromosomes and consist of tandem double-stranded 5′-TTAGGG-3′ repeats and a single-stranded G-rich 3′ overhang that are capped and protected by proteins named shelterin, such as TRF2 and POT1, which can sequester telometric DNA and prevent it from being recognized as DNA damage (Hayashi et al. 2012). Telomeric DNA is gradually lost with each S phase because DNA polymerases are unidirectional and require a liable primer, but the ends of linear DNA molecules cannot satisfy this requirement and thus cannot be completely replicated, thus resulting in shortening with each cell division. Telomere erosion does not happen in cells with telomerase expression, the reverse transcriptase that can replenish the shortening telomeric DNA *de novo*. Most cells do not express telomerase, and introduction of telomerase expands the lifespan of normal cells, for example, overexpression of hTERT, the catalytic subunit of telomerase, can counteract telomere erosion, extend replicative potential, and is used in many cell models to prevent replicative senescence. Upon disruption of shelterin components, telomeres are recognized as DNA damage sites, leading to phosphorylation of Histone H2AX (γ-H2AX) and subsequently recruiting DNA repair proteins such as 53BP1, which can be observed as telomere dysfunction-induced foci (TIF) and intimately linked to senescence (Levy et al. 1992; Hayashi et al. 2012).

DNA damage, irrespective of genomic localization, is also able to promote senescence such as DNA double-stranded breaks (DSBs) induced by ionizing radiation, topoisomerase inhibitors, and other chemotherapies. Moreover, other DNA lesions such as DNA base damage or single-stranded breaks induced by oxidative stress or other agents can also drive cells into senescence, since these lesions can be converted to DSBs during DNA replication or DNA excision repair (Parrinello et al. 2003; Sedelnikova et al. 2010). p53 is activated downstream of the DNA damage checkpoint cascade in response to genomic damage or other stresses, which consist of several typical checkpoint kinase complexes such as DNA-dependent protein kinase (DNA-PK), ataxia telangiectasia mutated (ATM), rad-3 related (ATR), checkpoint kinase 1 (CHK1), checkpoint kinase 2 (CHK2), and MAPK-activated protein kinase 2 (MK2) that finally merge on activation of p53. Upon activation, p53 will transcriptionally induce a set of target genes that promote transient cell cycle arrest to allow enough time for DNA damage repair. Otherwise, persistent DNA damage will drive cells into senescence, an irreversible cell cycle arrest, or activate several apoptotic proteins such as the BH3-only protein PUMA (p53 upregulated modulator of apoptosis), the BH domain proteins Bcl-2-associated protein X (BAX), and Bcl-2 antagonist/killer (BAK), promoting cells to undergo apoptosis (Reinhardt and Schumacher 2012).

Cellular senescence can also be triggered by chronic, strong, or unbalanced mitogenic signals such as senescence induced by oncogenes. The first report about oncogene-induced senescence showed that activated H-RAS (H-RAS^{G12V}) could provoke senescence through chronic activation of the MAPK signaling pathway. Likewise, overexpression of growth factor receptors such as ErbB2, continuous proliferation stimulation by cytokines such as interferon-β, and loss of tumor suppressor PTEN that stimulates mitogenic signals can also induce senescence. Senescence converges on either or both of p53/p21 and p16^{INK4a}/pRB signaling pathways with various upstream regulators, downstream effectors, and modifying side branches (Campisi 2013; Macheret and Halazonetis 2015). A recently published study demonstrated that cross talk between Rb pathway and AKT signaling can act as a quiescence–senescence switch (Imai et al. 2014). Quiescence is the state of reversible cell cycle arrest required for tissue homeostasis in multicellular organisms with the Rb pathway also involved in quiescence regulation. According to the study, cross talk between the Rb-AKT signaling pathways exist by controlling the overlapping functions of FoxO3a and FoxM1 transcription factors in cultured fibroblasts. In the absence of mitogenic stimuli, although FoxM1 expression was repressed by the Rb pathway, FoxO3a prevented reactive oxygen species (ROS) production by maintaining SOD2 expression, leading to quiescence. However, when the Rb pathway was activated in the presence of mitogenic signals as well as FoxO3a inactivation by AKT, SOD2 expression was reduced, consequently allowing extensive ROS production. This situation induced senescence as a consequence of irreparable DNA damage (Imai et al. 2014).

As mentioned above, cellular senescence is mainly dependent on two major tumor-suppressive pathways: the p53/p21 and p16^{INK4a}/pRB pathways. In both humans and mice, most cancers harbor mutations in one or two of the pathways. Furthermore, defects in either pathways attenuated the ability to undergo senescence, leading to an increased incidence of cancer. In-depth studies from human tissues and mouse models strongly support the opinion that senescence can suppress cancers *in vivo*. Premalignant human nevi and colon adenomas contained cells that express senescence markers such as positive SA-β-gal staining and activation of DDR signaling. However, as these lesions gradually develop to become malignant melanomas and adenocarcinomas, senescent cells dramatically diminish. Likewise, in mouse models harboring oncogenic RAS or PTEN deletion, senescent cells were abundantly present in premalignant lesions, but became scarce at the fully developed cancer stage. In addition, some tumor cells still retain the ability to undergo senescence in response to conventional anticancer therapies, which can be regarded as an alternative approach for cancer intervention. For example, most of the currently used anticancer agents can lead to DNA damage and trigger senescence. These senescent cells can be subsequently cleared via the innate immune response, which partially increased the efficiency of anticancer therapies. As mentioned above, a host of oncogenic genes have the ability to drive senescence and this phenomenon is highly prevalent in many premalignant lesions at the early age of tumor development, which can be considered as a potential therapeutic target similar to agents targeting tumor-specific mutations. Although it is known that activated oncogenes can induce senescence, the mechanisms determining this outcome are hardly known and warrants further investigation (Campisi 2013; Perez-Mancera et al. 2014). According to work from Amati's group, Cdk2 (cyclin-dependent kinase 2) plays a critical role in suppressing Myc-induced senescence and was considered a potential therapeutic target in Myc-driven tumors (Campaner et al. 2010). In addition, the Schmitt group has shown that metabolic pathway alteration in senescent cells could be used for combinatorial cancer therapy (Dorr et al. 2013). They established both therapy-induced senescent (TIS)-competent and incompetent transgenic mouse lymphoma models and treated them with the DNA-damaging drug cyclophosphamide (CTX), broadly used in the treatment of lymphomas and leukemia. TIS-competent cells displayed higher glucose avidity and uptake of glucose, higher glycolysis, as well as accelerated oxygen consuming rate (OCR), which made them more vulnerable to inhibition of energy-generating catabolic pathways. These metabolic differences have provided a way to combinatorially treat TIS cells through synthetic lethal targeting of metabolic pathways (Dorr et al. 2013).

4.5 AUTOPHAGY-SENESCENCE CROSS TALK IN CANCER: CLINICAL RELEVANCE

In this section, we explore the emerging cross talk between autophagy and cellular senescence. As described in the previous section, senescence is an extremely stable form of cell cycle arrest activated in response to stress. Likewise, autophagy, a catabolic pathway that targets cellular organelles and cytoplasmic constituents to the lysosomes for degradation, can also be triggered by cellular stress. Both autophagy and senescence show either cytoprotective or cytotoxic response to stress.

A study by Young et al. (2009) first posited that autophagy can contribute to establishment of senescence and modulate functional activity of senescent cells (Young et al. 2009). The authors observed autophagosome accumulation during oncogene-induced senescence and that overexpression of ULK3 induced both autophagy and senescence. Furthermore, autophagy inhibition was shown to delay senescence and accumulation of SASP. Another study demonstrated that ASPP2 (N-terminal Apoptosis-stimulating of p53 protein 2), which inhibits autophagy, is able to dictate cell fate response to RAS toward either proliferation or senescence by modulating autophagy activity (Wang et al. 2012).

Senescence has also been shown to regulate autophagy mainly through the autocrine or paracrine effects of SASP, which consist of various cytokines, chemokines, and growth factors. Senescent cells are metabolically active and produce large amount of proteins such as the SASP components. The error-prone production and processing of SASP components may result in proteotoxic stress, leading to autophagic activation (Young and Narita 2009; Dorr et al. 2013). Certain components of SASP were found to regulate autophagy in neighboring cells such as CSF-2, which modulate the bystander effect of irradiation-induced senescent cells via inhibition of autophagy (Huang et al. 2014).

The relationship between autophagy and senescence has also been studied in the chemotherapeutic context. Studies using pseudolaric-acid B (PAB) show that ROS-mediated autophagic induction could lead to senescence, as a means of cell survival or resistance to PAB (Yu et al. 2016). Additionally, the Gerwitz group showed that inhibition of autophagy only transiently repressed senescence and did not abrogate it in MCF-7 breast tumor and HCT116 colon cancer cell lines treated with doxorubicin and camptothecin (Goehe et al. 2012). Their studies suggest that autophagy and senescence are related but senescence is not directly dependent on autophagy. Another study describing the autophagy–senescence transition from the Lisanti group showed that in hTERT-BJ telomerase-immortalized human foreskin fibroblasts stably expressing BNIP3, CTSB, or ATG16L1, constitutive autophagic expression also showed features of senescence (Capparelli et al. 2012).

In summary, autophagy and senescence can be reciprocally regulated and play multifaceted roles in health and disease, calling for further in-depth investigation. It remains to be seen whether autophagy and senescence in therapy occur in parallel, independent pathways or through interlinked pathways. Understanding this cross talk better will provide insights into the cytotoxic and cytoprotective functions of autophagy and senescence, important considerations for treatment.

4.6 METHODOLOGY

Since a large body of evidence has demonstrated vast and critical roles of autophagy and senescence in diverse biological processes and diseases, evaluation of autophagy and senescence status *in vitro* or *vivo* systems is critical.

In 2008, Daniel J. Klionsky, an expert in the field of autophagy, gathered thousands of researchers studying autophagy worldwide together to publish guidelines for standardizing the research in autophagy. This consortium compiled a large database of typical and relatively new cutting-edge methods to monitor autophagy in different systems, simultaneously pointing out the merits and demerits of each method. These guidelines are updated every few years based on newly developed technologies and knowledge to aid in an accurate study of autophagy (Klionsky et al. 2008, 2012; Mizushima et al. 2010; Gump and Thorburn 2014). Notably, it is generally recommended that scientists use a combination of various assays to evaluate autophagic flux.

Methods for detection of senescence are relatively invariable depending on the definition and major hallmarks of senescent cells, although markers to be detected are dependent on cell context and stress condition (Cho et al. 2004; Coppe et al. 2010; Carnero 2013; Itahana et al. 2013).

Here, we discuss some classical and commonly used methods utilized to detect autophagy and senescence level in mammalian cells.

4.6.1 EVALUATION OF AUTOPHAGY

1. LC3 (microtubule-associated protein 1 light chain 3) detection and quantification

   LC3 proteins are the mammalian homologs of yeast Atg8 and consist of eight family members, which are LC3A, LC3B, LC3B2, LC3C, GABARAP, GABARAPL1 (GABA$_A$ receptor–associated protein like 1), GABARAPL2 (GABA$_A$ receptor–associated protein like 2), and GABARAPL3 (GABA$_A$ receptor–associated protein like 3). Here, we only discuss detection of LC3B according to its abundance and major role in autophagy induction (hereafter referred to as LC3). LC3-I, the nascent form of LC3, is initially synthesized in an unprocessed form and mainly localized evenly across the cytoplasm. Upon autophagy induction, LC3-I is proteolytically processed, leading to a lack of C-terminal amino acids, and then conjugated to PE (phosphatidyl-ethanolamine), finally resulting in the formation of PE-modified LC3 and LC3-II. LC3-II is inserted to the mature autophagosome and can be used as a reliable marker of autophagosome, followed by hydrolase-mediated degradation of autolysosome. In mammalian cells, we usually use the conversion of LC3-I to LC3-II as a gold standard marker for autophagy induction, but the conclusion should be based on careful interpretation, as there are two possible causes of LC3-II accumulation. On the one hand, it makes sense that an increase of LC3-II is due to elevated autophagy, thus leading to increased autophagosome formation. However, there is a caveat: accumulation of LC3-II can also be observed as a consequence of interrupted autophogosome–lysosome fusion or autolysosome degradative activities. Hence, one cannot conclude autophagic induction based solely on increase of LC3-II levels. It is recommended that one also uses a combination of autophagic inhibitors such as Bafilomycin A1 to block autolysosome activity and/or autophagosome–lysosome fusion step to confirm accumulation of LC3-II is mainly derived from increased autophagosome formation.

   In addition, GFP-LC3 puncta formation is considered another powerful tool to analyze autophagy via fluorescence microscope-based observation and quantitation. GFP-LC3 I is distributed evenly inside the cell. Upon autophagic induction, LC3 I is converted to LC3 II and inserted into autophagosomes, thus leading to GFP "dots" formation, known as GFP puncta. Likewise, it is also important to carefully clarify the reasons behind these accumulated GFP puncta as per the Western blotting method for LC3 II described above.

2. SQSTM1/P62 and other LC3 binding proteins detection

   SQSTM1/P62 serves as an adaptor linking LC3 and ubiquitinated substrates that are subsequently delivered to the autophagosomes for lysosomal degradation and is thus considered a specific indicator of autophagic degradation. In addition to p62, there are several other adaptor proteins such as NBR1 (Neighbor of Brca1 gene), an adaptor protein in the Wnt signaling pathway, dishevelled (Dvl), which interacts with LC3 through their LIR(LC3 interacting region) and is then degraded by autophagy. However, adaptor proteins such as p62 are not only associated with the autophagy pathway but are also regulated at different layers. For example, p62 was shown to be transcriptionally upregulated in response to autophagic induction (Nakaso et al. 2004; He and Klionsky 2009). As such, its degradation in protein levels should not be the only reliable maker for evaluating autophagy.

3. Transmission electron microscope

The phenomenon of autophagy was first observed by transmission electron microscope (TEM) since more than 50 years ago and was termed autophagy by Christian de Duve in the 1960s. TEM visualization is believed to be one of the most reliable methods to evaluate autophagic status which depends on specific vesicular structures during autophagic processing. Autophagosomes, expanded from sequestered phagophores, have double membranes. The cytoplasm within is not degraded and has the same electron density as the cytoplasm outside the autophagosome. At times, only one electron-dense membrane is present or the autophagosomes seem to have no limiting membrane at all. In contrast, autolysosomes have only one layer membrane containing cytosol and/or organelles at different stages of degradation. Since autophagy is a dynamic process and undergoes formation of autophagosome and autolysosomes subsequently, it is recommended that researchers consider the overall formation of autophagosomes and autolysosomes to monitor autophagy.

4.6.2 EVALUATION OF SENESCENCE

1. Morphological alteration in senescent cells

Morphological changes are direct evidence of senescence both at the cellular and organismal level. Senescent cells usually become flattened and enlarged. This is relative to the expression level and activities of several cell structure determinants include integrin, focal adhesion complexes, and small Rho GTPases.

2. Senescence-associated β-galactosidase staining assay

In 1995, senescence-associated β-galactosidase (SA-β-gal) was initially identified as a biomarker of senescent cells in cells and in aging skin *in vivo*, and was conspicuously absent in pre-senescent, quiescent, or terminally differentiated cells (Dimri et al. 1995). Senescent cells express β-galactosidase which can be detected by histochemical staining cells with the artificial substrate X-gal at pH 6.0, forming a local blue precipitate upon cleavage of the substrate. To date, there have been thousands of publications which have cited this method as one of the more typical senescence markers because of its simplicity and reliability.

3. Lack of proliferation markers and activation of cell cycle arrest machinery

Due to its permanent cell cycle arrest feature, senescent cells are usually accompanied by lack of proliferation markers such as BrdU or Ki67, and accumulation of cell cycle arrest regulators such as p53 and its transcriptional target p21, the cyclin-dependent kinase (CDK) inhibitor. In addition, nonphosphorylated pRB is stabilized and binds to E2F factors, which lead to inactivation of E2F and inhibition of DNA replication. Thus, besides detection of p53 and p21 by Western blotting, nonphosphorylated pRB also serves as an additional marker.

4. Senescence-associated secretory phenotype

The SASP represents the property of senescent cells in that they secrete a plethora of soluble/insoluble factors that execute autocrine or paracrine functions, originally proposed by the Campisi group (Coppe et al. 2010). SASP components can be divided into several subsets including signaling factors (interleukins, chemokines, and growth factors), secreted proteases, and secreted insoluble factors/extracellular matrix (ECM) components. Through the SASP, senescent cells possess a tumor-suppressive role by preventing genomic instability or facilitate tumor progression by affecting the tumor microenvironment or neighboring cells. ELISA and SILAC-based quantitative proteomics are the most common methods utilized for detection of SASP factors (Acosta et al. 2013; Rodier 2013).

REFERENCES

Acosta, J. C., and J. Gil. 2012. Senescence: A new weapon for cancer therapy. *Trends Cell Biol* 22 (4):211–19.

Acosta, J. C., A. P. Snijders, and J. Gil. 2013. Unbiased characterization of the senescence-associated secretome using SILAC-based quantitative proteomics. *Methods Mol Biol* 965:175–84.

Alers, S., A. S. Loffler, S. Wesselborg, and B. Stork. 2012. Role of AMPK-mTOR-Ulk1/2 in the regulation of autophagy: Cross talk, shortcuts, and feedbacks. *Mol Cell Biol* 32 (1):2–11.

Bensaad, K., A. Tsuruta, M. A. Selak, et al. 2006. TIGAR, a p53-inducible regulator of glycolysis and apoptosis. *Cell* 126 (1):107–20.

Buckbinder, L., R. Talbott, S. Velasco-Miguel, et al. 1995. Induction of the growth inhibitor IGF-binding protein 3 by p53. *Nature* 377 (6550):646–9.

Budanov, A. V., and M. Karin. 2008. p53 target genes sestrin1 and sestrin2 connect genotoxic stress and mTOR signaling. *Cell* 134 (3):451–60.

Campaner, S., M. Doni, P. Hydbring, et al. 2010. Cdk2 suppresses cellular senescence induced by the c-myc oncogene. *Nat Cell Biol* 12 (1):54–9; sup. pp 1–14.

Campisi, J. 2013. Aging, cellular senescence, and cancer. *Annu Rev Physiol* 75:685–705.

Capparelli, C., C. Guido, D. Whitaker-Menezes, et al. 2012. Autophagy and senescence in cancer-associated fibroblasts metabolically supports tumor growth and metastasis via glycolysis and ketone production. *Cell Cycle* 11 (12):2285–302.

Carnero, A. 2013. Markers of cellular senescence. *Methods Mol Biol* 965:63–81.

Chan, E. Y., S. Kir, and S. A. Tooze. 2007. siRNA screening of the kinome identifies ULK1 as a multidomain modulator of autophagy. *J Biol Chem* 282 (35):25464–74.

Chang, Y. Y., and T. P. Neufeld. 2009. An Atg1/Atg13 complex with multiple roles in TOR-mediated autophagy regulation. *Mol Biol Cell* 20 (7):2004–14.

Cho, K. A., S. J. Ryu, Y. S. Oh, et al. 2004. Morphological adjustment of senescent cells by modulating caveolin-1 status. *J Biol Chem* 279 (40):42270–8.

Cisowski, J., V. I. Sayin, M. Liu, C. Karlsson, and M. O. Bergo. 2016. Oncogene-induced senescence underlies the mutual exclusive nature of oncogenic KRAS and BRAF. *Oncogene* 35 (10):1328:33.

Coppe, J. P., P. Y. Desprez, A. Krtolica, and J. Campisi. 2010. The senescence-associated secretory phenotype: The dark side of tumor suppression. *Annu Rev Pathol* 5:99–118.

Crighton, D., S. Wilkinson, J. O'Prey, et al. 2006. DRAM, a p53-induced modulator of autophagy, is critical for apoptosis. *Cell* 126 (1):121–34.

Dewaele, M., H. Maes, and P. Agostinis. 2010. ROS-mediated mechanisms of autophagy stimulation and their relevance in cancer therapy. *Autophagy* 6 (7):838–54.

Dimri, G. P., X. Lee, G. Basile, et al. 1995. A biomarker that identifies senescent human cells in culture and in aging skin in vivo. *Proc Natl Acad Sci U S A* 92 (20):9363–7.

Dorr, J. R., Y. Yu, M. Milanovic, et al. 2013. Synthetic lethal metabolic targeting of cellular senescence in cancer therapy. *Nature* 501 (7467):421–5.

Feng, Y., Z. Yao, and D. J. Klionsky. 2015. How to control self-digestion: Transcriptional, post-transcriptional, and post-translational regulation of autophagy. *Trends Cell Biol* 25 (6):354–63.

Feng, Z. 2010. p53 regulation of the IGF-1/AKT/mTOR pathways and the endosomal compartment. *Cold Spring Harb Perspect Biol* 2 (2):a001057.

Feng, Z., W. Hu, E. de Stanchina, et al. 2007. The regulation of AMPK beta1, TSC2, and PTEN expression by p53: Stress, cell and tissue specificity, and the role of these gene products in modulating the IGF-1-AKT-mTOR pathways. *Cancer Res* 67 (7):3043–53.

Funderburk, S. F., Q. J. Wang, and Z. Yue. 2010. The Beclin 1-VPS34 complex–at the crossroads of autophagy and beyond. *Trends Cell Biol* 20 (6):355–62.

Galluzzi, L., F. Pietrocola, J. M. Bravo-San Pedro, et al. 2015. Autophagy in malignant transformation and cancer progression. *EMBO J* 34 (7):856–80.

Ganley, I. G., H. Lam du, J. Wang, X. Ding, S. Chen, and X. Jiang. 2009. ULK1.ATG13. FIP200 complex mediates mTOR signaling and is essential for autophagy. *J Biol Chem* 284 (18):12297–305.

Goehe, R. W., X. Di, K. Sharma, et al. 2012. The autophagy-senescence connection in chemotherapy: Must tumor cells (self) eat before they sleep? *J Pharmacol Exp Ther* 343 (3):763–78.

Gozuacik, D., and A. Kimchi. 2004. Autophagy as a cell death and tumor suppressor mechanism. *Oncogene* 23 (16):2891–906.

Gozuacik, D., and A. Kimchi. 2006. DAPk protein family and cancer. *Autophagy* 2 (2):74–9.

Green, D. R., and J. E. Chipuk. 2006. p53 and metabolism: Inside the TIGAR. *Cell* 126 (1):30–2.

Green, D. R., L. Galluzzi, and G. Kroemer. 2011. Mitochondria and the autophagy-inflammation-cell death axis in organismal aging. *Science* 333 (6046):1109–12.

Gump, J. M., and A. Thorburn. 2014. Sorting cells for basal and induced autophagic flux by quantitative ratiometric flow cytometry. *Autophagy* 10 (7):1327–34.

Guo, J. Y., B. Xia, and E. White. 2013. Autophagy-mediated tumor promotion. *Cell* 155 (6):1216–9.

Hardie, D. G. 2007. AMP-activated/SNF1 protein kinases: Conserved guardians of cellular energy. *Nat Rev Mol Cell Biol* 8 (10):774–85.

Harrison, B., M. Kraus, L. Burch, et al. 2008. DAPK-1 binding to a linear peptide motif in MAP1B stimulates autophagy and membrane blebbing. *J Biol Chem* 283 (15):9999–10014.

Hawley, S. A., J. Boudeau, J. L. Reid, et al. 2003. Complexes between the LKB1 tumor suppressor, STRAD alpha/beta and MO25 alpha/beta are upstream kinases in the AMP-activated protein kinase cascade. *J Biol* 2 (4):28.

Hawley, S. A., D. A. Pan, K. J. Mustard, et al. 2005. Calmodulin-dependent protein kinase kinase-beta is an alternative upstream kinase for AMP-activated protein kinase. *Cell Metab* 2 (1):9–19.

Hayashi, M. T., A. J. Cesare, J. A. Fitzpatrick, E. Lazzerini-Denchi, and J. Karlseder. 2012. A telomere-dependent DNA damage checkpoint induced by prolonged mitotic arrest. *Nat Struct Mol Biol* 19 (4):387–94.

He, C., and D. J. Klionsky. 2009. Regulation mechanisms and signaling pathways of autophagy. *Annu Rev Genet* 43:67–93.

Herrero-Martin, G., M. Hoyer-Hansen, C. Garcia-Garcia, et al. 2009. TAK1 activates AMPK-dependent cytoprotective autophagy in TRAIL-treated epithelial cells. *EMBO J* 28 (6):677–85.

Hoshino, A., S. Matoba, E. Iwai-Kanai, et al. 2012. p53-TIGAR axis attenuates mitophagy to exacerbate cardiac damage after ischemia. *J Mol Cell Cardiol* 52 (1):175–84.

Hosokawa, N., T. Hara, T. Kaizuka, et al. 2009. Nutrient-dependent mTORC1 association with the ULK1-Atg13-FIP200 complex required for autophagy. *Mol Biol Cell* 20 (7):1981–91.

Hou, J., Z. P. Han, Y. Y. Jing, et al. 2013. Autophagy prevents irradiation injury and maintains stemness through decreasing ROS generation in mesenchymal stem cells. *Cell Death Dis* 4:e844.

Huang, Y. H., P. M. Yang, Q. Y. Chuah, et al. 2014. Autophagy promotes radiation-induced senescence but inhibits bystander effects in human breast cancer cells. *Autophagy* 10 (7):1212–28.

Imai, Y., A. Takahashi, A. Hanyu, et al. 2014. Crosstalk between the Rb pathway and AKT signaling forms a quiescence-senescence switch. *Cell Rep* 7 (1):194–207.

Itahana, K., Y. Itahana, and G. P. Dimri. 2013. Colorimetric detection of senescence-associated beta galactosidase. *Methods Mol Biol* 965:143–56.

Itakura, E., C. Kishi, K. Inoue, and N. Mizushima. 2008. Beclin 1 forms two distinct phosphatidylinositol 3-kinase complexes with mammalian Atg14 and UVRAG. *Mol Biol Cell* 19 (12):5360–72.

Jung, C. H., C. B. Jun, S. H. Ro, et al. 2009. ULK-Atg13-FIP200 complexes mediate mTOR signaling to the autophagy machinery. *Mol Biol Cell* 20 (7):1992–2003.

Kang, M. R., M. S. Kim, J. E. Oh, et al. 2009. Frameshift mutations of autophagy-related genes ATG2B, ATG5, ATG9B and ATG12 in gastric and colorectal cancers with microsatellite instability. *J Pathol* 217 (5):702–6.

Kaushik, S., and A. M. Cuervo. 2012. Chaperone-mediated autophagy: A unique way to enter the lysosome world. *Trends Cell Biol* 22 (8):407–17.

Kim, W. Y., and N. E. Sharpless. 2006. The regulation of INK4/ARF in cancer and aging. *Cell* 127 (2):265–75.

Kim, Y. C., and K. L. Guan. 2015. mTOR: A pharmacologic target for autophagy regulation. *J Clin Invest* 125 (1):25–32.

Kimata, M., S. Matoba, E. Iwai-Kanai, et al. 2010. p53 and TIGAR regulate cardiac myocyte energy homeostasis under hypoxic stress. *Am J Physiol Heart Circ Physiol* 299 (6):H1908–16.

Klionsky, D. J., F. C. Abdalla, H. Abeliovich, et al. 2012. Guidelines for the use and interpretation of assays for monitoring autophagy. *Autophagy* 8 (4):445–544.

Klionsky, D. J., H. Abeliovich, P. Agostinis, et al. 2008. Guidelines for the use and interpretation of assays for monitoring autophagy in higher eukaryotes. *Autophagy* 4 (2):151–75.

Kouroku, Y., E. Fujita, I. Tanida, et al. 2007. ER stress (PERK/eIF2alpha phosphorylation) mediates the polyglutamine-induced LC3 conversion, an essential step for autophagy formation. *Cell Death Differ* 14 (2):230–9.

Kundu, M., T. Lindsten, C. Y. Yang, et al. 2008. Ulk1 plays a critical role in the autophagic clearance of mitochondria and ribosomes during reticulocyte maturation. *Blood* 112 (4):1493–502.

Lee, E. J., and C. Tournier. 2011. The requirement of uncoordinated 51-like kinase 1 (ULK1) and ULK2 in the regulation of autophagy. *Autophagy* 7 (7):689–95.

Levy, J. M., and A. Thorburn. 2011. Targeting autophagy during cancer therapy to improve clinical outcomes. *Pharmacol Ther* 131 (1):130–41.

Levy, M. Z., R. C. Allsopp, A. B. Futcher, C. W. Greider, and C. B. Harley. 1992. Telomere end-replication problem and cell aging. *J Mol Biol* 225 (4):951–60.

Liang, C., J. S. Lee, K. S. Inn, et al. 2008. Beclin1-binding UVRAG targets the class C Vps complex to coordinate autophagosome maturation and endocytic trafficking. *Nat Cell Biol* 10 (7):776–87.

Liang, X. H., S. Jackson, M. Seaman, et al. 1999. Induction of autophagy and inhibition of tumorigenesis by beclin 1. *Nature* 402 (6762):672–6.

Liu, E. Y., and K. M. Ryan. 2012. Autophagy and cancer—Issues we need to digest. *J Cell Sci* 125 (Pt 10):2349–58.

Lorin, S., A. Borges, L. Ribeiro Dos Santos, et al. 2009. c-Jun NH2-terminal kinase activation is essential for DRAM-dependent induction of autophagy and apoptosis in 2-methoxyestradiol-treated Ewing sarcoma cells. *Cancer Res* 69 (17):6924–31.

Lorin, S., G. Pierron, K. M. Ryan, P. Codogno, and M. Djavaheri-Mergny. 2010. Evidence for the interplay between JNK and p53-DRAM signalling pathways in the regulation of autophagy. *Autophagy* 6 (1):153–4.

Ma, B., W. Cao, W. Li, et al. 2014a. Dapper1 promotes autophagy by enhancing the Beclin1-Vps34-Atg14L complex formation. *Cell Res* 24 (8):912–24.

Ma, X. H., S. F. Piao, S. Dey, et al. 2014b. Targeting ER stress-induced autophagy overcomes BRAF inhibitor resistance in melanoma. *J Clin Invest* 124 (3):1406–17.

Macheret, M., and T. D. Halazonetis. 2015. DNA replication stress as a hallmark of cancer. *Annu Rev Pathol* 10:425–48.

Maiuri, M. C., L. Galluzzi, E. Morselli, O. Kepp, S. A. Malik, and G. Kroemer. 2010. Autophagy regulation by p53. *Curr Opin Cell Biol* 22 (2):181–5.

Manchado, E., M. Guillamot, and M. Malumbres. 2012. Killing cells by targeting mitosis. *Cell Death Differ* 19 (3):369–77.

Matsunaga, K., T. Saitoh, K. Tabata, et al. 2009. Two Beclin 1-binding proteins, Atg14L and Rubicon, reciprocally regulate autophagy at different stages. *Nat Cell Biol* 11 (4):385–96.

Mihaylova, M. M., and R. J. Shaw. 2011. The AMPK signalling pathway coordinates cell growth, autophagy and metabolism. *Nat Cell Biol* 13 (9):1016–23.

Mizushima, N., and M. Komatsu. 2011. Autophagy: Renovation of cells and tissues. *Cell* 147 (4):728–41.

Mizushima, N., B. Levine, A. M. Cuervo, and D. J. Klionsky. 2008. Autophagy fights disease through cellular self-digestion. *Nature* 451 (7182):1069–75.

Mizushima, N., T. Yoshimori, and B. Levine. 2010. Methods in mammalian autophagy research. *Cell* 140 (3):313–26.

Morselli, E., S. Shen, C. Ruckenstuhl, et al. 2011. p53 inhibits autophagy by interacting with the human ortholog of yeast Atg17, RB1CC1/FIP200. *Cell Cycle* 10 (16):2763–9.

Munoz-Espin, D., and M. Serrano. 2014. Cellular senescence: From physiology to pathology. *Nat Rev Mol Cell Biol* 15 (7):482–96.

Nakaso, K., Y. Yoshimoto, T. Nakano, et al. 2004. Transcriptional activation of p62/A170/ZIP during the formation of the aggregates: Possible mechanisms and the role in Lewy body formation in Parkinson's disease. *Brain Res* 1012 (1–2):42–51.

Park, K. J., S. H. Lee, C. H. Lee, et al. 2009. Upregulation of Beclin-1 expression and phosphorylation of Bcl-2 and p53 are involved in the JNK-mediated autophagic cell death. *Biochem Biophys Res Commun* 382 (4):726–9.

Parkhitko, A. A., C. Priolo, J. L. Coloff, et al. 2014. Autophagy-dependent metabolic reprogramming sensitizes TSC2-deficient cells to the antimetabolite 6-aminonicotinamide. *Mol Cancer Res* 12 (1):48–57.

Parrinello, S., E. Samper, A. Krtolica, J. Goldstein, S. Melov, and J. Campisi. 2003. Oxygen sensitivity severely limits the replicative lifespan of murine fibroblasts. *Nat Cell Biol* 5 (8):741–7.

Peng, X., F. Gong, Y. Chen, et al. 2014. Autophagy promotes paclitaxel resistance of cervical cancer cells: Involvement of Warburg effect activated hypoxia-induced factor 1-alpha-mediated signaling. *Cell Death Dis* 5:e1367.

Perez-Mancera, P. A., A. R. Young, and M. Narita. 2014. Inside and out: the activities of senescence in cancer. *Nat Rev Cancer* 14 (8):547–58.

Qin, L., Z. Wang, L. Tao, and Y. Wang. 2010. ER stress negatively regulates AKT/TSC/mTOR pathway to enhance autophagy. *Autophagy* 6 (2):239–47.

Qu, X., J. Yu, G. Bhagat, et al. 2003. Promotion of tumorigenesis by heterozygous disruption of the beclin 1 autophagy gene. *J Clin Invest* 112 (12):1809–20.

Reinhardt, H. C., and B. Schumacher. 2012. The p53 network: cellular and systemic DNA damage responses in aging and cancer. *Trends Genet* 28 (3):128–36.

Rodier, F. 2013. Detection of the senescence-associated secretory phenotype (SASP). *Methods Mol Biol* 965:165–73.

Rodier, F., and J. Campisi. 2011. Four faces of cellular senescence. *J Cell Biol* 192 (4):547–56.

Rosenfeldt, M. T., J. O'Prey, J. P. Morton, et al. 2013. p53 status determines the role of autophagy in pancreatic tumour development. *Nature* 504 (7479):296–300.

Rosenfeldt, M. T., and K. M. Ryan. 2009. The role of autophagy in tumour development and cancer therapy. *Expert Rev Mol Med* 11:e36.

Sedelnikova, O. A., C. E. Redon, J. S. Dickey, A. J. Nakamura, A. G. Georgakilas, and W. M. Bonner. 2010. Role of oxidatively induced DNA lesions in human pathogenesis. *Mutat Res* 704 (1–3):152–9.

Soussi, T. 2007. p53 alterations in human cancer: More questions than answers. *Oncogene* 26 (15):2145–56.

Soussi, T., and G. Lozano. 2005. p53 mutation heterogeneity in cancer. *Biochem Biophys Res Commun* 331 (3):834–42.

Sui, X., R. Chen, Z. Wang, et al. 2013. Autophagy and chemotherapy resistance: A promising therapeutic target for cancer treatment. *Cell Death Dis* 4:e838.

Sun, Q., W. Westphal, K. N. Wong, I. Tan, and Q. Zhong. 2010. Rubicon controls endosome maturation as a Rab7 effector. *Proc Natl Acad Sci U S A* 107 (45):19338–43.

Tang, J., J. Di, H. Cao, J. Bai, and J. Zheng. 2015. p53-mediated autophagic regulation: A prospective strategy for cancer therapy. *Cancer Lett* 363 (2):101–107.

Tasdemir, E., M. C. Maiuri, L. Galluzzi, et al. 2008. Regulation of autophagy by cytoplasmic p53. *Nat Cell Biol* 10 (6):676–87.

Turcotte, S., and A. J. Giaccia. 2010. Targeting cancer cells through autophagy for anticancer therapy. *Curr Opin Cell Biol* 22 (2):246–51.

Vaseva, A. V., N. D. Marchenko, K. Ji, S. E. Tsirka, S. Holzmann, and U. M. Moll. 2012. p53 opens the mitochondrial permeability transition pore to trigger necrosis. *Cell* 149 (7):1536–48.

Veldhoen, R. A., S. L. Banman, D. R. Hemmerling, et al. 2013. The chemotherapeutic agent paclitaxel inhibits autophagy through two distinct mechanisms that regulate apoptosis. *Oncogene* 32 (6):736–46.

Wang, Y., X. D. Wang, E. Lapi, et al. 2012. Autophagic activity dictates the cellular response to oncogenic RAS. *Proc Natl Acad Sci U S A* 109 (33):13325–30.

Wei, Y., S. Pattingre, S. Sinha, M. Bassik, and B. Levine. 2008. JNK1-mediated phosphorylation of Bcl-2 regulates starvation-induced autophagy. *Mol Cell* 30 (6):678–88.

Wesselborg, S., and B. Stork. 2015. Autophagy signal transduction by ATG proteins: From hierarchies to networks. *Cell Mol Life Sci* 72 (24): 4721–57.

White, E. 2012. Deconvoluting the context-dependent role for autophagy in cancer. *Nat Rev Cancer* 12 (6):401–10.

White, E. 2015. The role for autophagy in cancer. *J Clin Invest* 125 (1):42–6.

Woods, A., S. R. Johnstone, K. Dickerson, et al. 2003. LKB1 is the upstream kinase in the AMP-activated protein kinase cascade. *Curr Biol* 13 (22):2004–8.

Wullschleger, S., R. Loewith, and M. N. Hall. 2006. TOR signaling in growth and metabolism. *Cell* 124 (3):471–84.

Xiao, B., M. J. Sanders, E. Underwood, et al. 2011. Structure of mammalian AMPK and its regulation by ADP. *Nature* 472 (7342):230–3.

Xie, Z., and D. J. Klionsky. 2007. Autophagosome formation: Core machinery and adaptations. *Nat Cell Biol* 9 (10):1102–9.

Yang, S., X. Wang, G. Contino, et al. 2011. Pancreatic cancers require autophagy for tumor growth. *Genes Dev* 25 (7):717–29.

Yang, Z., and D. J. Klionsky. 2009. An overview of the molecular mechanism of autophagy. *Curr Top Microbiol Immunol* 335:1–32.

Yang, Z., and D. J. Klionsky. 2010. Mammalian autophagy: Core molecular machinery and signaling regulation. *Curr Opin Cell Biol* 22 (2):124–131.

Young, A. R., and M. Narita. 2009. SASP reflects senescence. *EMBO Rep* 10 (3):228–30.

Young, A. R., M. Narita, M. Ferreira, et al. 2009. Autophagy mediates the mitotic senescence transition. *Genes Dev* 23 (7):798–803.

Yu, J., C. Chen, T. Xu, et al. 2016. Pseudolaric acid B activates autophagy in MCF-7 human breast cancer cells to prevent cell death. *Oncol Lett* 11 (3):1731–7.

Zalckvar, E., H. Berissi, M. Eisenstein, and A. Kimchi. 2009. Phosphorylation of Beclin 1 by DAP-kinase promotes autophagy by weakening its interactions with Bcl-2 and Bcl-XL. *Autophagy* 5 (5):720–2.

Zhang, L., F. Zhou, and P. Ten Dijke. 2013. Signaling interplay between transforming growth factor-beta receptor and PI3K/AKT pathways in cancer. *Trends Biochem Sci* 38 (12):612–20.

Zilfou, J. T., and S. W. Lowe. 2009. Tumor suppressive functions of p53. *Cold Spring Harb Perspect Biol* 1 (5):a001883.

Section III

Autophagy in Immunity
and Metabolism

5 Autophagy and Regulation of Immune Response

Rut Valdor
University of Murcia-Biomedical Research
Institute of Murcia (IMIB-Arrixaca)
Murcia, Spain

CONTENTS

5.1 INTRODUCTION

Autophagy is a constitutive process in most eukaryotic cells. It was initially characterized as a cellular mechanism for the turnover of cellular components through lysosomal degradation, implicating it as a key process in quality control and in providing an alternative source of energy during adaptive response to metabolic stress (e.g., starvation). In the last few years, since autophagy was discovered in the 1960s, we could find up to 17,300 articles in Pubmed, showing a better understanding of the molecular mechanisms in charge of the activation and regulation of several forms of autophagy and the advance of new tools to assess autophagic activity. The acquisition of knowledge on this highly conserved mechanism of protein degradation by lysosomes has led to elucidation of its wide array of functions in many tissues and organs. These include the modulation of cell death and survival, regulation of organ development and cell differentiation, cell defense against several types of stressors, and protective role in modulating age-associated pathologies (Levine 2005; Yang and Klionsky 2010; Cuervo and Macian 2014; Cuervo and Wong 2014). In the last years, lysosomal degradation of proteins via autophagy has been shown to play a key role in maintaining proper cell homeostasis by reducing the accumulation of damaged proteins and recycling amino acids for new protein synthesis. Additionally, it also plays a crucial role in providing cells with an alternative source of energy and modifying protein levels in response to extracellular signals (Hubbard et al. 2012; Valdor and Macian 2012). Autophagy is critical for proper cell function and its failure is seen in several diseases, leading to the intracellular accumulation of abnormal proteins, defective regulation of many cellular processes, and altered response to stress (Hubbard et al. 2010; Kon et al. 2011; Cannizzo et al. 2012). In addition, the importance of autophagy is further supported by the observations that failure of one or more forms of autophagy leads to different pathologies, such as neurodegenerative diseases, myopathies, cancer, and inflammatory processes (Sridhar et al. 2012; Guo et al. 2013; Nixon 2013; Cuervo and Wong 2014). In this chapter, we will focus on the latest findings that demonstrate different forms of autophagy, which are active in the immune system and their roles in regulating the innate and adaptive immune response, with particular focus on adaptive immune system and T cells.

Three main types of autophagy have been described in mammalian cells: macroautophagy (MA), chaperone-mediated autophagy (CMA), and microautophagy (Cuervo et al. 2004; Klionsky 2005; Mizushima et al. 2008). MA is involved in the degradation of cytosolic proteins and whole organelles and involves sequestering of cargo into a *de novo*-formed double-membrane vesicle termed "autophagosome." Eventually, autophagosomes fuse with lysosomes and breakdown of the cargo occurs. Two ubiquitin-like conjugation systems are involved in the biogenesis of the autophagosome: light chain 3 protein (LC3)-phosphatidyl-ethanolamine and autophagy-related protein (Atg) Atg12-Atg5. Genetic deletion of the proteins involved in the conjugation processes (e.g., Atg5 or Atg7) leads to MA inhibition (Mizushima et al. 2011; Subramani and Malhotra 2013). However, MA has been shown to progress without some of the Atg proteins previously considered essential (i.e., Beclin, Atg5, or Atg7) (Chu et al. 2007; Nishida et al. 2009; Kroemer et al. 2010), and it is still a question whether these different forms are alternative to classic autophagy or if they might coexist, converging from multiple pathways to coordinate and integrate signals that modulate autophagy (Kroemer et al. 2010). Furthermore, Atg proteins are identified to have independent functions to autophagy process (Subramani and Malhotra 2013). MA activity shows constitutive basal levels in almost all cell types and is induced in response to several stressors being able to selectively recognize and target specific molecules, organelles, and pathogens to the autophagic vacuoles through adaptor proteins such as p62 and Neighbor of BRCA1 gene 1 (NBR1). These adaptors interact with LC3 protein or LC3-related proteins by identification of ubiquitinated substrates in the nascent autophagic vacuole (Cadwell et al. 2008). Autophagosome formation is regulated by two PI3K complexes: the PI3K class III enzyme, Vps34, and Beclin 1 form part of a complex that stimulates autophagosome nucleation, whereas PI3K class I enzyme activates the mammalian target of rapamycin (mTOR), which has an inhibitory effect on MA (Feng et al. 2014).

Microautophagy is another form of autophagy that also caters to bulk degradation. However, in this case, the cargo is not engulfed in cytosolic vacuoles but instead, is sequestered directly into the lysosomal lumen through invaginations, protrusions, or septation of the lysosomal membrane. Microautophagy is induced by starvation and rapamycin, as in MA, and is regulated by the Target of rapamycin (TOR) and the exit from rapamycin-induced growth arrest (EGO) signaling complexes that lead to direct invagination and degradation of the vacuolar limiting membrane (Uttenweiler et al. 2007). Initially, a vacuolar transporter chaperone (VTC complex), which is present on the endoplasmic reticulum, vacuoles, and close to the cell membrane, was identified as essential for microautophagy. Removal of VTC complex impairs cargo uptake into vacuoles, and therefore microautophagic activity. Although it seems that this second form of autophagy is implicated in the organelle size homeostasis and membrane composition under nutrient deprivation, the regulation mechanism of this process, which might be promoted by any signal or stressor, remains to be elucidated (Ahlberg and Glaumann 1985).

The third form of autophagy is CMA that degrades soluble cytosolic proteins. In CMA, a chaperone cytosolic complex that contains the heat shock cognate protein of 70 kDA (Hsc70) binds to protein substrates containing a sequence biochemically related to the pentapeptide KFERQ. The chaperone complex carries the substrate to the lysosomal membrane, where it is unfolded and translocated through the lysosomal receptor for CMA, lysosome-associated membrane protein type 2A (LAMP-2A), into the lysosomal lumen and degraded. CMA activity is directly dependent on the levels of LAMP-2A at the lysosomal membrane, because the binding of substrate proteins to LAMP-2A is the limiting step of this pathway (Kaushik et al. 2011). Lysosomal levels of LAMP-2A are usually regulated through changes in its turnover and intralysosomal distribution and do not usually involve *de novo* protein synthesis. However, under conditions requiring maximal activation of this autophagic process, such as in response to oxidative stress, activation of CMA occurs through upregulation of the expression of *Lamp2a* transcription and the synthesis of new LAMP-2A protein (Kiffin et al. 2004; Kaushik et al. 2006; Valdor et al. 2014). The compensatory effect between these three forms of autophagy has been described as a cross talk to maintain the lysosomal balance and cellular homeostasis (Massey et al. 2006). Impaired MA activity has been shown to induce constitutive activation of CMA to compensate for macroautophagic failure (Kaushik et al. 2008). Cross talk also exists between MA and apoptosis (Marino et al. 2014), whereas the activity of

CMA has been associated with a protective role from death cell and to maintain the metabolic alterations characteristic of malignant cells (Kon et al. 2011). In the last few years, the regulation and functions of these autophagic variants have been shown in different types of immune cells, such as macrophages, dendritic cells (DCs), and lymphocytes. Their roles in the immune system are described as a general function in maintenance of cell homeostasis; however, specific functions, such as deletion of pathogens, antigen presentation, or maintaining of a proper T cell activation, are reported to be tightly regulated by these process as well.

5.2 AUTOPHAGY REGULATES THE INNATE IMMUNE RESPONSE AGAINST PATHOGENS

5.2.1 TARGETING AND KILLING OF PATHOGENS

Autophagy has initially been reported to be a non-selective process in all cell types to maintain cellular homeostasis in response to starvation (Kristensen et al. 2008). However, in recent years, selective form of autophagy participating in the selective recognition, capturing, and elimination of intracellular pathogens known as xenophagy has been identified (Levine 2005). The first evidence on the clearance of pathogens within non-phagocytic cells was demonstrated by the presence of some bacterial pathogenic group A Streptococcus being enveloped by the autophagosome and degraded upon lysosome fusion (Nakagawa et al. 2004). It is well established now that specific types of poly-ubiquitination on the substrates can serve as a recognition signal for recruitment of autophagic adaptor/receptor proteins to selectively target the substrate to the inner autophagosome membrane. The autophagic adaptor/receptor proteins possess both ubiquitin-binding domains (UBA or UBZ) and LC3-interacting motifs (WXXXL or WXXI) to facilitate tethering of the poly-ubiquitinated substrate to the autophagosome membrane for specific engulfment (Bjorkoy et al. 2005). There are four different adaptor/receptor proteins currently known in mammalian cells. These include p62/sequestosome 1 (Pankiv et al. 2007), NBR1 (Kirkin et al. 2009), Nuclear dot protein 52 KDa (NDP52) (Thurston et al. 2009), and optineurin (Wild et al. 2011). NDP52 has been shown to be important for the autophagy of intracellular *Salmonella typhimurium* by recognizing poly-ubiquitinated proteins on the bacterial surface to facilitate recruitment of TANK-binding kinase (TBK1) and p62 to suppress the bacterial growth (Thurston et al. 2009; Zheng et al. 2009). p62 has also been shown to participate in the autophagy of Shigella and *Lysteria monocytogenes* (Dupont et al. 2009; Yoshikawa et al. 2009), as well as in the autophagy of virus by linking the virus capsid proteins to the autophagosome to promote viral protein clearance and cell survival (Orvedahl et al. 2010). Optineurin is phosphorylated by TBK1 to induce LC3 binding to the cytosolic ubiquitinated Salmonella to enhance bacterial clearance (Wild et al. 2011).

Phagocytosis is also an ancient and highly conserved process involved in the removal of extracellular organisms and the destruction of organisms in the cytosol. Macroutophagy has been shown to cross talk with phagocytic pathways to kill pathogens in infected cells or phagocytic cells by different mechanisms (Figure 5.1). Protein levels have been shown to be modulated by autophagic inducers in the phagosomes of macrophages, enhancing pathogens colocalization with the autophagosome to be killed in the

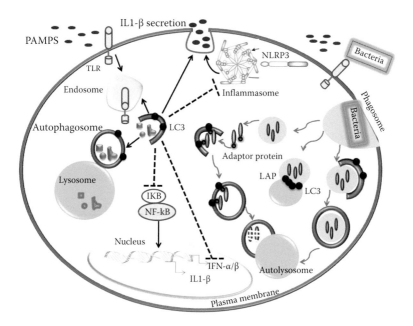

FIGURE 5.1 Autophagy regulates the innate immune system: inflammation and elimination of intracellular pathogens. Autophagy enhances delivery of PAMPS and induces formation of TLR endosomes through TLR activation. Under basal levels of inflammasome activation in response to NLRP (Nucleotide-binding oligomerization domain, Leucine reach repeat and Pyrin domain containing) autophagy inducers, autophagy cooperates with inflammasome to secret Interleukin1-β (IL1-β). Autophagy inhibits excessive inflammasome activation and regulates inflammation, downregulating type I Interferon (IFN) responses and inhibiting nuclear factor-kappaB (NF-κB) activation through the Inhibitor of NF-κB (IκB). On the contrary, autophagy degrades intracellular pathogens through several mechanisms. Phagocytic pathways promote the extracellular pathogens engulfment into phagosomes that can be fully engulfed by the autophagosome to fuse with the lysosome and form the autolysosome. Phagosomes can also be directly fused to the lysosome, forming the autophagolysosome through the process named LC3-associated phagocytosis. Pathogens can escape from phagosomes and be ubiquitinated to bind adaptor proteins such as optineurin or p62 that target the pathogens to nascent autophagosomes for delivery to lysosomes for degradation.

lysosomes (Xu et al. 2007; Shui et al. 2008). One mechanism through which MA participates in the elimination of pathogens is a process known as LC3-associated phagocytosis (LAP), where toll-like receptor (TLR) engagement on murine macrophages activates phagocytosis, which triggers the autophagosome marker LC3 to be recruited to the phagosome membrane to result in phagosome maturation and fusion with the lysosome (Sanjuan et al. 2009). Other mechanism, where autophagy and phagocytosis are complementary, is based on the engulfment of the whole phagosome by the nascent autophagosome upon MA induction, sequestering and controlling pathogens that are infecting the cell for degradation by fusion with the lysosome (Figure 5.1). This is seen in the case of *Mycobacterium tuberculosis* which resides in the phagocytic vesicles of macrophages and is cleared upon MA induction where phagosomes are triggered for engulfment by autophagosomes by enhanced levels of autophagic markers such as LC3 and Beclin 1, leading to subsequent fusion with the lysosomes to form phagolysosome (Gutierrez et al. 2004; Xu et al. 2007). Several *in vivo* models have detected that failed MA in macrophages impairs clearance of *M. tuberculosis*, therefore demonstrating that MA has an important defensive role in the phagocytic pathway of pathogens (Intemann et al. 2009; Kumar et al. 2010; Watson et al. 2012). MA can also regulate the clearance of intracellular bacteria, such as *Streptococcus* A, by autophagosome formation through selective autophagy without the existence of phagocytic pathways (Nakagawa et al. 2004). MA has also been demonstrated to eliminate virion and virus proteins to prevent viral accumulation in a mechanism dependent on the regulation by protein kinase R and the elongation factor 2 alpha (Talloczy et al. 2006). MA regulates viral replication, virus-induced cell death programs, viral pathogenesis, and antiviral responses (Lin et al. 2010). Besides protection against bacterial and viral infections, MA also offers cellular protection against parasites. That is seen in the case of the protozoan *Toxoplasma gondii*, which infects macrophages by phagocytic pathway, and after disruption of the vesicle, can be targeted by autophagosomes or engulfed in autophagosomes-like vacuoles for delivery to the lysosomes (Andrade et al. 2006; Ling et al. 2006). A different mechanism of cross talk between phagocytic pathways and MA is demonstrated in bacteria-infected macrophages, where the pathogen is destroyed inside the phagosome by several peptides with antimicrobial properties that are produced in the autophagosomes (Alonso et al. 2007; Ponpuak et al. 2010).

Although MA protects against pathogens in the cells, it is well known that bacteria have also developed mechanisms to escape from the autophagic process to survive inside the infected cells. Bacteria are able to escape from phagosomes by making cholesterol-dependent cytolysin listeriolysin-O pores in the phagosome membrane. However, these escaped cytosolic bacteria can still be targeted to the autophagosome to be cleared by lysosome degradation (Py et al. 2007; Meyer-Morse et al. 2010). Furthermore, bacteria are able to present actin A protein on their surface to avoid recognition by the autophagic machinery to escape from autophagosome engulfment (Yoshikawa et al. 2009). Bacteria can also evade autophagosome by protein IcsB presentation on the surface, which blocks the interaction between VirG protein and Atg5 autophagic marker (Ogawa et al. 2005). Also, many viruses have been found to develop escape strategies to avoid autophagic destuction. The targeting of autophagic protein Beclin 1 by virulence gene products has been shown to block autophagy activity and leads to viral pathogenesis (Leib et al. 2009).

5.2.2 MA REGULATION IN INNATE IMMUNE RESPONSE

Autophagy also regulates innate immune responses by promoting the engulfment and delivery of viral nucleic acids, known as cytosolic pathogens-associated molecular patterns (PAMPS), to the endosomal TLRs, which have been shown to induce an IFN-type proinflammatory response in plasmocytoids DCs (Lee et al. 2007). After activation of endosomal TLRs (TLR7 and TLR9), and then inflammasome activation, MA regulates inflammatory response with the induction of an excess of IL-1β level as well as IL-18 and high-mobility group box 1 (Dupont et al. 2011; Ohman et al. 2014). However, autophagy inhibits inflammasome activation when this pathway is highly activated and required to be regulated to control cytokines secretion (Figure 5.1). Autophagy prevents excessive tissue damaging through degradation of several inflammasome components, including NLPR3 and the IFN-inducible protein AIM2. Activation of the NLPR3 inflammasome is induced by mitochondrial reactive oxygen species (ROS) production and oxidized mitochondrial DNA releasing, which are regulated by autophagy activity (Nakahira et al. 2011; Zhou et al. 2011; Shimada et al. 2012). Autophagy also interferes and downregulates the signaling pathways of Type I IFN responses and inhibits NF-κB activation (Deretic et al. 2015) (Figure 5.1). In addition, MA is regulated in a loop of activation and inhibition; where it can also be promoted by extracellular IL-1β upon inflammasome activation, but the subsequent activation of caspase-1 suppresses its activity, through degradation of required adaptor proteins for autophagy initiation (Pilli et al. 2012; Jabir et al. 2014).

Toll-like receptors are the first identified autophagy regulators in the innate immune system. TRLs such as Pam3CSK4 (TLR1), Pam2CSK4 (TLR2), poly (I:C)(TLR3), LPS (TLR4), flagellin (TLR5), MALP-2 (TLR6), ssRNA (TLR7/8), and CpG oligodeoxy nucleotides (TLR9) have been shown to induce autophagy upon TLR engagement in several cell types including macrophages, DCs, and neutrophils (Delgado et al. 2008; Shi and Kehrl 2008). The activation of both MyD88- and TRIF (TIR-domain-containing adapter-inducing interferon-β) mediated pathways is enhanced by autophagy induction through TLRs (Alonso et al. 2007; Deretic et al. 2015). MyD88 and TRIF act as adaptor molecules interacting with Beclin 1, which dissociates from anti-apoptotic proteins Bcl-2 and Bcl-XL, leading to autophagosome formation and autophagy induction (Shi and Kehrl 2008). The TLR4 stimulation activates E3 ligase TRAF6 to enhance the ubiquitination of Beclin 1 at Lys117, and thereby promotes the dissociation from Bcl-2 and Bcl-XL, and to trigger autophagy. In an opposing manner, the enzyme A20 promotes the deubiquitination of Beclin 1 to inhibit autophagy (Shi and Kehrl 2010). TRAF6 is also an intermediate in the TLR-induced MyD88-dependent activation of the IκB kinase (IKK) complex in an independent manner from NF-κB signaling pathway activation (Criollo et al. 2010). On the contrary, the TLRs are shown to promote LAP, uptaking the cargos containing TLR ligands and facilitating LC3 recruitment onto the phagosome surface, to induce lysosomal degradation without formation of autophagosomes (Sanjuan et al. 2009). The uptake of the NOD-like receptors has also a protective antibacterial role, enhancing autophagy in response to bacterial muramyl dipeptide. NOD1 and NOD2 recruit ATG16L1 to the bacterial entry point at the cell plasma membrane, to interact with it and to promote an efficient autophagy induction (Travassos et al. 2010). A proper major histocompatibility complex (MHC) II antigen presentation in DCs has been shown when NOD2 recognizes the bacterial peptides and induces autophagosome formation by interaction with ATG16L1 (Cooney et al. 2010). Single nucleotide polymorphisms in several autophagy genes including, among others, ATG16L1 and IRGM, are linked to genetic predisposition to Crohn's disease (Valdor and Macian 2012). Interaction between NOD receptors and ATG16L1 is very important for autophagy induction to prevent onset of inflammatory bowel disease (Cadwell et al. 2010; Hubbard-Lucey et al. 2014). Other immune receptors such as CD40 are also able to activate autophagy, and the vacuole/lysosome fusion through autophagy is dependent on TRAF6 signaling, acting downstream of CD40 and upon TNF-alpha stimulation (Subauste et al. 2007). In addition, some tripartite motif (TRIM) family proteins can act as autophagy receptors. For example, TRIM5α is a retroviral restriction factor, which directly recognizes the retroviral capsid without any ubiquitination target and delivers it to autophagosome (Mandell et al. 2014). The regulation of autophagy by cytokines has also been found to induce autophagy. IFN-γ, in an IFN-γ-inducible GTPase Irgm1/IRGM (murine/humans)–dependent manner, is known to induce autophagy in murine and human macrophages (Singh et al. 2006; Harris et al. 2007). While TNF-α and IL-1β induce autophagy (Keller et al. 2011; Mostowy et al. 2011; Pilli et al. 2012), IL4 and IL-13 inhibit autophagy in bacteria-infected murine macrophages, depending on the types of signaling pathways activated in response to the stimuli (Harris et al. 2007). IL-10 also inhibits starvation-induced autophagy via the Akt pathway (Park et al. 2011), as well as IL-6 (Shen et al. 2012).

5.3 AUTOPHAGY REGULATES THE ADAPTIVE IMMUNE RESPONSES

5.3.1 AUTOPHAGY REGULATES ANTIGEN PRESENTATION TO ADAPTIVE IMMUNE CELLS

Autophagy has a role in the antigen presentation to adaptive immune cells like CD4+ and CD8+ T lymphocytes. This was one of the first few functions of autophagy implicated in the regulation of the adaptive immune response by DCs, macrophages, and B cells (Figure 5.2). These cells deliver intracellular proteins to class II loading compartments to promote the presentation of intracellular antigens on MHC class II molecules. MA inducers, such as starvation, have been shown to lead to an increase in peptides presentation from cytosol and lysosomes on MHC class II molecules (Dengjel et al. 2005). MA has been found to regulate not only the presentation of cellular endogenous proteins but also plays a role in the delivery of different viral and bacterial antigens to MHC class II and I antigen-presenting molecules, leading to the activation of immune response in T lymphocytes (English et al. 2009; Hayward and Dinesh-Kumar 2010; Lee et al. 2010; Fiegl et al. 2013; Jin et al. 2014). Autophagy induces antigen presentation to both CD4+ and CD8+ T lymphocytes in DCs, and specifically in bacterial peptides presentation. MA is regulated by stimulation of nucleotide-binding oligomerization domain-containing protein 2 (NOD2) (Cooney et al. 2010; Ravindran et al. 2014). As we have seen in this chapter previously, autophagy enhances presentation of extracellular antigens through phagocytic pathways where pathogens are often targeted by several adaptor proteins, such as p62, NDP52, or optineurin (Ponpuak et al. 2010; Wild et al. 2011). Constitutive MA works under basal conditons to deliver LC3-tagged viral proteins to MHC class II-associated endosomal compartments to promote antigen presentation to T cells (Schmid et al. 2007). In addition, during the antigen processing in viral and bacterial compartments, autophagy modulates MHC class II antigen presentation with a protective role in antimicrobial defense, which stabilizes phagosomes that contain pathogens (Paludan et al. 2005; Jagannath et al. 2009; Romao et al. 2013). On the contrary, autophagy can regulate both immune responses of tolerance and autoimmunity during antigen presentation. MA enhances self-antigen presentation by thymic epithelial cells during positive and negative selection of CD4+ T cells, modulating the establishment of central tolerance in the immune system through

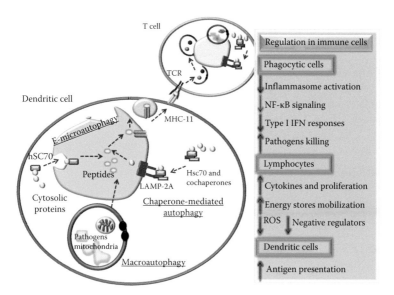

FIGURE 5.2 Different types of autophagy regulating specific functions of innate and adaptive immune cells. In dendritic cells, MA engulfs pathogens and endogenous proteins in addition to organelles, such as mitochondria (mitophagy). CMA and E-microautophagy specifically degrade cytosolic proteins that contain a KFERQ-like motif. E-microautophagy is facilitated by engulfment of selected cytosolic proteins bound to the Hsc70 chaperone by endosomes, while CMA requires the lysosomal receptor (LAMP-2A) to uptake the unfolding cytosolic proteins that are recognized and delivered to the membrane of the MHC-II-loading compartment, to be translocated and degraded into peptides. MHC-II-loading compartments receive peptides from all types of autophagy for antigen presentation. In T cells, basal MA modulates T cell homeostasis, whereas MA in response to T cell activation, through the TCR engagement, degrades cytosolic components and substrates proteins to increase the required energetic metabolism and to regulate specific pathways. CMA degrades protein substrates involved in regulating and maintaining T cell activation. Therefore, autophagy activity regulates specific functions in immune cells and if it is altered or inhibited (aging, diseases), these functions are deregulated. Box on the right summarized how autophagy activates or inhibits specific functions in immune cells.

the negative selection of thymocytes when thymic epithelial cells express low antigen concentration (Nedjic et al. 2008; Aichinger et al. 2013). MA is also implicated in peripheral tolerance by promoting autoimmunity through MA-mediated, autoreactive self-antigen presentation by DCs and B cells (Ireland and Unanue 2011). In addition, MA affects lymphocyte responses through antigen presentation such as T cell polarization. MA-defective macrophages, which produce excessive IL-1α and IL-1β levels, promote polarization and persistence of Th17 responses (Castillo et al. 2012). MA-deficient DCs destabilizes synapse formation with T cells, allowing abnormal duration of immunologic synapses and consecutively enhancing Th17 responses (Wildenberg et al. 2012). MA also promotes efficient tumor antigen presentation and induces stronger antitumor T cell responses. Delivery of tumor intracellular proteins is achieved through the generation of LC3 fusions to increase antigen presentation of these proteins in antigen-presenting cells (APCs) directly on class II MHC or via a process termed "cross-presentation" on class I MHC (Munz 2013). MA is required to produce an effective antitumor T cell response upon chemotherapy. MA modulates chemotherapy-induced cell death, making tumor cells release ATP and influencing them to become immunogenic (Michaud et al. 2011). Modulation of MA by several inducers in tumor cell lines has shown to promote tumor antigen cross-presentation, leading to stronger T cells responses (Li et al. 2011, 2012). Furthermore, regulation of MA has been found to improve vaccination, as induction of MA in bacteria-infected DCs results in successful presentation of pathogenic antigens, and thereby increases the efficiency of the adaptive immune response (Jagannath et al. 2009; Uhl et al. 2009). Other important type of autophagy for antigen presentation to adaptive immune cells is known as endosomal-microautophagy (E-microautophagy), which has the characteristics of canonical microautophagy, but occurs in late endosomes of DCs instead of plasma membrane (Sahu et al. 2011). E-microautophagy involves the invaginations of the endosomal membrane to engulf specific proteins with a specific KFERQ motif that is previously recognized by Hsc70 chaperone. Contrarily to CMA, the proteins are engulfed by the endosomal compartment without requirement to be unfolded, as seen in the case of CMA (Figure 5.2). Cytosolic proteins are internalized in an endosomal sorting complex required for transport (ESCRT)-dependent manner (Sahu et al. 2011). Besides E-microautophagy and MA, CMA is also implicated in antigen presentation on MHC molecules. Increased expression levels of LAMP-2A, the lysosomal receptor for CMA, also induces antigen presentation on MHC class II of lymphoblastoid B cells, thereby enhancing stronger T cell responses (Zhou et al. 2005). In conclusion, all forms of autophagy are implicated in the processing or delivery of proteins to antigen-loading compartments for antigen presentation to the adaptive immune cells (Figure 5.2).

5.3.2 MA REGULATES T CELL FUNCTION

We have seen that autophagy can modulate T cell responses by its involvement in the presentation of antigens in the thymus and peripheral APCs. In addition to regulating the cells from the adaptive immune system through the innate immune system, MA also plays specific roles in the T and B cells. In particular, MA is important for the maintenance of cellular homeostasis in T cells and T cell activation.

Basal MA activity that controls organelle homeostasis, such as mitochondrial and endoplasmic reticulum clearance, is implicated in the T cell development from thymocytes to mature T cells. MA-deficient T cells show higher levels of ROS and deregulation of calcium mobilization due to impaired mitochondria and endoplasmic reticulum degradation (Pua et al. 2009; Jia et al. 2011). Perturbation in mitochondria homeostasis caused by a failure to initiate autophagosome formation has been shown to decrease the number of peripheral T cells (Pua et al. 2007; Willinger and Flavell 2012; Parekh et al. 2013). T cells with this phenotype promote inflammatory responses, which are associated with a reduced suppression function by alterations of peripheral maintenance and function of CD4(+) Foxp3(+) regulatory T cells (Parekh et al. 2013). Furthermore, basal levels of MA activity are required for thymocyte progenitor survival at all stages of development, including from double negative to single positive cells. Malfunction of MA activity causes significant reduction in total number of thymocytes (Stephenson et al. 2009; Arsov et al. 2011). However, MA is upregulated from basal levels in response to T cell receptor (TCR) engagement (Figure 5.2) in peripheral T cells and excludes mitochondria from autophagosomes (Hubbard et al. 2010). MA activity has two specific functions that regulate T cell activation: first, to degrade organelles to obtain substrates to increase energy metabolism required during T cell activation and second, to selectively degrade Bcl10, a component of the IκκB-activating complex, to regulate NFκB activation, inducing a proper T cell activation (Hubbard et al. 2012; Paul et al. 2012). Upon TCR stimulation, Bcl10 is ubiquitinated and bound to the adaptor protein p62 to be recruited to the autophagosome for degradation (Paul et al. 2012). Reduced mitochondrial turnover in response to TCR engagement allows the T cell to maintain these organelles for high energy metabolism required during T cell activation. Impaired MA in T cells leads to poor T cell activation through TCR engagement, with a decrease in the levels of T cell proliferation and cytokine production (Hubbard et al. 2010; Yang et al. 2013). How this process is regulated is still not completely understood, but as far as we know, MA is increased from basal levels through mTOR inhibition, which is required for T cell survival and proliferation (Li et al. 2006; Pua et al. 2007). During T cell activation, the expression level of Beclin 1 is upregulated through NF-κB activation in response to TCR stimulation (Copetti et al. 2009). Formation of Vps34/Beclin 1 complex, an essential complex to activate MA, is induced in response to TCR engagement and is shown to be required for T cell survival (McLeod et al. 2011; Willinger and Flavell 2012). However, it remains unclear whether MA can act upon T cell activation through Vps34/Beclin 1-dependent canonical pathway or Vps34-independent non-canonical pathways.

MA also modulates cell death in T cells. Peripheral populations of MA-deficient T cells are reduced in number as a consequence of excessive cell death promoted in response to increased activation. Beclin 1 plays a role in the modulation of the intracellular levels of proapoptotic proteins such as caspase 3, caspase 8, and Bim. Therefore, failure to trigger MA activation in Beclin 1–deficient T cells enhances high accumulation of proapoptotic proteins and prevents the protective effect of MA against apoptosis (Kovacs et al. 2012). This process is tightly regulated by the Fas-associated protein with death domain FADD and caspase 8. These proteins inhibit MA activity upon T cell activation and prevent deregulated autophagosome degradation (Bell et al. 2008). Excessive MA activity induces T-cell cell death in response to some pathogens that destabilize autophagosomes degradation. That is the case for envelope glycoprotein (ENV) released by HIV 1 virus, which enhances deregulated MA to promote T-cell cell death (Espert et al. 2006).

5.3.3 MA Regulates B Cell Function

Macroautophagic activity regulates B cell development and function. Failed autophagosome formation impairs transition between pro- and pre-B cells, leading to reduced numbers of B cell precursors and defective populations of B1 cells (Miller et al. 2008). MA regulates B cell activation as in T cells, which takes place upon the engagement of the antigen receptor, the B cell receptor (BCR). Autophagosomes are localized to the BCR after B cell stimulation and assist through fusion with endosomes, to recruit the toll-like receptor TLR9 and induce specific signaling pathways of B cell activation (Chaturvedi et al. 2008; Monroe and Keir 2008; Weindel et al. 2015). Furthermore, MA activity modulates the maintenance of a proper antibody response from B cells and is essential for plasma cell homeostasis. MA is required for the energy metabolism of activated plasma cells and is critical for preventing cell death of short- and long-lived plasma cells that promote humoral immune response (Pengo et al. 2013).

5.3.4 CMA Regulates T Cell Function

CMA is also implicated in the regulation of T cell responses and is induced in response to T cell activation (Valdor et al. 2014). The expression of the lysosomal receptor for CMA, LAMP-2A, is upregulated upon TCR engagement, which induces CMA activity in different subsets of CD4+ T cells, including naïve and effector cells. In addition to TCR engagement, the kinetics of LAMP-2A expression seems to be accelerated upon CD28 engagement. However, CMA is not dependent on costimulation, it only appears to require TCR stimulation. During T cell activation, there is an increase in the lysosome number as a result of Transcription factor EB activation, the master transcription factor controlling lysosomal biogenesis (Settembre et al. 2011). Mitochondrial and cytosolic ROS production is known to be essential for efficient calcium signaling pathways induced by TCR engagement (Kwon et al. 2010; Sena et al. 2013). Such ROS generation regulates CMA activity through the upregulation of the *Lamp2a* gene and subsequent LAMP-2A expression upon TCR stimulation in T cells. Furthermore, ROS production also activates nuclear factor of activated T cells transcription factor, which binds to the *Lamp2a* gene promoter and enhances the expression of *Lamp2a* gene (Figure 5.3). The role of CMA activity in T cell is to maintain a proper T cell activation and to promote T cell functions such as cytokine production and T cell proliferation. CMA is required for the degradation of negative regulators of T cell activation such as the ubiquitin ligase Itch and the calcium inhibitor Rcan-1, the deficient degradation with age is in part responsible for the age-related diminished adaptive immune responses. T cells from old mice and humans show impaired degradation of negative regulator by CMA as a result of a reduction in LAMP-2A expression levels under T cell stimulation.

5.4 MATERIAL AND METHODS TO MONITOR AUTOPHAGY IN IMMUNE SYSTEM

5.4.1 Innate Immune System

5.4.1.1 Induction of Autophagy

Induction of autophagy can be carried out by amino acid starvation or treatment with the mTOR inhibitor, rapamycin, or with IFN-γ or LPS. However, the rate of autophagosome formation may be different depending on the stimuli. Starvation usually is very fast inducer, resulting in rapid formation of LC3-positive autophagic vesicles after 30 minutes treatment. Treatments with LPS (1–100 ng/ml) and IFN-γ (100–500 U/ml) will require between 2 and 48 hours to observe autophagosome

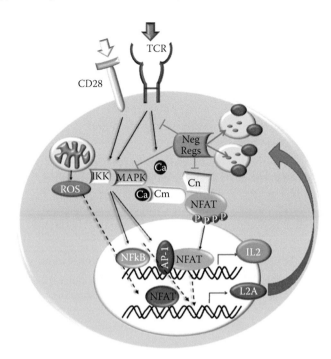

FIGURE 5.3 Mechanism of CMA regulation and its role in T cell function. During T cell activation through the TCR receptor and costimulatory molecules, there is activation of transcription factors that promote high reactive oxygen species (ROS) levels release. High ROS production induces NFAT transcription factor-dependent upregulation of *Lamp2a* gene. NFAT binds to the *Lamp2a* promoter to induce its gene expression, and thereby induces lysosomes to present high levels of CMA receptor LAMP-2A at the lysosomal membrane, which is essential for CMA activity. Increased CMA activity degrades Itch and RCAN-1, negative regulators of the T cell activation that need to be eliminated to maintain the IL-2 production and T cell proliferation and in conclusion, to keep a proper T cell activation.

formation in the cells (Harris et al. 2009). Macrophage cell lines, such as murine RAW 264.7 and J774 monocytes/macrophages, or murine bone marrow–derived macrophages, or human peripheral blood–derived macrophages, are useful cell models for studying autophagy in immune cells. Cell density of $0.1–0.5 \times 10^6$ cells/ml is recommended for immunofluorescence studies, whereas higher concentrations are used to obtain protein lysates for LC3-flux detection by immunoblotting studies. Rapamycin (Sigma) treatment at 2–50 μM for 10 minutes to 48 hours can be used to induce MA in cell culture. To induce MA by starvation, culture media is replaced with Earle's balanced salt solution (Sigma) for 10 minutes to 4 hours.

5.4.1.2 Immunofluorescence Analysis of Autophagosomes

Autophagosomes can be visualized in fixed cells by immunofluorescence staining with anti-LC3 antibody (Sigma #L8918) and further localized to different cellular compartments through colocalization studies using different organelle markers. Specific dyes highlighting different organelles include mitotracker, ER-tracker, and lysotracker (Invitrogen) that stain for mitochondria, ER, and lysosomes, respectively. Alternatively, monodansylcadaverine (MDC) staining of live cells at 50 μM is useful for tracking autophagic vacuoles in live cells. Macroautophagic activity can be quantified by percentage of cells with MDC-positive vesicles or number of MDC-positive vesicles per cell (Harris et al. 2009).

5.4.1.3 Analysis of Autophagosome Formation and Turnover by Immunoblotting

Immunoblotting assay aids in the detection of autophagosome formation in innate immune cells through the analysis of the conversion of cytosolic LC3-I to lipidated autophagosome membrane-bound LC3-II. Autophagosome turnover can be measured by quantifying the magnitude of LC3-II levels accumulated upon lysosomal inhibition treatment as compared to the untreated control condition.

5.4.1.4 Dissecting the Role of Autophagy Function in Immune Functions

To investigate the possible role of autophagy in immune system, autophagy can be ablated using small interference RNA against essential autophagy proteins (e.g., ATG5 or LC3), overexpression of dominant negative mutant forms of autophagy proteins (e.g., mutant ATG4), or using innate immune cells from autophagy-deficient mice (e.g., ATG5, ATG7, or LC3 tissue-specific "Knock out" mice). Due to the impaired autophagic degradation, accumulation of autophagic substrates can be detected and analyzed by proteomics analysis of the isolated phagosomes from control cells and autophagy-deficient cells to identify substrate proteins modulated by autophagy (Shui et al. 2008). For example, the failed clearance of a pathogen observed upon autophagy ablation is an indicator that autophagy regulates the infection of that specific pathogen (Harris et al. 2007, 2009).

5.4.2 Adaptive Immune System

5.4.2.1 MA and MHC Class II Presentation of Antigens

Intracellular antigens presented on MHC class II molecules can be detected after degradation via MA. Antigen targeting for autophagic degradation via fusion with the ATg8/LC3 protein is incorporated on the autophagosome to be degraded by autophagy and loaded in MHC class II compartments (Schmid and Munz 2008). The localization of Atg8/LC3 fusion antigens in MHC class II molecules can be detected by immunofluorescence assay as previously described for macrophages. In this protocol, Hacat keratinocyte cell line and human breast carcinoma cell line MDAMC, which express MP1-LC3 fusion construct, are treated with 200 U/ml IFN-γ for 24 or 36 hours to induce expression of MHC class II compartments. During the last 6 hours of culturing, one control set of cells is treated with chloroquine at 50 μM, to inhibit degradation of MP1-LC3 by lysosomes proteases. Detection and colocalization of antigen cargo, and autophagy in MHCII compartment, can be performed by immunofluorescence using anti-HLA-DR/DP/DQ antibody and anti-LC-3 antibody, and/or anti-MP1 antibody.

5.4.2.2   Antigen Processing for MHC Presentation via MA

1. Cell culture and isolation of antigen-presenting cells

    Antigen-processing studies can be done in cell lines such as human embryonic kidney epithelial cell line (HLA-DR4+ HEK 293), human lung epithelium A549, and human melanoma cell line M199. Cell lines are cultured in Dulbeco's Modified Eagle's medium (DMEM, Whittaker) with 10% fetal bovine serum (FBS), 2 mM glutamine, and 110 μg/ml sodium pyruvate and gentamicin 2 μg/ml or penicillin/streptomycin at 100 μg/ml (Gibco). Induction of MHC class II antigen-processing machinery expression can be done by 200 U/ml of IFN-γ (Prepotech) treatment for 48 hours. Furthermore, human or murine DCs can be purified from PMBCs isolated by leukocytes concentration from peripheral blood using density gradient centrifugation in Ficoll-Hypaque. CD14+ monocytes/macrophages are purified by magnetic beads isolation (Miltenyi Biotec), and monocytes-derived DCs are obtained by differentiation for 5 days, with 20 ng/ml for both recombinant human interleukin-4 (IL-4, Prepotech) and granulocyte-macrophage colony-stimulating factor (Invitrogen), in RPMI 1640 medium supplemented with 2% FBS (Invitrogen). After 5 days, immature DCs are collected and separated by pipetting up and down several times from adherent cells in culture plates. Matured DCs can be obtained after 6 days with a stimulus such as pathogen infection. Maturation of DCs can be confirmed by Fluorescence-Activated Cell Sorting (FACS) analysis with antibodies against CD11c, CD83, CD86, HLA-DR, and HLA-ABC (BD Biosciences).

2. Analysis of autophagosome flux and MHCII compartment by immunocytochemistry

    Detection of autophagosomes or MA activity can be done by expression of MA reporters based on fusion of fluorescent proteins with autophagic marker LC3 such as mRFP-GFP-LC3 tandem fluorescent construct (to monitor fusion of autophagosomes with lysosomes), pEGFP-LC3 construct (to monitor autophagosome formation and flux) (Addgene), or specific antigen protein fused to pEGFP-LC3 construct (to monitor antigen processing through MA) (Gannage et al. 2013). Transient overexpression of the various constructs for 24 hours is sufficient to check LC3 flux by fluorescence microscopy or for use in MHC presentation assays. To avoid degradation of autophagosomes by lysosomal proteases and to visualize the LC3 flux by immunofluorescence, treatment with chloroquine at 50 μM or with a combination of NH4Cl (20 mM) and leupeptin (100 μM) can be added at least 6 hours before cell fixation with 4% paraformaldehyde. Pathogens infections or TLR stimulation treatments for 24 hours are enough to see colocalization of LC3 and MHCII compartments by immunofluorescence costaining.

3. MHC class II presentation assays with antigen-specific CD4+ T cells

    An antigen-specific human CD4+ T cell clone (Gannage et al. 2013) is cultured in RPMI 1640 with 10% FBS, with 450 U/ml of human recombinant IL-2 (Peprotech), gentamicin or penicillin/streptomycin, and 2 mM glutamine and then seeded in a 96-well plate at 10^5 cells/well in ratios 2:1, 1:1, or 1:2 with APCs for 18–24 hours. As a positive control, T cells are stimulated with the specific peptide (1–5 µM) or phyto-hemagglutinin A-L (1 µg/ml, Sigma). Negative controls for resting condition of T cells are performed with just the cell culture media. Analysis of IFN-γ secretion by T cells, as a T cell effector functions against the specific antigen presentation by APCs through MA, can be done by ELISA, following the manufacturer's instructions of ELISA kit for human IFN-γ (R&D).

4. Silencing of MA in APCs

Immature monocyte-derived DCs at day 4 and cell lines are electroporated with Amaxa nucleofactor electroporator according to the manufacturer's instructions (Lonza), using 1 nM of siRNA-targeting Atg16L1 (sense-5'-GAG UUG UCU UCA GCC CUG AUG GCA G-3' and antisense-5'CUG CCA UCA GGG CUG AAG ACA ACU-3) (Gannage et al. 2013), or Atg5 or scrambled siRNA as negative control. In the case of immature DCs, maturation is carried out the following day after electroporation by stimuli of interest such as pathogens. At day 6, impaired MA can be detected as absence of LC3 autophagic flux by immunofluorescence microscopy or immunoblotting. Maturation phenotype and viability of the cells must be checked.

5.4.2.3   Analysis of E-Microautophagy in Antigen Presentation by Dendritic Cells to Adaptive Immune Cells

Detection of cytosolic protein substrates that are delivered to late endosomes is based on the fact that KFERQ-containing specific proteins are degraded by E-microautophagy. To study E-microautophagy degradation, late endosomes need to be isolated using centrifugation in Percoll gradient (Castellino and Germain 1995). A DCs line (Jaws), which is cultured in DMEM medium with FBS, glutamine, and antibiotics, can be used following the material and methods as described by Sahu et al. (2011). Methotrexate (MTX), which prevents protein unfolding, and chemical U18666A, which inhibits cholesterol trafficking (Sahu et al. 2011), can be added to the culture media for 48 and 16 hours, respectively. Detection of antigen presentation through E-microautophagy to adaptive immune cells, such as T cells, can be monitored as described for MA previously, or following the methods to measure T cell responses against antigen presentation through E-microautophagy as described by Cannizzo et al. (2012). To confirm if a KFERQ-containing protein is being degraded by E-microautophagy, the following criteria can be checked (Klionsky et al. 2012): (1) Inhibition of lysosomal proteolysis (e.g., with NH4Cl and leupeptin) blocks degradation. (2) Knockdown of LAMP-2A does not inhibit E-microautophagy. (3) Knockdown of components of ESCRT I and II (e.g., VPS4 and TSG101) inhibits E-microautophagy. (4) Interfering with the capability to unfold the substrate protein does not block degradation by E-microautophagy of the protein. Therefore, soluble proteins, oligomers, and protein aggregates can undergo E-microautophagy. (5) *In vitro* uptake of E-microautophagy substrates can be reconstituted using isolated late endosomes.

5.4.2.4   Measure of MA in B and T Cells

1. Cell culture and isolation of T/B cells

Primary CD4+ T cells and B cells are isolated from lymph nodes and spleen of mice between 6 and 8 weeks old or from human peripheral blood obtained with leukopaks. Isolation for T cells can be done by positive depletion with anti-CD4-coupled magnetic beads (Life Technologies), and for B cells with Dynabeads® Mouse Pan B (B220) (ThermoFisherscientific). Isolated T cells are stimulated with 0.5 µg/ml plate-bound anti-CD3 and 0.5 µg/ml anti-CD28 antibodies (clones 2C11 and 37.51: BD Biosciences). CD4+ T cells can be differentiated to T helper cells 1 (T_H1) by adding IL-12 (10 ng/ml) (Cell Sciences) and anti-IL-4 (10 µg/ml) for 6 days and 10 U/ml recombinant human IL-2, or IL-4 (10 ng/ml) and anti-IFNγ (10 µg/ml) (BD Biosciences) to differentiate to T helper cell 2 (T_H2). For stimulation of B cells, plastic-immobilized anti-IgM Ab is prepared by incubating Phosphate-buffered saline (PBS) containing 10 µg/ml F(ab')2 fragments of goat anti-mouse IgM Ab (Cappel, Aurora, OH, USA) in wells of plastic dishes at 37°C for 2 hours, followed by washing with PBS. Cells are stimulated with 2 µg/ml F(ab')2 fragments of anti-IgM Ab or immobilized anti-IgM Ab in the presence or absence of 2 µg/ml anti-mouse CD40 monoclonal Ab (mAb) HM40-3 (Abnova).

2. Measurement of MA

   After induction of autophagy by plate-bound antibodies stimulation as indicated above, cells are treated with NH4Cl and leupeptin to analyze the autophagic activity (Hubbard et al. 2010). LC3 flux is detected by immunoblotting or immunofluorescence microscopy with anti-LC3 antibody as described previously in this chapter for other cell types. Autophagic vacuoles can be identified by electron microscopy using previously established criteria (Hubbard et al. 2010).

3. Inhibition of MA in T and B cells

   To study effects of impaired MA, T cells can be isolated from Atg7[F/F-Lck-cre] mice, which have deficient MA specifically in T cells. Furthermore, T and B cells can be isolated from Atg3[F/F] Rosa26-Cre-ER or Ub-Cre-ER bone and treated with 4-hydroxytamoxifen (4OHT, 0.5 µM in ethanol, Sigma) to activate Cre recombinase and excise the ATG7 gene *in vitro*. MA can also be inhibited by knocking down Atg7 with siRNA against Atg7, and also pharmacologically inhibited by 3-methyladenine (10 mM) for 4 hours prior to the cell stimulation.

4. Measurement of T and B cell immune responses

   T and B cell function can be measured after MA induction by kinetics of stimulation to correlate immune response time course with the autophagosome degradation (Chaturvedi et al. 2008; Hubbard et al. 2010; Pengo et al. 2013). Measurement of cytokine production and cell proliferation can be performed as previously described in this chapter for T cell responses. Cell death (Roche), levels of ATP (Roche), and lactate (MBL) can be measured per the manufacturer's instructions.

5.4.2.5 Measurement of CMA in T Cells

1. Inducing or inhibiting CMA activity

   CMA activity is induced in T cells upon T cell activation. As described above for isolation of T cells and induction of MA, CMA is upregulated under the TCR engagement with anti-CD3 and anti-CD28 antibodies that stimulate

T cell activation. To assay CMA induced by ROS production, 5 µM CellROX-green (Life Technologies) is added to the cells for the last 30 minutes of stimulation with anti-CD3 and anti-CD28 antibodies for 12 hours incubation. CMA can be blocked in murine T cells, by lentiviral-mediated knockdown of *Lamp2a* using one of the three sequences targeting exon 8a of the *Lamp2a* gene 5'GACTGCAGTGCAGATGAAG-3', 5'-CTGCAATCTGATTGATTA-3' or 5'-TAAACACTGCTTGACCACC-3' (Massey et al. 2006). Measurement of the T cell immune response correlated to CMA activity is analyzed as described above for measurement of MA.

2. Monitor CMA activity

The level of LAMP-2A is indicative of CMA activity, and LAMP-2A mRNA levels can be analyzed by real-time PCR using specific primers for *Lamp2a* gene and compared with controls (Valdor et al. 2014). LAMP-2A protein levels can be analyzed by immunoblotting with anti-LAMP-2A antibody (Invitrogen). Functional LAMP-2A expression for CMA activity is identified by microscopic colocalization with lysosomal markers such as lysotracker or anti-LAMP1 (Abcam). Changes in CMA activity can be tracked in T cells using an artificial CMA-fluorescent substrate probe that contains the KFERQ sequence in frame with a photoswitchable fluorescent protein. Induction of CMA leads to lysosomal localization of the CMA substrate probe, which is detected as a change from a diffused fluorescence distribution to a punctate pattern. CMA activity is measured as the number of fluorescent puncta per cell or as the decay in fluorescence activity over a time course (Valdor et al. 2014).

3. Detection of CMA substrates

CMA is the only type of autophagy that does not need formation of intermediate autophagic vacuoles (autophagosomes or microvesicles) for the uptake of cargo into lysosomes. Instead, the CMA-soluble protein substrates are translocated across the lysosomal membrane through the lysosome membrane protein receptor LAMP2A assisted by chaperones HSPA8/HSC70 located in the cytosol and lysosome lumen. The following requirements (Klionsky et al. 2012) need to be checked for consideration as a CMA substrate (Valdor et al. 2014): (1) Identify the presence of a KFERQ-related motif by analysis of the amino acid sequence of the protein. (2) Identify colocalization of the protein, at least partially, with the lysosome by lysosomal markers (typically LAMP-2A and/or LysoTracker). In addition, to support the hypothesis that the protein is a substrate of CMA, the protein substrate must be enriched in the lysosomal compartment under stimuli-activating CMA such as anti-CD3/CD28 stimulation, or after blocking of lysosomal proteases. The lysosomes active for CMA are those containing HSPA8 in their lumens in addition to the presence of LAMP-2A. (3) Detection of the protein substrate interacting with HSPA8 chaperone by coimmunoprecipitation assay. (4) Detection of the protein substrate interacting with LAMP-2A by coimmunoprecipitation assay. Due to the only specific antibody for the LAMP-2A variant made against the cytosolic tail of LAMP-2A protein, which also binds to the substrate proteins, it is recommended to purify the protein of interest and then detect the presence of the LAMP-2A receptor. (5) Demonstrate that the protein substrate is preferentially degraded by selective upregulation of CMA or is enriched by blocking CMA. Overexpression of LAMP-2A upregulates CMA activity in T cells, and the most specific method to block CMA is by knockdown of LAMP-2A, the limiting factor for CMA activity (Kaushik et al. 2010). (6) The most conclusive method to demonstrate that a protein is a CMA substrate in T cells is by showing translocation of the substrate protein into the lysosomes for degradation. An enriched fraction of the protein must be detected in lysosomal fractions incubated with inhibitors of lysosomal proteases when compared with the untreated lysosomal fractions, which is indicative of active uptake and degradation of the protein of interest by the lysosomes. Known CMA substrates such as glyceraldehyde-3-phosphate dehydrogenase, Itch, or RCAN1 can be used as control CMA substrates. Isolation of lysosomes from T cells is performed by discontinuous metrizamide density gradient as described by Cuervo et al. (1997).

ACKNOWLEDGMENTS

Valdor R. is supported by grants from MINECO SAF2015-73923-JIN- AEI/FEDER/ UE. Part of this chapter was elaborated, thanks to a mobility Seneca Foundation grant in Spain ("Programa Jimenez de la Espada, 19667/EE/14"). I would like to gratefully acknowledge Fernando Macian and Ana Maria Cuervo for their help in suggestions and discussions on autophagy.

REFERENCES

Ahlberg, J. and H. Glaumann (1985). Uptake—Microautophagy—And degradation of exogenous proteins by isolated rat liver lysosomes. Effects of pH, ATP, and inhibitors of proteolysis. *Exp Mol Pathol* **42**(1): 78–88.

Aichinger, M., C. Wu, et al. (2013). MA substrates are loaded onto MHC class II of medullary thymic epithelial cells for central tolerance. *J Exp Med* **210**(2): 287–300.

Alonso, S., K. Pethe, et al. (2007). Lysosomal killing of Mycobacterium mediated by ubiquitin-derived peptides is enhanced by autophagy. *Proc Natl Acad Sci U S A* **104**(14): 6031–6.

Andrade, R. M., M. Wessendarp, et al. (2006). CD40 induces macrophage anti-Toxoplasma gondii activity by triggering autophagy-dependent fusion of pathogen-containing vacuoles and lysosomes. *J Clin Invest* **116**(9): 2366–77.

Arsov, I., A. Adebayo, et al. (2011). A role for autophagic protein beclin 1 early in lymphocyte development. *J Immunol* **186**(4): 2201–9.

Bell, B. D., S. Leverrier, et al. (2008). FADD and caspase-8 control the outcome of autophagic signaling in proliferating T cells. *Proc Natl Acad Sci U S A* **105**(43): 16677–82.

Bjorkoy, G., T. Lamark, et al. (2005). p62/SQSTM1 forms protein aggregates degraded by autophagy and has a protective effect on huntingtin-induced cell death. *J Cell Biol* **171**(4): 603–14.

Cadwell, K., J. Y. Liu, et al. (2008). A key role for autophagy and the autophagy gene Atg16l1 in mouse and human intestinal Paneth cells. *Nature* **456**(7219): 259–63.

Cadwell, K., K. K. Patel, et al. (2010). Virus-plus-susceptibility gene interaction determines Crohn's disease gene Atg16L1 phenotypes in intestine. *Cell* **141**(7): 1135–45.

Cannizzo, E. S., C. C. Clement, et al. (2012). Age-related oxidative stress compromises endosomal proteostasis. *Cell Rep* **2**(1): 136–49.

Castellino, F. and R. N. Germain (1995). Extensive trafficking of MHC class II-invariant chain complexes in the endocytic pathway and appearance of peptide-loaded class II in multiple compartments. *Immunity* **2**(1): 73–88.

Castillo, E. F., A. Dekonenko, et al. (2012). Autophagy protects against active tuberculosis by suppressing bacterial burden and inflammation. *Proc Natl Acad Sci U S A* **109**(46): E3168–76.

Chaturvedi, A., D. Dorward, et al. (2008). The B cell receptor governs the subcellular location of Toll-like receptor 9 leading to hyperresponses to DNA-containing antigens. *Immunity* **28**(6): 799–809.

Chu, C. T., J. Zhu, et al. (2007). Beclin 1-independent pathway of damage-induced mitophagy and autophagic stress: Implications for neurodegeneration and cell death. *Autophagy* **3**(6): 663–6.

Cooney, R., J. Baker, et al. (2010). NOD2 stimulation induces autophagy in dendritic cells influencing bacterial handling and antigen presentation. *Nat Med* **16**(1): 90–7.

Copetti, T., F. Demarchi, et al. (2009). p65/RelA binds and activates the beclin 1 promoter. *Autophagy* **5**(6): 858–9.

Criollo, A., L. Senovilla, et al. (2010). The IKK complex contributes to the induction of autophagy. *EMBO J* **29**(3): 619–31.

Cuervo, A. M., J. F. Dice, et al. (1997). A population of rat liver lysosomes responsible for the selective uptake and degradation of cytosolic proteins. *J Biol Chem* **272**(9): 5606–15.

Cuervo, A. M. and E. Wong (2014). Chaperone-mediated autophagy: Roles in disease and aging. *Cell Res* **24**(1): 92–104.

Cuervo, A. M. and F. Macian (2014). Autophagy and the immune function in aging. *Curr Opin Immunol* **29**: 97–104.

Cuervo, A. M., L. Stefanis, et al. (2004). Impaired degradation of mutant alpha-synuclein by chaperone-mediated autophagy. *Science* **305**(5688): 1292–5.

Delgado, M. A., R. A. Elmaoued, et al. (2008). Toll-like receptors control autophagy. *EMBO J* **27**(7): 1110–21.

Dengjel, J., O. Schoor, et al. (2005). Autophagy promotes MHC class II presentation of peptides from intracellular source proteins. *Proc Natl Acad Sci U S A* **102**(22): 7922–7.

Deretic, V., T. Kimura, et al. (2015). Immunologic manifestations of autophagy. *J Clin Invest* **125**(1): 75–84.

Dupont, N., S. Jiang, et al. (2011). Autophagy-based unconventional secretory pathway for extracellular delivery of IL-1beta. *EMBO J* **30**(23): 4701–11.

Dupont, N., S. Lacas-Gervais, et al. (2009). Shigella phagocytic vacuolar membrane remnants participate in the cellular response to pathogen invasion and are regulated by autophagy. *Cell Host Microbe* **6**(2): 137–49.

English, L., M. Chemali, et al. (2009). Autophagy enhances the presentation of endogenous viral antigens on MHC class I molecules during HSV-1 infection. *Nat Immunol* **10**(5): 480–7.

Espert, L., M. Denizot, et al. (2006). Autophagy is involved in T cell death after binding of HIV-1 envelope proteins to CXCR4. *J Clin Invest* **116**(8): 2161–72.

Feng, Y., D. He, et al. (2014). The machinery of MA. *Cell Res* **24**(1): 24–41.

Fiegl, D., D. Kagebein, et al. (2013). Amphisomal route of MHC class I cross-presentation in bacteria-infected dendritic cells. *J Immunol* **190**(6): 2791–806.

Gannage, M., R. B. da Silva, et al. (2013). Antigen processing for MHC presentation via MA. *Methods Mol Biol* **960**: 473–88.

Guo, J. Y., B. Xia, et al. (2013). Autophagy-mediated tumor promotion. *Cell* **155**(6): 1216–9.

Gutierrez, M. G., S. S. Master, et al. (2004). Autophagy is a defense mechanism inhibiting BCG and *Mycobacterium tuberculosis* survival in infected macrophages. *Cell* **119**(6): 753–66.

Harris, J., O. Hanrahan, et al. (2009). Measuring autophagy in macrophages. *Curr Protoc Immunol* Chapter 14: Unit 14 14.

Harris, J., S. A. De Haro, et al. (2007). T helper 2 cytokines inhibit autophagic control of intracellular *Mycobacterium tuberculosis*. *Immunity* **27**(3): 505–17.

Hubbard, V. M., R. Valdor, et al. (2010). MA regulates energy metabolism during effector T cell activation. *J Immunol* **185**(12): 7349–57.

Hubbard, V. M., R. Valdor, et al. (2012). Selective autophagy in the maintenance of cellular homeostasis in aging organisms. *Biogerontology* **13**(1): 21–35.

Hubbard-Lucey, V. M., Y. Shono, et al. (2014). Autophagy gene Atg16L1 prevents lethal T cell alloreactivity mediated by dendritic cells. *Immunity* **41**(4): 579–91.

Hayward, A. P. and S. P. Dinesh-Kumar (2010). Special delivery for MHC II via autophagy. *Immunity* **32**(5): 587–90.

Intemann, C. D., T. Thye, et al. (2009). Autophagy gene variant IRGM-261T contributes to protection from tuberculosis caused by *Mycobacterium tuberculosis* but not by M. africanum strains. *PLoS Pathog* **5**(9): e1000577.

Ireland, J. M. and E. R. Unanue (2011). Autophagy in antigen-presenting cells results in presentation of citrullinated peptides to CD4 T cells. *J Exp Med* **208**(13): 2625–32.

Jabir, M. S., N. D. Ritchie, et al. (2014). Caspase-1 cleavage of the TLR adaptor TRIF inhibits autophagy and beta-interferon production during *Pseudomonas aeruginosa* infection. *Cell Host Microbe* **15**(2): 214–27.

Jagannath, C., D. R. Lindsey, et al. (2009). Autophagy enhances the efficacy of BCG vaccine by increasing peptide presentation in mouse dendritic cells. *Nat Med* **15**(3): 267–76.

Jia, W., H. H. Pua, et al. (2011). Autophagy regulates endoplasmic reticulum homeostasis and calcium mobilization in T lymphocytes. *J Immunol* **186**(3): 1564–74.

Jin, Y., C. Sun, et al. (2014). Regulation of SIV antigen-specific CD4+ T cellular immunity via autophagosome-mediated MHC II molecule-targeting antigen presentation in mice. *PLoS One* **9**(3): e93143.

Kaushik, S., A. C. Massey, et al. (2006). Lysosome membrane lipid microdomains: Novel regulators of chaperone-mediated autophagy. *EMBO J* **25**(17): 3921–33.

Kaushik, S., A. C. Massey, et al. (2008). Constitutive activation of chaperone-mediated autophagy in cells with impaired MA. *Mol Biol Cell* **19**(5): 2179–92.

Kaushik, S., U. Bandyopadhyay, et al. (2011). Chaperone-mediated autophagy at a glance. *J Cell Sci* 124(Pt 4): 495–9.

Keller, C. W., C. Fokken, et al. (2011). TNF-alpha induces MA and regulates MHC class II expression in human skeletal muscle cells. *J Biol Chem* **286**(5): 3970–80.

Kiffin, R., C. Christian, et al. (2004). Activation of chaperone-mediated autophagy during oxidative stress. *Mol Biol Cell* **15**(11): 4829–40.

Kirkin, V., D. G. McEwan, et al. (2009). A role for ubiquitin in selective autophagy. *Mol Cell* **34**: 259–69.

Klionsky, D. J. (2005). Autophagy. *Curr Biol* **15**(8): R282–3.

Klionsky, D. J., F. C. Abdalla, et al. (2012). Guidelines for the use and interpretation of assays for monitoring autophagy. *Autophagy* **8**(4): 445–544.

Kon, M., R. Kiffin, et al. (2011). Chaperone-mediated autophagy is required for tumor growth. *Sci Transl Med* **3**(109): 109ra117.

Kovacs, J. R., C. Li, et al. (2012). Autophagy promotes T-cell survival through degradation of proteins of the cell death machinery. *Cell Death Differ* **19**(1): 144–52.

Kristensen, A. R., S. Schandorff, et al. (2008). Ordered organelle degradation during starvation-induced autophagy. *Mol Cell Proteomics* **7**(12): 2419–28.

Kroemer, G., G. Marino, et al. (2010). Autophagy and the integrated stress response. *Mol Cell* **40**(2): 280–93.

Kumar, D., L. Nath, et al. (2010). Genome-wide analysis of the host intracellular network that regulates survival of *Mycobacterium tuberculosis*. *Cell* **140**(5): 731–43.

Kwon, J., K. E. Shatynski, et al. (2010). The nonphagocytic NADPH oxidase Duox1 mediates a positive feedback loop during T cell receptor signaling. *Sci Signal* **3**(133): ra59.

Lee, H. K., J. M. Lund, et al. (2007). Autophagy-dependent viral recognition by plasmacytoid dendritic cells. *Science* **315**(5817): 1398–401.

Lee, H. K., L. M. Mattei, et al. (2010). In vivo requirement for Atg5 in antigen presentation by dendritic cells. *Immunity* **32**(2): 227–39.

Leib, D. A., D. E. Alexander, et al. (2009). Interaction of ICP34.5 with Beclin 1 modulates herpes simplex virus type 1 pathogenesis through control of CD4+ T-cell responses. *J Virol* **83**(23): 12164–71.

Levine, B. (2005). Eating oneself and uninvited guests: Autophagy-related pathways in cellular defense. *Cell* **120**(2): 159–62.

Li, C., E. Capan, et al. (2006). Autophagy is induced in CD4+ T cells and important for the growth factor-withdrawal cell death. *J Immunol* **177**(8): 5163–8.

Li, H., Y. Li, et al. (2011). Alpha-alumina nanoparticles induce efficient autophagy-dependent cross-presentation and potent antitumour response. *Nat Nanotechnol* **6**(10): 645–50.

Li, Y., T. Hahn, et al. (2012). The vitamin E analogue alpha-TEA stimulates tumor autophagy and enhances antigen cross-presentation. *Cancer Res* **72**(14): 3535–45.

Lin, L. T., P. W. Dawson, et al. (2010). Viral interactions with MA: A double-edged sword. *Virology* **402**(1): 1–10.

Ling, Y. M., M. H. Shaw, et al. (2006). Vacuolar and plasma membrane stripping and autophagic elimination of *Toxoplasma gondii* in primed effector macrophages. *J Exp Med* **203**(9): 2063–71.

Mandell, M. A., A. Jain, et al. (2014). TRIM proteins regulate autophagy and can target autophagic substrates by direct recognition. *Dev Cell* **30**(4): 394–409.

Marino, G., M. Niso-Santano, et al. (2014). Self-consumption: The interplay of autophagy and apoptosis. *Nat Rev Mol Cell Biol* **15**(2): 81–94.

Massey, A. C., S. Kaushik, et al. (2006). Consequences of the selective blockage of chaperone-mediated autophagy. *Proc Natl Acad Sci U S A* **103**(15): 5805–10.

McLeod, I. X., X. Zhou, et al. (2011). The class III kinase Vps34 promotes T lymphocyte survival through regulating IL-7Ralpha surface expression. *J Immunol* **187**(10): 5051–61.

Meyer-Morse, N., J. R. Robbins, et al. (2010). Listeriolysin O is necessary and sufficient to induce autophagy during *Listeria monocytogenes* infection. *PLoS One* **5**(1): e8610.

Michaud, M., I. Martins, et al. (2011). Autophagy-dependent anticancer immune responses induced by chemotherapeutic agents in mice. *Science* **334**(6062): 1573–7.

Miller, B. C., Z. Zhao, et al. (2008). The autophagy gene ATG5 plays an essential role in B lymphocyte development. *Autophagy* **4**(3): 309–14.

Mizushima, N., B. Levine, et al. (2008). Autophagy fights disease through cellular self-digestion. *Nature* **451**(7182): 1069–75.

Mizushima, N., T. Yoshimori, et al. (2011). The role of Atg proteins in autophagosome formation. *Annu Rev Cell Dev Biol* **27**: 107–32.

Monroe, J. G. and M. E. Keir (2008). Bridging Toll-like- and B cell-receptor signaling: Meet me at the autophagosome. *Immunity* **28**(6): 729–31.

Mostowy, S., V. Sancho-Shimizu, et al. (2011). p62 and NDP52 proteins target intracytosolic Shigella and Listeria to different autophagy pathways. *J Biol Chem* **286**(30): 26987–95.

Munz, C. (2013). Antigen processing via autophagy—Not only for MHC class II presentation anymore? *Curr Opin Immunol* **22**(1): 89–93.

Nakagawa, I., A. Amano, et al. (2004). Autophagy defends cells against invading group A Streptococcus. *Science* **306**(5698): 1037–40.

Nakahira, K., J. A. Haspel, et al. (2011). Autophagy proteins regulate innate immune responses by inhibiting the release of mitochondrial DNA mediated by the NALP3 inflammasome. *Nat Immunol* **12**(3): 222–30.

Nedjic, J., M. Aichinger, et al. (2008). Autophagy in thymic epithelium shapes the T-cell repertoire and is essential for tolerance. *Nature* **455**(7211): 396–400.

Nishida, Y., S. Arakawa, et al. (2009). Discovery of Atg5/Atg7-independent alternative MA. *Nature* **461**(7264): 654–8.

Nixon, R. A. (2013). The role of autophagy in neurodegenerative disease. *Nat Med* **19**(8): 983–97.

Ogawa, M., T. Yoshimori, et al. (2005). Escape of intracellular Shigella from autophagy. *Science* **307**(5710): 727–31.

Ohman, T., L. Teirila, et al. (2014). Dectin-1 pathway activates robust autophagy-dependent unconventional protein secretion in human macrophages. *J Immunol* **192**(12): 5952–62.

Orvedahl, A., S. MacPherson, et al. (2010). Autophagy protects against Sindbis virus infection of the central nervous system. *Cell Host Microbe* **7**(2): 115–27.

Paludan, C., D. Schmid, et al. (2005). Endogenous MHC class II processing of a viral nuclear antigen after autophagy. *Science* **307**(5709): 593–6.

Pankiv, S., T. H. Clausen, et al. (2007). p62/SQSTM1 binds directly to Atg8/LC3 to facilitate degradation of ubiquitinated protein aggregates by autophagy. *J Biol Chem* **282**(33): 24131–45.

Parekh, V. V., L. Wu, et al. (2013). Impaired autophagy, defective T cell homeostasis, and a wasting syndrome in mice with a T cell-specific deletion of Vps34. *J Immunol* **190**(10): 5086–101.

Park, H. J., S. J. Lee, et al. (2011). IL-10 inhibits the starvation induced autophagy in macrophages via class I phosphatidylinositol 3-kinase (PI3K) pathway. *Mol Immunol* **48**(4): 720–7.

Paul, S., A. K. Kashyap, et al. (2012). Selective autophagy of the adaptor protein Bcl10 modulates T cell receptor activation of NF-kappaB. *Immunity* **36**(6): 947–58.

Pengo, N., M. Scolari, et al. (2013). Plasma cells require autophagy for sustainable immunoglobulin production. *Nat Immunol* **14**(3): 298–305.

Pilli, M., J. Arko-Mensah, et al. (2012). TBK-1 promotes autophagy-mediated antimicrobial defense by controlling autophagosome maturation. *Immunity* **37**(2): 223–34.

Ponpuak, M., A. S. Davis, et al. (2010). Delivery of cytosolic components by autophagic adaptor protein p62 endows autophagosomes with unique antimicrobial properties. *Immunity* **32**(3): 329–41.

Pua, H. H., I. Dzhagalov, et al. (2007). A critical role for the autophagy gene Atg5 in T cell survival and proliferation. *J Exp Med* **204**(1): 25–31.

Pua, H. H., J. Guo, et al. (2009). Autophagy is essential for mitochondrial clearance in mature T lymphocytes. *J Immunol* **182**(7): 4046–55.

Py, B. F., M. M. Lipinski, et al. (2007). Autophagy limits Listeria monocytogenes intracellular growth in the early phase of primary infection. *Autophagy* **3**(2): 117–25.

Ravindran, R., N. Khan, et al. (2014). Vaccine activation of the nutrient sensor GCN2 in dendritic cells enhances antigen presentation. *Science* **343**(6168): 313–7.

Romao, S., N. Gasser, et al. (2013). Autophagy proteins stabilize pathogen-containing phagosomes for prolonged MHC II antigen processing. *J Cell Biol* **203**(5): 757–66.

Sahu, R., S. Kaushik, et al. (2011). Microautophagy of cytosolic proteins by late endosomes. *Dev Cell* **20**(1): 131–9.

Sanjuan, M. A., S. Milasta, et al. (2009). Toll-like receptor signaling in the lysosomal pathways. *Immunol Rev* **227**(1): 203–20.

Schmid, D. and C. Munz (2008). Localization and MHC class II presentation of antigens targeted for MA. *Methods Mol Biol* **445**: 213–25.

Schmid, D., M. Pypaert, et al. (2007). Antigen-loading compartments for major histocompatibility complex class II molecules continuously receive input from autophagosomes. *Immunity* **26**(1): 79–92.

Sena, L. A., S. Li, et al. (2013). Mitochondria are required for antigen-specific T cell activation through reactive oxygen species signaling. *Immunity* **38**(2): 225–36.

Settembre, C., C. Di Malta, et al. (2011). TFEB links autophagy to lysosomal biogenesis. *Science* **332**(6036): 1429–33.

Shen, S., M. Niso-Santano, et al. (2012). Cytoplasmic STAT3 represses autophagy by inhibiting PKR activity. *Mol Cell* **48**(5): 667–80.

Shi, C. S. and J. H. Kehrl (2008). MyD88 and Trif target Beclin 1 to trigger autophagy in macrophages. *J Biol Chem* **283**(48): 33175–82.

Shi, C. S. and J. H. Kehrl (2010). TRAF6 and A20 regulate lysine 63-linked ubiquitination of Beclin-1 to control TLR4-induced autophagy. *Sci Signal* **3**(123): ra42.

Shimada, K., T. R. Crother, et al. (2012). Oxidized mitochondrial DNA activates the NLRP3 inflammasome during apoptosis. *Immunity* **36**(3): 401–14.

Shui, W., L. Sheu, et al. (2008). Membrane proteomics of phagosomes suggests a connection to autophagy. *Proc Natl Acad Sci U S A* **105**(44): 16952–7.

Singh, S. B., A. S. Davis, et al. (2006). Human IRGM induces autophagy to eliminate intracellular Mycobacteria. *Science* **313**(5792): 1438–41.

Sridhar, S., Y. Botbol, et al. (2012). Autophagy and disease: Always two sides to a problem. *J Pathol* **226**(2): 255–73.

Stephenson, L. M., B. C. Miller, et al. (2009). Identification of Atg5-dependent transcriptional changes and increases in mitochondrial mass in Atg5-deficient T lymphocytes. *Autophagy* **5**(5): 625–35.

Subauste, C. S., R. M. Andrade, et al. (2007). CD40-TRAF6 and autophagy-dependent antimicrobial activity in macrophages. *Autophagy* **3**(3): 245–8.

Subramani, S. and V. Malhotra (2013). Non-autophagic roles of autophagy-related proteins. *EMBO Rep* **14**(2): 143–51.

Talloczy, Z., H. W. T. Virgin, et al. (2006). PKR-dependent autophagic degradation of herpes simplex virus type 1. *Autophagy* **2**(1): 24–9.

Thurston, T. L., G. Ryzhakov, et al. (2009). The TBK1 adaptor and autophagy receptor NDP52 restricts the proliferation of ubiquitin-coated bacteria. *Nat Immunol* **10**(11): 1215–21.

Travassos, L. H., L. A. Carneiro, et al. (2010). Nod1 and Nod2 direct autophagy by recruiting ATG16L1 to the plasma membrane at the site of bacterial entry. *Nat Immunol* **11**(1): 55–62.

Uhl, M., O. Kepp, et al. (2009). Autophagy within the antigen donor cell facilitates efficient antigen cross-priming of virus-specific CD8+ T cells. *Cell Death Differ* **16**(7): 991–1005.

Uttenweiler, A., H. Schwarz, et al. (2007). The vacuolar transporter chaperone (VTC) complex is required for microautophagy. *Mol Biol Cell* **18**(1): 166–75.

Valdor, R., E. Mocholi, et al. (2014). Chaperone-mediated autophagy regulates T cell responses through targeted degradation of negative regulators of T cell activation. *Nat Immunol* **15**(11): 1046–54.

Valdor, R. and F. Macian (2012). Autophagy and the regulation of the immune response. *Pharmacol Res* **66**(6): 475–83.

Watson, R. O., P. S. Manzanillo, et al. (2012). Extracellular *M. tuberculosis* DNA targets bacteria for autophagy by activating the host DNA-sensing pathway. *Cell* **150**(4): 803–15.

Weindel, C. G., L. J. Richey, et al. (2015). B cell autophagy mediates TLR7-dependent autoimmunity and inflammation. *Autophagy* **11**(7): 1010–24.

Wild, P., H. Farhan, et al. (2011). Phosphorylation of the autophagy receptor optineurin restricts Salmonella growth. *Science* **333**(6039): 228–33.

Wildenberg, M. E., A. C. Vos, et al. (2012). Autophagy attenuates the adaptive immune response by destabilizing the immunologic synapse. *Gastroenterology* **142**(7): 1493–503 e6.

Willinger, T. and R. A. Flavell (2012). Canonical autophagy dependent on the class III phosphoinositide-3 kinase Vps34 is required for naive T-cell homeostasis. *Proc Natl Acad Sci U S A* **109**(22): 8670–5.

Xu, Y., C. Jagannath, et al. (2007). Toll-like receptor 4 is a sensor for autophagy associated with innate immunity. *Immunity* **27**(1): 135–44.

Yang, Z. and D. J. Klionsky (2010). Mammalian autophagy: Core molecular machinery and signaling regulation. *Curr Opin Cell Biol* **22**(2): 124–31.

Yang, Z., H. Fujii, et al. (2013). Phosphofructokinase deficiency impairs ATP generation, autophagy, and redox balance in rheumatoid arthritis T cells. *J Exp Med* **210**(10): 2119–34.

Yoshikawa, Y., M. Ogawa, et al. (2009). Listeria monocytogenes ActA-mediated escape from autophagic recognition. *Nat Cell Biol* **11**(10): 1233–40.

Zheng, Y. T., S. Shahnazari, et al. (2009). The adaptor protein p62/SQSTM1 targets invading bacteria to the autophagy pathway. *J Immunol* **183**(9): 5909–16.

Zhou, D., P. Li, et al. (2005). Lamp-2a facilitates MHC class II presentation of cytoplasmic antigens. *Immunity* **22**(5): 571–81.

Zhou, R., A. S. Yazdi, et al. (2011). A role for mitochondria in NLRP3 inflammasome activation. *Nature* **469**(7329): 221–5.

6 Cilia-Related Autophagy in Nutrient Sensing

Olatz Pampliega
Institut des Maladies Neurodégéneratives, UMR 5293
CNRS, Université de Bordeaux
Bordeaux, France

Patrice Codogno
Université Paris Descartes-Sorbonne
Paris, France

CONTENTS

6.1 INTRODUCTION

Nutrients, also known as macronutrients, are required for energy production and synthesis of cellular components, and therefore, maintaining a correct nutrient homeostasis is vital in all living organisms. Under basal conditions, nutrients are synthesized *de novo*, whereas during high metabolic demands and stress conditions, the cell obtains these macromolecules from extracellular sources. Indeed, nutrient scarcity is a strong selective pressure that has resulted in the selection of efficient sensing mechanisms in all organisms. As a result, the cell provides a plethora of mechanisms to sense the abundance of glucose, related sugars, amino acids, and lipids within cell, but also in the extracellular space, from where the information has to be transduced into the intracellular machineries. Although extracellular and intracellular nutrient-sensing mechanisms are essential for an optimal nutrient homeostasis, their working mechanisms are understudied compared with their physiological importance [1].

Nutrient availability in the extracellular milieu is transduced into intracellular signaling pathways by receptors located in the plasma membrane. Their relative location on the cell surface will determine their availability to the upcoming signals, as well as the regulatory and switch-off mechanisms. In order to maximize these sensory functions, most cell types on their surface contain microtubule-based organelles, known as cilia, engulfed in a continuous but yet specialized portion of the plasma membrane. The ciliary membrane is enriched in specific receptors, and thus, cilia can be recognized as signaling hubs that compartmentalize the transmission of extracellular signals into effector cascades.

In recent years, growing number of cilia-dependent signaling cascades have been described, which in turn has broadened the connection between this organelle and other cellular events [2].

Inside the cell, different catabolic pathways act simultaneously to provide energy and macromolecules during nutrient starvation. Among them, autophagy mobilizes nutrient stores such as glycogen, lipid droplets, proteins, and organelles. The principal function of autophagy is to provide nutrients and essential amino acids through degradation of proteins, allowing cell survival during starvation. Nevertheless, recent data show that autophagy mobilizes intracellular stores also in basal conditions, which is tightly linked to the maintenance of cellular homeostasis. The nutrient, growth factor, and hormonal status of the cell regulate autophagy through the integration of different signaling pathways that are mainly coordinated by mTOR, a major autophagy inhibitor. Nutrient scarcity is sensed by the adenosine monophosphate–activated protein kinase (AMPK), which in turn inactivates mTORC1, one of the two mTOR complexes [3]. In fact, regulation of nutrient sensing by the two mTOR complexes occurs at the lysosome, the ultimate organelle in the autophagy pathway where degradation occurs. During nutrient-rich conditions, mTORC1 senses amino acids, and it gets translocated to the lysosomal surface to be activated, which in turn represses macroautophagy (here referred as autophagy) [4]. Additionally, mTORC2 through its kinase Akt1 inhibits chaperone-mediated autophagy—the selective type of autophagy that degrades single cytosolic proteins through the recognition of a pentapeptide motif [5,6]. Thus, nutrient-sensing mechanisms are intrinsically linked to autophagy at different levels, which highlights the importance of this catabolic pathway in the maintenance of cell homeostasis.

Recently, it has been uncovered that the regulation of autophagy through the primary cilium (PC) in periods of nutrient scarcity, linking the extracellular sensing mechanisms with the intracellular effector pathways such as autophagy, ultimately provides the amino acids and energy during starvation. Here, we extensively describe the links between autophagy and the PC. We first describe the major structural and functional components of the PC, as we understand that autophagy components are described in other chapters within this book. We then revise the signaling pathways that require the presence of the cilium to function, and which also regulate autophagy. Finally, we provide evidence of the link between autophagy and ciliogenesis, as well as the potential implications of this bidirectional interrelation in pathological conditions.

6.2 THE DYNAMIC ROLE OF CILIARY COMPONENTS

Although for many years the PC was visualized as a static and vestigial organelle, the exhaustive molecular characterization of its components in the latest years has changed this vision. More importantly, the description of the cilium as a prominent signaling hub, where several key signaling pathways converge, has finally depicted this organelle as a highly dynamic structure in continuous regulation. Thus, here, we describe the structure and functional roles of the main ciliary components as a way to better understand the dynamics of this organelle (Figure 6.1).

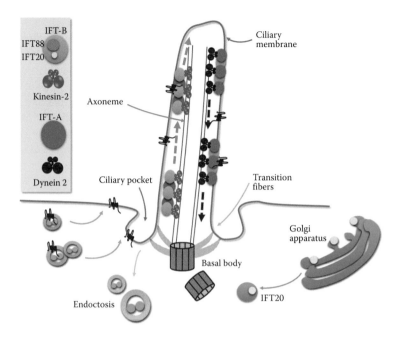

FIGURE 6.1 Structure of primary cilia. Microtubule-based ciliary axoneme is rooted to the basal body, which is anchored to the plasma membrane by the transition fibers. Cilia elongation is driven by anterograde transport of IFT-B complexes through kinesin-2 motors, while IFT-A complexes participate in the retrograde transport through dynein-2 motors. Some ciliary proteins such as IFT20 participate in vesicular trafficking from the Golgi apparatus. The ciliary pocket is an invagination of the plasma membrane, closely related to the ciliary membrane with high activity in endocytosis.

6.2.1 THE BASAL BODY

The central axis of the cilium—the axoneme—emerges from the basal body, a structure enriched in gamma-tubulin. The mechanism for basal body generation differs in PC and motile cilium (MC). PC axoneme is derived from the mother centriole, which after cell division generates a daughter centriole that remains associated in the same structure through interconnecting fibers. Only the mature mother centriole of the PC can nucleate the ciliary axoneme, and proteins present in the appendages, such as OFD2, are required for cilia formation. At the distal end of the mother centriole, the transition zone arises, which is continuous with the ciliary axoneme [7]. By contrast, multiciliated cells require *de novo* centriole multiplication preceding ciliogenesis, which occurs within a single cell and not during cell division. Axonemes will be associated to a single centriole in this case. In both PC and MC formations, the centrioles migrate toward the cell surface during maturation, where they acquire accessory structures and get surrounded by the electron-dense pericentriolar material (PCM) [8]. In a last step, mature basal bodies migrate to the cell surface, associating with trafficking vesicles that subsequently fuse with the plasma membrane, giving rise to the ciliary membrane. The basal bodies are anchored to the actin-rich base of the ciliary membrane transition fibers [9].

6.2.2 THE CILIARY AXONEME

The main structure of cilia comprises a central axoneme of tubulin polymers enriched in acetylated tubulin, the composition of which defines two types of cilia, PC and MC. Primary cilia are nonmotile and solitary organelles with nine outer microtubule doublets and an additional central pair in their axoneme. Motile cilia instead are beating organelles that lack the central pair of microtubules, and are found on the epithelial surfaces of the oviduct, the airways, and of the brain ependyma, being responsible for gamete and zygote transport, mucociliary clearance, and ependymal flow, respectively [10]. Motile and primary cilia axoneme share an equal molecular structure, where each of the outer pair of microtubules is composed by an A tubule and a B tubule. The nine pairs of microtubules are held together by nexin linkages. A tubule connects to the B tubule in the next pair of microtubules through outer and inner dynein arms, which are the main driving forces for axonemal movement in motile cilia. Elongation of the axoneme occurs by addition of $\alpha\beta$-tubulin subunits to the distal (+) end.

It is important to remember that the ciliary axoneme grows rooted to the basal body in a highly dynamic way. New tubulin molecules are incorporated at the tip of the axoneme, while at the same time it is also disassembling through a mechanism that involves molecular motors. Thus, this feature implies that not only the growth but also the maintenance of ciliary length is an active process that requires the cilia assembly machinery [11].

6.2.3 INTRAFLAGELLAR TRANSPORT

Axonemal growth is mainly maintained by the intraflagellar transport (IFT) machinery, due to the fact that the PC lacks the protein synthesis machinery. IFT is a regulated transport system essential for ciliogenesis, as its disruption or blockage severely disrupts ciliary architecture and function, leading to alterations incompatible with life [12]. Partial loss of function of IFT genes results in severe developmental defects that are categorized as ciliopathies. At the molecular level, the IFT system is composed of IFT-B and IFT-A complexes. IFT-B is responsible for the anterograde transport of cargo to the ciliary tip, which is based on kinesin (kinesin-2) motors, and is essential for the assembly and maintenance of cilia. The IFT-B complex contains 16 core proteins, of which IFT88 and IFT20 have been extensively studied. Indeed, genetic blockage of either of these proteins leads to blockage of ciliogenesis and cilia-related pathologies.

IFT particles are predominantly located in the cilium or centrosomal region, although during steady state, 80% of IFT complexes locate in the cytosol and are enriched in the basal body [13]. IFT20 protein is the exception to this rule, as it also localizes in the Golgi apparatus, and it shuttles from this organelle to the ciliary base in trafficking vesicles. Some studies suggest that IFT20 has a role in vesicular trafficking, participating in the sorting of proteins from the Golgi apparatus. In this proposed model, cargo proteins that need to be targeted to the ciliary membrane would traffic in IFT20-containing vesicles tethered by Golgi proteins such as GMAP120 [14]. Other extra-ciliary functions of IFT20 include the formation of the immune synapse, a structure that serves as a signaling platform between antigen-presenting cells and nonciliated T cells. Components of the ciliary transport machinery, including IFT52 and IFT57, participate in the formation and function of the immune synapse. Indeed, IFT20 also traffics the autophagy protein ATG16 toward the basal body [15], revealing that the IFT machinery could be a more general trafficking system to polarized structures [14,16].

However, most IFT-B proteins have cilia-specific functions. The most prominent and studied one is IFT88, which is homologous to the mouse and human *Tg737* gene. Mice with defective *Tg737* present severe kidney and liver defects similar to autosomal recessive polycystic kidney disease (ARPKD) in humans, which in fact represented the first evidence that defects in primary cilia structure and function were involved in human pathologies. Several other IFT-B proteins have a strong effect on ciliogenesis, such as IFT-172, -80, -52, and -46, whereas mutations on IFT-81, 74, -70, -57, and -27 have only a moderate effect on ciliogenesis. The relationship between the different IFT particles, as well as their association to the different subunits, is not well understood, and highlights the complex biogenesis of this signaling organelle [13]. On the contrary, IFT-A complex transports IFT trains and cargo back from the tip of the cilium through dynein-2 motors. Although IFT-A complex is not necessary for cilia assembly, mutations in proteins of this complex or in these motors result in a short or swollen cilia that accumulate IFT particles [11].

Besides its role in the physical assembly of cilia, IFT transport is required for the functioning of Hedgehog (Hh) pathway. Vertebrates rely on the Hh signaling for correct organogenesis, and alterations of this pathway are involved in tumorigenesis. Interestingly, the activity of Hh in vertebrates is dependent on the primary cilium, whereas invertebrates such as the fruit fly present a cilia-independent signaling mechanism [17]. The core proteins from the Hh pathway are enriched in the primary cilium. Indeed, upon Hh ligand binding to the Patched1 receptor, Smoothened (Smo) receptor (Smo) is released from intracellular vesicles and traffics to the ciliary axoneme in an IFT-dependent manner. However, it has been probed that Smo can be recruited to the cilium in a ligand-independent manner, and that indeed Hh binding solely increases ciliary accumulation of Smo. Upon recruitment of Smo to the cilium, Gli repressor forms (GliRs) are processed into Gli transcription factors (GliAs) that activate the expression of Hh target genes. IFT is indeed required for processing GliRs into GliAs [18]. More detail studies showed the role of the different IFT proteins in Hh signaling processing. It is known, for example, that IFT-B components such as IFT88, IFT172, or the motor protein Kif3a are required for Hh transduction downstream of Patched1 receptor, and upstream of the direct targets of Hh signaling [17]. While IFT-B components are required for trafficking of Hh components to cilium, it has been suggested that IFT-A proteins are required for localization of specific proteins into the ciliary membrane [19]. Although the Hh pathway is closely related to primary cilia, intriguingly other developmental signaling pathways such as transforming growth factor-β (TGF-β), Notch, and fibroblast growth factor (Fgf) do not seem to require primary cilia for a proper signaling activity [18].

6.2.4 THE BBSOME

In addition to the IFT system, other transport machineries participate in protein recruitment and trafficking into the cilium [20]. The BBSome is a transport mechanism critical for the delivery of proteins to the ciliary membrane, although it is not required for ciliary assembly itself. The exact function of Bardet–Biedl syndrome (BBS) proteins is yet to be elucidated, but it is known to be composed of seven core components that stabilize IFT-B and IFT-A during the anterograde transport. The BBSome is responsible for the assembly of IFT complexes at the base of the cilia, after which it traffics along the axoneme together with the IFT-B machinery. Therefore, BBS proteins can be found both at the ciliary tip and at the basal body. Once at the ciliary tip, the BBSome reorganizes the IFT complex for retrograde transport, leading to the formation of IFT-A [21]. Still, further studies are needed to better understand the functional role of the BBSome, as well as its interaction with the IFT transport machinery.

6.2.5 OTHER CILIARY TRAFFICKING MECHANISMS

Some studies have suggested that proteins involved in the formation and maintenance of epithelial polarity, such as Crumbs3, Par3, Par6, and aPKC, may have a role in cilia assembly. Nevertheless, these studies have to be taken carefully, because only Crumbs3 has been localized to the cilium, and it needs to be better understood whether polarity proteins directly affect cilia formation or if it is a secondary effect [21].

In addition, the Ras superfamily of small GTPases, such as Rab, Arf, and Arl subfamilies, are implicated in ciliary processes such as the budding of rhodopsin carriers from Golgi and their fusion to the connecting cilium in the retina [20].

6.2.6 THE TRANSITION FIBERS AND THE TRANSITION ZONE

Entry of proteins into the cilium is a regulated process that occurs through the transition fibers and the transition zone. Transition fibers form early in ciliogenesis, and are considered a functional region of the basal body. On one side, they represent the membrane attachment point for the basal body. On the contrary, they function as a transport blocking system, and therefore, it can be said that they contribute to compartmentalization. Indeed, the small inter-fiber spaces do not allow cytoplasmic vesicles to enter the cilium, and therefore, transition fibers limit the fusion of vesicles with the plasma membrane. They are also an important docking and assembly site for IFT proteins; therefore, they contribute to a correct cilia formation [9].

Another component that regulates cargo entry into the cilium is the transition zone, formed during the early steps of ciliogenesis. Once it matures, the transition zone works as a ciliary gatekeeper, acting as a membrane diffusion barrier as well as a modulator of IFT. Interestingly, the entry of cytoplasmic proteins into the cilium through the transition zone is not a passive mechanism, and requires an import signal similar to the nuclear localization signal (NLS). Its functioning is regulated by specific membrane-associated proteins, although at the molecular level, the transition zone shares components with the nuclear pore complex, the highly specialized and selective complex to regulate the entry of macromolecules into the cell nucleus [9]. Indeed, recent discoveries have led to the description of a ciliary pore in this region, which is described below.

6.2.7 THE CILIARY PORE

Although the transition zone has been known for many years in the cilia field, only recent discoveries have described the presence of a ciliary pore complex, a highly complex, specialized structure sited at the base of the cilium, and that is part of the transition zone. At the molecular level, the ciliary pore shares components with the nuclear import machinery, and therefore, cytosolic proteins require an import signal similar to the NLS to enter the cilium. It has also been described that several nucleoporins at the base of cilia function in a similar manner to their functioning at the cell nucleus. Therefore, it is recognized that the ciliary pore complex acts as a size-dependent diffusion barrier, determining the entry of proteins into the cilium [22].

The molecular and functional analogy of the nuclear and the ciliary pore complexes highlights the importance of the cilium as a signaling hub, where the unique molecular interactions that occur within its compartmentalized domains require an exquisite control of the entry of proteins. It also suggests that the signaling and cellular processes that occur within the ciliary domains are highly specialized and unique events that require tight control mechanisms [9].

6.2.8 THE CILIARY POCKET

The ciliary pocket is a membrane domain found at the base of the cilium, which serves as a platform for cilia-related vesicular trafficking. The ciliary pocket is not present in all types of cilia, as it is formed only during one type of ciliogenesis. During early ciliogenesis, an intracellular vesicle positioned on top the basal body fuses with the plasma membrane while the axoneme is growing from the mother centriole, giving rise to the ciliary pocket. This membrane invagination is enriched in budding clathrin–coated vesicles; therefore, it has been proposed that the ciliary pocket is a specialized hot spot for endocytic events [23]. It has to be noted that the ciliary pocket is not the preferential spot for total endocytosis. On the contrary, some authors have suggested that this region would be specialized in the internalization of ciliary components that arise during the recycling of membrane components. In addition, other functions would include the exocytosis of cargo, which would help in the clearance of proteins that are either not properly targeted to the cilia or that are restricted to enter the cilia through the selective action of the ciliary pore. The ciliary pocket also determines the interactions of the cilium with the actin cytoskeleton, which could serve as a structural element or participate in vesicle trafficking [24].

6.2.9 THE CILIARY MEMBRANE

Although continuous with the plasma membrane, the ciliary membrane contains a unique protein composition, and likely also, a unique lipidic fingerprint. In most cell types, the ciliary membrane is physically close to the axoneme, and therefore, it acquires a cylindrical shape. However, in certain specialized cell types, such as retinal photoreceptors, the ciliary membrane is expanded to form stacks that will allow the detection of light. Additional studies show that upon certain sensory inputs, the shape of the ciliary membrane is remodeled, adapting the structure to the functional needs [20].

It is also well described that the ciliary membrane is enriched in signaling receptors, as for example, the components of the Hh signaling pathway described above. Studies of *Paramecium* cilia proteome show that the ciliary membrane is enriched in Glycophosphatidylinositol (GPI)-anchored proteins [25], whereas ciliary membranes from olfactory sensory neurons contain mostly components from the olfactory signaling transduction pathway, as well as anexins, and calcium and chloride transporters [26]. These studies reinforce the idea that ciliary membranes are enriched in specific receptor sets that vary depending on the cell type. This enrichment can be accomplished through the action of diffusion barriers described here, such as the transition fibers and the ciliary pore complex. These protein selection mechanisms are ultimately closed related to the protein composition of the ciliary membrane.

The functions of the ciliary membrane are closely related to the general roles of primary cilia. One of the best described is the regulation of ciliary beat frequencies. In fact, calcium channels and pumps located within the ciliary membrane determine the calcium concentration within this organelle, which has been proved to be distinct from that of the cytosol. Calcium concentration within the cilium is regulated without affecting the overall calcium concentration in the intracellular milieu [27,28]. The influx of calcium causes a change in the direction of movement of motile cilia and flagella. The ciliary membrane has also adhesive properties that coined the term "haptocilia" to describe it. This property is used by organisms like *Chlamydomonas* for the initiation of the mating process. In addition, several pathogens use the haptocilia to interact with their hosts, as it has been described for *Trypanosoma* or *Mycoplasma pneumoniae* [20]. However, the most prominent role of the ciliary membrane is its role in transducing sensory functions such as mechanoreception, chemoreception, and photoreception, which are tightly associated to the downstream signaling functions [29]. The measurement of fluid flow in blood vessels and kidney tubules occurs through the mechanoreceptor properties of the mammalian primary cilia in endothelial and kidney epithelial cells, respectively, which in turn are mediated through calcium. Osteoclasts and osteocytes also respond to fluid flow in cilia-dependent but calcium-independent manner, whereas primary cilia in the node of developing embryos are involved in the establishment of the left–right asymmetry [20]. Retinal rod and cone cells possess highly specialized and modified primary cilia for photoreception, which occur through the trafficking of rhodopsin molecules to the outer segment of this organelle [30]. Examples of chemoreception functions by the cilium include the sperm of sea urchin toward the egg, chemosensory neurons in *Caenorhabditis elegans*, as well as specialized bipolar neurons in the olfactory epithelium of mammals' nose.

6.3 REGULATION OF CILIOGENESIS

Ciliogenesis is referred to as the assembly of primary cilia, a process in continuous dynamic regulation. Cilia are permanently present on the cell surface of terminally differentiated cells like neurons, kidney epithelial cells, or multiciliated cells. On the contrary, cycling cells typically form cilia during G0 or G1 phase, and disassemble it prior to enter mitosis. Although it is not clear why cilia need to be reabsorbed before cell division, one plausible explanation is that basal bodies need to be recycled for the formation of the mitotic spindle.

6.3.1 TWO MODELS OF CILIOGENESIS

Sorokin early proposed two models for ciliogenesis based on the relative position of the basal body within the cell [31]. While beta cells in the pancreas and secretory cells have an axoneme that is deeply rooted in the cell cytoplasm, kidney cells, and other epithelial cells dock the basal body directly to the plasma membrane, from which they protrude directly. In the extracellular pathway, the basal body docks to the plasma membrane prior to the growth of the axoneme. On the contrary, during the intracellular pathway, a cytosolic vesicle sites on top of the mother centriole, from where the axoneme starts growing. The fusion of the incipient cilia with the plasma membrane results in the formation of the ciliary pocket.

6.3.2 THE CONTROL OF CILIARY LENGTH

Ciliary length is controlled by the same mechanisms that are in charge of ciliogenesis, which in turn include the synthesis of precursors, intraflagellar transport, and turnover of components at the ciliary tip. Still, the genetic basis that controls ciliary length is unknown, which has prompted to propose theoretical models that would explain it.

These models are based on the observation that ciliary assembly at the tip continues even after reaching the final ciliary length. Therefore, ciliary length can only be maintained when the rate of cilia assembly equals the disassembly rate. In short cilia, assembly predominates over disassembly, leading to growth, with large IFT trains carrying more cargo to the ciliary tip. In long cilia, on the contrary, disassembly is the primary phenomenon, which leads to shrinkage. Other theoretical models suggest that a length sensor may exist, which by unknown molecular mechanisms would modulate axoneme length [11].

6.4 CILIA-MEDIATED AUTOPHAGY AND CILIOPHAGY

Numerous stimuli are known to activate autophagy, such as starvation, oxidative stress, DNA damage, or proteotoxicity, but how these extracellular events are transduced and linked to the autophagy initiation machinery has remained elusive until very recently. Recent reports indicate that external signals, such as nutrient sensing by the PC, are coordinately transduced to the autophagy machinery.

Initially, it was identified that activation of Hh pathway triggers autophagy in a cilia-dependent manner. Under starvation conditions, where ciliogenesis is maximal, Hh activation recruits autophagy initiation factors to the ciliary base. Specifically, ATG16, one of the autophagy-related proteins involved in the early steps of the autophagosome formation and a marker of the pre-autophagosomal structures, is trafficked to the basal body in IFT20-containing cytosolic vesicles. Other ATG proteins are also present in the periciliary region, such as ATG5 or LC3, indicating a close relation between the autophagy and IFT machinery [15]. Thus, this work suggests that the periciliary region could function as an autophagosome formation platform upon certain stimuli. This idea indicates that niches of autophagosome formation within the cell would be closely related to the type of stimuli that activates autophagy (Figure 6.2).

Additionally, autophagy regulates ciliogenesis in a more complex way. During basal conditions, where ciliogenesis is repressed, autophagy helps controlling IFT20 levels, and limiting the growth of the ciliary axoneme [15]. On the contrary, during starvation, autophagy degrades the centrosomal protein OFD1, which is repressing ciliogenesis, and therefore, activation of autophagy leads to longer primary cilia [32].

Although not through direct evidence, experimental data suggest that other ciliary pathways are tightly linked to autophagy. Some of these pathways are discussed in Section 6.5.

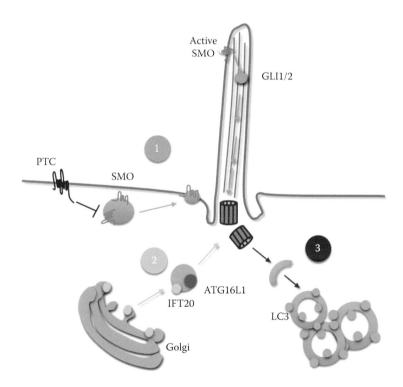

FIGURE 6.2 Interplay between autophagy and ciliogenesis: Hedgehog pathway activates autophagy. (1) Upon Hedgehog ligand binding to the Patched receptor, Smoothened is enriched in the ciliary axoneme, where it activates Gli transcription factors. (2) During active ciliogenesis, the autophagy protein ATG16L1 is recruited to the basal body in IFT20-containing cytosolic vesicles. (3) The arrival of autophagy proteins to the ciliary base increases autophagosome formation, leading to an overall activation of the autophagic capacity of the cell.

6.5 CILIA AND AUTOPHAGY-RELATED SIGNALING PATHWAYS IMPLICATED IN NUTRIENT SENSING

As an organelle in charge of sensing the extracellular milieu, the ciliary membrane is enriched in a variety of signaling receptors that are compartmentalized in this organelle. Here, we describe the main signaling pathways involved in nutrient sensing that rely on the PC for their functioning.

6.5.1 HEDGEHOG SIGNALING PATHWAY

6.5.1.1 Hedgehog Signaling and Primary Cilia

The Hh pathway is a key signaling pathway during development as it is involved in the determination of the left–right asymmetry and organogenesis. Other roles of this pathway include tumorigenesis and the maintenance of homeostasis in adult cells [10]. Vertebrate Hh signaling depends on the PC, and more importantly, it has been proved that it requires the IFT transport machinery for correct functioning [17].

The Hh pathway is a complex signaling mechanism, with several ligands that include Sonic Hh, Indian Hh, and Desert Hh. These secreted lipoproteins will bind the 12-transmembrane receptor Patched (Ptc) sited in the plasma membrane. In the absence of ligand, Ptc represses the Smo receptor, a 7-transmembrane receptor located in intracellular vesicles near the basal body. The effector proteins of the Hh pathway are indeed the Gli transcription factors, which include Gli1, Gli2, and Gli3 proteins, and which function as activators (GliA) or repressors (GliR) of the transcription of downstream target genes. Gli1 will always act as an activator, while Gli2 and Gli3 can be either activators or repressors. In the absence of Hh ligand, axonemal Sufu protein represses the activation of GliA, while proteasomes located at the base of the cilia process GliR. Upon Hh ligand binding to Ptc receptor, Smo is released and trafficked to the ciliary axoneme where it activates the processing of the GliA. Gli activators traffic to the nucleus in an IFT-dependent manner, where they activate the transcription of the Hh target genes, which include Gli1, Ptc1, and Gli2.

In *in vitro* cell cultures, nutrient starvation results in cell cycle arrest that activates ciliogenesis, with a parallel activation of the Hh pathway. Although still needs to be clearly probed, Hh signaling is likely to participate in nutrient sensing under these conditions. A recent work on flies described Hh as a metabolic hormone that links nutrient response in the growing larvae [33]. Despite the fact that invertebrate Hh signaling is cilia independent, it will be of great interest to decipher if mammalian Hh exerts a similar function.

6.5.1.2 Autophagy and Hedgehog Signaling

In *Drosophila* and HeLa cells, which are mostly non-ciliated cells, Hh activation inhibits autophagosome formation by a mechanism that involves Gli2 and its *Drosophila* ortholog Ci [34]. However, in terminally differentiated cells and tissues, activation of Hh pathway promotes autophagy. For example, Shh activation in vascular smooth muscle cells (SMCs) increases autophagosome formation and Akt phosphorylation by an mTOR-independent mechanism [35]. In addition, in hippocampal neurons, in which primary cilia are required for the neurogenic properties of the dentate gyrus, Shh activation increases the presence of autophagosomes in the synaptic terminals, an effect that is mediated by the class III phosphatidylinositol 3-kinase complex (PI3K-III) [36]. Therefore, Hh pathway has differentiated effects over autophagy, which could be linked to the requirement of the PC to mediate Hh signaling events.

6.5.2 PDGF-R Alpha

Platelet-derived growth factors (PDGF) are chemosensory regulators during developmental migration events, as well as during wound healing in adult tissues.

6.5.2.1 Primary Cilia and PDGF

While platelet-derived growth factor receptor β (PDGFRβ) is localized at the plasma membrane, PDGFRα expression is upregulated during fibroblast growth arrest, and relocalizes to the primary cilium. Its ligand PDGF-AA acts over the ciliary PDGRFα homodimer, activating Akt1 and MEK-Erk pathways, and leading to cell cycle reentry and ciliary resorption [37]. Similarly, PDGF-AA induces migration of cultured fibroblasts in a cilia-dependent manner, as cilia mutants fail to respond to PDGF-AA and do not show chemotaxis [38]. A recent work shows that PDFG-AA signaling is regulated by IFT transport through a mechanism involving increased mTOR signaling and diminished PP2A activity [39], reinforcing the role of this organelle as a signaling hub.

6.5.2.2 Autophagy and PDGF

Similar to Hh signaling, activation of PDGF pathway has different effects over autophagy. On one side, inhibition of this proliferation pathway activates autophagy in glioma cells, which are mostly cycling and have very low percentage of ciliated cells [40]. On the contrary, in vascular SMCs, PDGF is a potent autophagy inducer [41], whose effects can be inhibited by c-Ski, a versatile transcriptional regulator widely expressed [42]. Still, it is unclear whether the contradictory effects of PDGF over autophagy are related to the possible differential effects of PDGF-AA and PDGF-BB on autophagy. Thus, more detailed studies are needed to clarify these dissimilarities.

6.5.3 IGF-R

Insulin growth factor 1 (IGF-1) is secreted by the liver in response to growth hormone, participating in cell survival, proliferation, differentiation, and metabolism. Upon ligand binding and receptor auto-phosphorylation, IGF-1 receptor (IGF-1R) activates MAPK/ERK and PI3K pathways.

6.5.3.1 Primary Cilia and IGF-1

IGF-1 pathway requires the PC and basal body for its signaling events. 3T3-L1 preadipocytes need a functional IGF-1 signaling to induce growth arrest and differentiation to adipocytes. In addition to the plasma membrane, IGF-1R is also found in the

PC, where it is more sensitive to insulin. Ciliogenesis induces the recruitment of the IGF-1R downstream effector insulin receptor substrate 1 (IRS-1) together with Akt-1, to be phosphorylated at the basal body, which are cilia-dependent events [43]. In addition, IGF-1 is implicated in ciliary resorption and cell cycle entry, by a mechanism that activates IGF-1R, noncanonical Gβγ signaling, and phosphorylated (T94) Tctex-1 at the ciliary base. In a model proposed by Yeh et al., IGF-1R translocates to the ciliary base upon ligand binding, where the signaling cascade involving Gβγ results in ciliary resorption and priming to reenter the cell cycle [44]. The molecular components of this signaling platform include the Hsp90 chaperone, which is clustered at the periciliary base and it is also present at the ciliary neck, serving as a platform for IRS-1 to mediate IGF-1 receptor signaling [45].

6.5.3.2 Autophagy and IGF-1

Diverging effects have been described about the role of IGF-1 over autophagy, which can either inhibit or promote autophagy depending on the cell type. The inhibitory effects of IGF-1 upon autophagy can be explained by the stimulation of the IGF-1R/Akt/mTOR axis, and by increases in mitochondrial Ca^{2+} levels, oxygen consumption, and ATP levels that lead to AMPK inhibition in cultured cardiomyocytes [46]. Further mechanistic studies are needed to decipher these differences.

6.5.4 LEPTIN SIGNALING

Leptin is an adipocyte-derived cytokine that is modulated by the nutritional state, and with a major role in regulating energy balance. Leptin regulates body weight via leptin receptors that are expressed in several tissues, indicating that this cytokine has peripheral effects. Indeed, leptin is also produced in the CNS, where it has a prominent role in the hypothalamus by regulating food intake, and with additional paracrine or autocrine roles in the adipose tissue and pituitary secretion. Leptin acts on at least two neuronal populations: proopiomelanocortin (POMC) neurons, where it has a catabolic role and promotes the expression of the *Pomc* gene; and neuropeptide Y (NPY) neurons, with an anabolic function that represses the expression of *Npy* and *Agrp* genes [47].

6.5.4.1 Primary Cilia and Leptin Signaling

Leptin signaling has been linked to primary cilia from the observation that cilia-deficient mice are obese. Indeed, obesity is a common feature in several ciliary dysfunctions, such as BBS and Alström syndrome. Similarly, several conditional cilia mutants are obese, hyperphagic, and present high circulating levels of insulin, glucose, and leptin. Finally, hypothalamic neurons regulate satiety through the PC [48]. Later studies described that leptin receptor signaling requires BBS proteins [49], which mediate leptin receptor trafficking, and where BBS1 physically interacts with the leptin receptor. Mice lacking Bbs2-/-, Bbs4-/-, and Bbs6-/- showed decreased leptin signaling as well as reduced hypothalamic STAT3 and *Pomc* gene expression. In a previous work by the same group, it was described that *Bbs*-deficient mice are hyperphagic and have elevated circulating levels of leptin and decreased POMC expression in the hypothalamus [50].

Mechanistically, leptin controls the sensitivity of hypothalamic neurons to metabolic signals by regulating ciliary length. In terminally differentiated cells, such as hypothalamic neurons, the length of primary cilia is actively regulated by changes in the metabolic status, such as leptin treatments, starvation–feeding, and others. Long cilia are thought to be essential for the sensing of metabolic signals such as leptin, although the exact mechanism is still unknown. One hypothesis is based on the fact that ciliary signaling may amplify leptin signaling in basal bodies by a pathway involving cAMP produced by the ciliary adenylyl cyclase III (AC3) [51].

In fact, a recent work has shown that leptin promotes cilia elongation in hypothalamic neurons by acting on the transcriptional regulation of IFT genes as well as rearranging F-actin [52].

Parallel studies indicate that leptin resistance in cilia-mutant mice is secondary to the obese phenotype, because conditional *Ift88* mutants with different adiposity states showed leptin resistance only when mice are obese. This feature indicates that leptin signaling is not the primary event that leads to obesity in cilia-defective mice [53].

6.5.4.2 Autophagy and Leptin Signaling

As a nutrient-sensing adipokine, leptin regulates autophagy, although the physiological and pharmacological effects are diverse in a tissue-dependent manner. In peripheral tissues such as muscle, heart, liver, and kidney, leptin induces autophagy through the activation of AMPK, and the subsequent inhibition of mTOR [54]. By contrast, leptin inhibits AMPK in hypothalamic neurons, vascular SMCs, and peritoneal macrophages, which suggests that in these tissues, leptin could have an inhibitory effect over autophagy. In xenografts models, leptin induces tumor growth, which turns to be mediated by an activation of autophagy [55].

Interestingly, the interaction between leptin signaling and autophagy seems to be reciprocal, as loss of autophagy in the white adipose tissue increases leptin plasma levels, and impairs normal adipogenesis [56]. However, this dual interaction appears to be controversial, because loss of the autophagy protein ATG12 in POMC neurons leads to increased body weight, with disruption of energy balance and impaired leptin sensitivity. However, loss of ATG5 in the same POMC population has no effect over leptin signaling, suggesting that ATG12 could have autophagy-independent effects over leptin [57]. Overall, these studies highlight the complex interaction between autophagy and leptin signaling, suggesting an important role of this neuroendocrine axis in the control of autophagy.

6.5.5 мTOR

The mechanistic target of rapamycin, or mammalian target of rapamaycin (mTOR), is a serine/threonine protein kinase that regulates cell growth, cell proliferation, cell motility, cell survival, protein synthesis, autophagy, and transcription. It is indeed a master regulator of cell signaling, as mTOR integrates the input of upstream pathways that include insulin, growth factors, and amino acids. mTOR is the catalytic subunit of two structurally different complexes, mTORC1 and mTORC2, both of which can be inhibited by rapamycin with different affinities. Contrary to mTORC2, mTORC1 has been extensively studied, and it is composed by mTOR, GβL, and raptor, which control protein synthesis and cell growth. Mitogenic and nutrient availability signals are integrated via substrates such as p70 S6 kinase (S6K). Tuberous sclerosis complex (TSC) TSC1 (hamartin) and TSC2 (tuberin) proteins interact physically, forming a heterodimeric complex that inhibits mTORC1.

6.5.5.1   Primary Cilia and mTOR Pathway

Several recent evidences link ciliogenesis to the mTOR pathway. First, TSC proteins regulate ciliogenesis via rapamycin-insensitive and PC1-independent pathways. Indeed, TSC1 localizes at the basal body, which can be considered as a signal integration center. TSC1-/- and TSC2-/- fibroblasts have increased percentage of ciliated cells, and cilia are longer, indicating that TSC1 and TSC2 may have a role in ciliogenesis. Nevertheless, ciliogenesis-related phenotype in TSC mutants seems to be mTOR independent, as regulation of ciliary length is rapamycin insensitive [58]. However, in a report by Yuan et al., ciliary length and size are controlled by the TSC-mTOR pathway in a rapamycin-sensitive manner. Moreover, according to this study, regulation of ciliary length is an evolutionary-conserved mechanism that is dependent on the protein synthesis [59]. However, continuous exposure to amino acids shuts down mTORC1 signaling, in a mechanism dependent on Skp2 and which inhibits mTORC1 localization at the lysosome. Indeed, loss of Skp2 results in mTORC1 hyperactivation and increased ciliary length and cell size, whereas autophagy is reduced. Therefore, inhibition of mTORC1 signaling is critical for the regulation of cell size, ciliary length, and autophagy by Skp2 [60], which could have further implications in the pathophysiology of age-related disorders and cancer.

On the contrary, mTOR pathway is activated in autosomal dominant polycystic kidney disease (ADPKD) patients, where cilia are insensitive to urinary flow [61–63]. However, Hartman et al. do not find an overall basal activation of mTOR in ADPKD. Indeed, only 30–35% of ADPKD cysts have mTOR pathway activation [58]. A more recent work describes that, in fact, primary cilia regulate mTOR activity through Lkb1, a tumor suppressor that negatively regulates mTOR, resulting in the modulation of cell size [64]. Lkb1 localizes both at the cilium and basal body, together with AMPK. Upon ciliary bending by flow, phosphorylated AMPK accumulates at the ciliary base, indicating that activation of Lkb1 pathway leads to mTOR downregulation and control of cell size.

6.5.5.2   Autophagy and mTOR Pathway

mTOR is one of the major regulators of autophagic function, as it maintains autophagy repressed during nutrient-rich conditions, while inhibition of mTOR during nutrient starvation leads to autophagy induction. Consequently, numerous efforts have been focused on the development of mTOR inhibitors, mainly rapamycin analogs, with the aim to burst autophagy during pathological conditions.

At molecular level, mTORC1 prevents autophagy through the inhibition of autophagy-related proteins such as ULK1, ATG13, AMBRA, and ATG14L. In addition, mTORC1 prevents the nuclear translocation of TFEB, the principal transcriptional regulator of autophagy and lysosomal genes. More specifically, mTORC1 inhibits the autophagy initiator ULK-1 by a mechanism involving direct phosphorylation of this protein. The mTORC1 also destabilizes ULK1, and impairs the autophagy scaffold protein AMBRA1 by its phosphorylation. During nutrient scarcity, AMPK binds and activates ULK1 through phosphorylation, a molecular event that is amplified to further inhibit mTORC1. Nutrient availability is sensed by mTORC1 directly at the lysosomal membrane, where different molecular complexes—including AMPK—are physically sited. Upon amino acid deprivation, AMPK is recruited to the lysosome, while mTORC1 remains in the cytoplasm. However, in nutrient-rich conditions, amino acids are actively transported from the lysosomal lumen to the cytoplasm, where the regulatory complexes recruit mTORC1 to the lysosomal membrane, inhibiting autophagy by transcriptional suppression of lysosomal biogenesis genes [65].

As mentioned above, pharmacological modulation of autophagy through mTOR is an attractive landscape for treating diseases that are characterized by an altered autophagic function. Rapamycin is originally an antifungal and strong immunosuppressant that severely inhibits autophagy. In order to improve its pharmacological properties, several derivative compounds have been synthesized, which are commonly known as rapalogs. Rapamycin and rapalogs are extensively used to activate autophagy *in vitro* and *in vivo*, with the addition that the rapalog RAD001 is an FDA-approved compound for certain types of cancer [66]. Other pharmacological modulators that influence autophagy are the ATP-competitive mTOR inhibitors, such as torin 1, which are designed to directly target the kinase activity of both mTORC1 and mTORC2, and inhibit both mTOR and the PI3K. Some of these compounds are currently under clinical trials as anticancer agents. Finally, metformin, a widely used drug for the treatment of type II diabetes, activates the upstream AMPK and thereby inhibits mTOR downstream of AMPK, which has been demonstrated to be beneficial for several types of cancer *in vivo* and *in vitro*. Overall, pharmacological targeting of autophagy-related disorders through mTOR modulation has to be taken carefully, due to the potentially strong side effects of the competitive inhibitors, as well as the immunosuppressive effects of rapalogs that could limit their clinical application.

6.5.6 Wnt

The Wnt signaling pathways are activated by binding of extracellular Wnt-protein ligands to a Frizzled family receptor. Frizzled receptors are sited in the plasma membrane, and activate intracellular dishevelled (Dvl) protein, which is thought to switch between the canonical (β-catenin dependent) and noncanonical (β-catenin independent) Wnt pathways. In basal conditions, β-catenin is phosphorylated by several kinases and targeted for proteasomal degradation. Upon Wnt activation, β-catenin degradation complexes dissociate, and active β-catenin is translocated to the nucleus for the activation of transcription. On the contrary, the signaling effectors of noncanonical Wnt are mostly unknown. However, disruption of noncanonical Wnt—also known as planar cell polarity (PCP) pathway—regulates cytoskeleton [67].

6.5.6.1   Primary Cilia and Wnt Signaling

The first reports indicated that basal bodies participate in the regulation of Wnt signaling, while cilia favor the noncanonical wing of the pathway [68]. Indeed, primary
cilia compartmentalize Wnt signaling components, which reinforces the role of these
organelles in regulating cell signaling transduction. Cilia sequester the Wnt protein jouberin (Jbn), impeding its nuclear translocation, and also limiting β-catenin
access to the nucleus. As a result, Wnt signaling effects are discretely controlled by
a mechanism that is dependent on the IFT machinery. Abolishment of ciliogenesis,
thus, results in a potentiated Wnt response [69]. Further studies have demonstrated
that indeed, Dvl proteins are continuously required for PCP signaling in motile cilia,
and that disruption of Dvl leads to hydrocephalus [70]. Proper functioning of Dvl is
regulated by phosphatase and tensin homolog (PTEN) protein, which prevents Dvl-
mediated cilia disassembly, and therefore promotes ciliogenesis [71].

6.5.6.2   Autophagy Regulates Wnt Signaling

Autophagy negatively regulates Wnt signaling by promoting Dvl degradation during metabolic stress conditions, such as nutrient starvation. Dvl aggregation in the
cytoplasm leads to pVHL (von Hippel–Lindau tumor suppressor)-mediated ubiquitylation, and a subsequent targeting to the autophagy pathway. At molecular level, LC3
protein associates with Dvl through p62 to promote degradation in the autophagosomes [72]. Upstream Dapper 1 (Dpr1), which directly interacts with Dvl, promotes
the ubiquitylation of Dvl2 by pVHL [73].

6.6   CILIA-RELATED AUTOPHAGY IN PATHOLOGY

Although the link between primary cilia and autophagy is recent, there are already
several evidences that point to an important role of cilia-related autophagy during
pathophysiology. Defects in primary cilia structure and functioning lead to a broad
group of syndromes classified as ciliopathies. Ciliopathies affect almost all major
body systems, including brain, eyes, liver, kidney, skeleton, and limbs. These syndromes show a broad range of clinical features such as early fetal lethality, polydactyly, cardiac defects, cystic kidneys, hepatic fibrosis, obesity, and nervous system
defects that emphasize the importance of primary cilia in development and homeostasis. Despite the fact that more than 50 genes have been linked to ciliopathies, little
is known about the specific molecular mechanisms that occur behind. A great effort
is currently being made to link the phenotypic characterization of this broad range
of disorders with the identification of the causative genes. Overall, the high amount
of nervous system defects in ciliopathies suggests that cilia have important roles in
development and functioning of the CNS.

Recent data indicate that cilia may modulate neural circuity that monitors food
intake and appetite. Cilia loss in leptin-responsive neurons (POMC) leads to obesity,
and BBS proteins are implicated in the trafficking of the leptin receptor. Nevertheless,
other works suggest that leptin resistance in cilia-mutant mice is secondary to obesity.
The obese phenotype in ciliopathies could also result from the mistargeted melanin-
concentrating hormone receptor 1 (Mchr1) in hypothalamic neurons in BBS-mutant
mice, or from alterations of the mTOR activity due to cilia loss [74].

6.6.1 Chronic Obstructive Pulmonary Disease

Cilia are important players in the physiology of several organs and systems, as for example, the respiratory tract and kidney tubules. In the case of the lung airways, these tracts are covered by motile, multiciliated epithelial cells, which eliminate particles and pathogens trapped in the mucus. Impairment of mucociliary clearance occurs in severe pathologies such as chronic obstructive pulmonary disease (COPD), for which cigarette smoking is the most common cause. Cigarette smoking causes a reduction in the length of cilia of the airway epithelia, as well as cell death. As a consequence, there is a regrowth of the epithelial cells accompanied by an excess in mucus production.

At the molecular level, shortening of ciliary length in COPD is linked to human enhancer of filamentation protein 1 (HEF1), aurora A (AurA), histone deacetylase 6 (HDAC6), and glycogen synthase kinase β (GSK3β). Among these proteins, HDAC6 has a central role in controlling deacetylation of tubulin, and therefore participates in microtubule stabilization, which is essential in cilia formation. On the contrary, HDAC6 is an ubiquitin-binding protein that facilitates the retrograde transport of ubiquitinated proteins into aggresomes, and increases the fusion rate between autophagosomes and lysosomes.

It has been recently proved that cigarette smoking induces cilia shortening accompanied by an increase in autophagy. Indeed, lack of autophagy protected mice from cilia shortening after exposure to cigarette smoking, indicating that autophagy is controlling ciliary length. Cigarette smoking induces global protein ubiquitination and aggregation of ciliary proteins, which are recognized by p62 and cleared through HDAC6-dependent autophagy. Part of this ubiquitinated pool includes the ciliary IFT88 protein, which is ubiquitinated and degraded by the autophagy machinery [75]. Therefore, in the airway epithelia, where ciliogenesis does not depend on nutrient availability and cilia are required for a proper tissue physiology, autophagy helps controlling ciliary length by removing IFT proteins, with a mechanism similar to that observed during basal autophagy [15].

6.6.2 Huntington's Disease

Huntington's disease (HD) is an autosomal dominant disease caused by the expansion of CAG repeats within the first exon of the Huntingtin (*Htt*) gene. In HD, both autophagy and primary cilia are impaired, although until now there is not a direct link between the autophagy and ciliogenesis defects in HD. Encoded mutant Htt protein contains an extended cue of polyglutamines (polyQ) at the N-terminal that leads to toxic function of the protein, losing the neuroprotective role of the wild-type (wt) Htt. Wt-Htt participates in vesicular and organelle trafficking, for which it directly associates with microtubules through dynein. Indirectly, Htt is associated with cytoskeleton through the huntingtin-associated protein 1 (HAP1), which binds the p150-glued subunit of dynactin. The resulting Htt-HAP1-dynactin complex is altered by mutant Htt, resulting in a reduced vesicular transport of brain-derived neurotrophic factor (BDNF) in neurons [76].

Although the native function of Htt protein is not totally understood, wt-Htt participates as a scaffold for several ATG proteins during selective autophagy, where it interacts directly with p62, facilitating the association with LC3 and ubiquitinated substrates [77]. Selective autophagy is usually linked to the degradation of altered organelles upon stress, whereas nutrient-dependent or in bulk macroautophagy do not require wt-Htt function [77]. Additionally, autophagy is abnormally activated in HD, where mutant Htt inhibits mTORC1, but an inefficient mutant Htt clearance and a defective cargo recognition result in the accumulation of empty autophagosomes [78].

More recent is the link between ciliogenesis and HD. Indeed, HAP1, the interacting partner of Htt, interacts with proteins such as PCM1, a major component of the centriolar satellites that is required for ciliogenesis, indicating that ciliary growth also requires a functional trafficking mechanism. Htt localizes at the centrosome through a microtubule-dependent transport, forming a complex with PCM1 and HAP1. Moreover, polyQ expansion in Htt leads to PCM1 accumulation at the centrosome, which in turn results in an increased ciliogenesis both *in vitro* and in *in vivo*. These results are in agreement with the description of an impairment of autophagy in HD, in which nondegraded, toxic mutant Htt accumulates leading to neuronal death [76].

As described above, several studies point individually to alterations in autophagy and ciliogenesis in HD. Nevertheless, further studies are required to probe that indeed ciliary alterations lead to an aberrant autophagy function in HD.

6.6.3   POLYCYSTIC KIDNEY DISEASE

Polycystic kidney disease (PKD) was the first syndrome linked to defects in primary cilia, opening the door to the term "ciliopathy." PKD is characterized by defects in primary cilia structure and function, with an inappropriate activation of the mTOR pathway. Several mutations have been linked to PKD, which can be further classified according to the genetics background. Among the different PKDs, autosomal dominant polycystic kidney disease (ADPKD) is the most prevalent of the inherited forms, which is characterized by mutations in PKD1 gene, which represent the 85%, as well as in the gene PKD2.

The products of these genes, the PC1 and PC2 proteins respectively, localize at the primary cilia. Overall, with a prevalence of 1/1000, ADPKD is considered a common disease that courses with alterations in the growth of the renal epithelia, in fluid transport, and morphogenesis. It is characterized by renal cysts, neural tube defects (NTDs), retinal malformation, and polydactyly [79].

In addition to its cilium localization, PC1 protein can be found in the lateral domain of the plasma membrane and adhesion complexes of polarized epithelial cells, such as tight and adherent junctions, desmosomes, and focal adhesions. At the molecular level, PC1 has to be cleaved at the N-terminal to be totally functional. Due to its characteristics, PC1 could be a sensor of mechanical flow in the lumen of renal epithelia, and participate in cell–cell and cell–matrix interactions. On the contrary, PC2 localizes mostly in intracellular compartments that allow the regulation of calcium release from intracellular stores, being only a portion of PC2 localized at the cilium. In the cilium, PC1 and PC2 form a complex that act as a channel in response to ciliary bending, and that recently has been described as a calcium channel for the control of this ion concentration within the cilia [28].

Autosomal recessive PKD (ARPKD) is the first primary cilia–related syndrome described in the literature, which results from mutations in the PKHD1 gene.

PKHD1 gene undergoes alternative splicing of its 86 exons, and location of each mutation critically determines the pathology of each case. With a prevalence of 1/20000, it is classified as a rare disease. A hallmark report describes that the insertion mutation in the *Tg737* gene, the mouse homolog of *Ift88* that encodes polaris protein, results in a phenotype similar to the human ARPKD [80]. Later, description of Orpk mice (Oak Ridge polycyctic kidney disease) concluded that mice with mutations in *Tg737* gene result in shorter cilia in kidneys, indicating for the first time that IFT is linked to ciliary assembly, and that disturbance of ciliary structure leads to PKD [81].

Interestingly, mounting evidence points to an altered autophagic function in renal diseases. Although these autophagy defects are not directly linked to altered ciliary function, mTOR pathway is activated in ADPKD mouse models and human cells. Moreover, kidney cells with mutated PC1 cannot induce autophagy after glucose starvation, suggesting that autophagy induction defects could be linked to altered primary cilia [82]. More indepth studies will indicate the role of cilia-related autophagy in renal cystic diseases such as ADPKD and ARPKD.

6.6.4 Neurodevelopmental Disorders

Neurological defects are commonly found in many ciliopathies, which can be partially explained by the fact that the main ciliary pathways are key in the embryonic development of the CNS, such as Hh or canonical and noncanonical Wnt pathway. Specifically, Hh pathway has a critical role in CNS patterning and neuronal specification. Primary cilia modulate Wnt pathway by sequestering jouberin and β-catenin proteins in the cilium, limiting the nuclear entry of these proteins [83]. In addition, the noncanonical Wnt pathway, or PCP pathway, is implicated in the left–right patterning and neural tube closure during development, which is tightly linked to the primary cilia [21].

Other neurodevelopmental disorders like the Meckel–Gruber syndrome (MKS) and Joubert syndrome (JS) are characterized by mutations affecting proteins within the transition zone. This ciliary structure ensures the access of selected proteins within the cilium—such as Arl13b and Smoothened—while it prevents the fast diffusion of unselected proteins. The MKS is an autosomal recessive, inherited, pleiotropic, neonatal lethal disorder characterized by defects in the neural tube, considered as the most severe ciliopathy [84]. MKS is phenotypically characterized by occipital encephalocele, cystic kidneys, and fibrotic liver changes, while the neurodevelopmental abnormalities include syndromic forms of NTDs, and midline orofacial clefting anomalies, among others. All the six known disease genes in MKS (MKS1, MKS2/TMEM216, MKS3/TMEM6, RPGRIP1L, CEP290, and CC2D2A) encode proteins localized either in the cilium or in the basal bodies. Although the molecular pathogenesis is unclear, MKS is caused by dysfunction of the PC during early embryogenesis, with a disruption of Wnt signaling and hyperactivation of the PCP pathway.

The Joubert syndrome is an autosomal or X-linked recessive ciliopathy, with 22 mutations in different genes identified to date, all of which encode ciliary proteins. However, mutations in these genes account for only half of the cases, highlighting the genetic complexity of this syndrome. "Molar tooth sign" is the hallmark feature of this syndrome, which refers to a distinctive midbrain and hindbrain malformation [21].

Other neurodevelopmental ciliopathies include the BBS, an autosomal recessive, heterogeneous disorder with 17 mutations reported in genes encoding for BBSome proteins. BBS phenotype includes obesity, polydactyly, kidney abnormalities, and retinal dystrophy, with mild to moderate intellectual disability. Additionally, the orofaciodigital syndromes (OFDs) comprise a group of ciliopathies characterized by orofacial (tongue nodules, cleft palate) and digital (i.e., polydactyly) abnormalities, and are frequently associated with intellectual disability. A total of 13 different forms of OFDs have been described, all of them X-linked and caused by disruptions in OFD1 gene, which encodes a centriolar protein involved in ciliogenesis.

Although the role of primary cilia has been extensively studied, much less is known about the implications of autophagy in neurodevelopment. Growing evidence indicates that autophagy is critical for axonal and dendritic development and maturation, as deletion of essential autophagy genes such as *Atg7* or *Ulk1* impairs neurite outgrowth. In a similar way, TSC proteins, which inhibit mTOR, localize in axons and are required for axonal growth. Some neurodevelopmental disorders, including autism, X-fragile and tuberous sclerosis, course with alterations in mTOR and PI3K/Akt pathways, which in turn modulate autophagy. However, more detailed studies are needed to provide a direct role of autophagy in the pathophysiology of these neurodevelopmental disorders [85].

6.7 METHODS TO STUDY CILIA-RELATED AUTOPHAGY

6.7.1 METHODS TO STUDY THE ROLE OF THE CILIUM ON AUTOPHAGY

The role of primary cilia signaling on autophagy can be studied using cilia mutants such as *Ift88-/-* or *Ift20-/-* as models with impaired ciliogenesis. Using these mutants, autophagy can be studied by the classical biochemical LC3 flux assay and/or the by dual mCherry-GFP-LC3 fluorescent autophagy reporter. When possible, it is encouraged to use both methods to have a complete overview of the effect of cilia lose on autophagy. In addition, transmission electron microscopy allows to visualize the amount and content of autophagic vacuoles, and therefore, we encourage to use it as it is the only method with enough resolution to monitor structural changes in the autophagy compartments [86].

In addition, the function of ciliary signaling pathways over autophagy can be studied by using specific pharmacological agonists and antagonists. In this case, control and cilia mutants are stimulated with the pathway agonist/antagonist, and autophagic function is measured by the autophagy methods mentioned above. The results allow to discern between cilia-mediated and non-cilia-mediated signaling effects over autophagy.

6.7.2 METHODS TO STUDY THE FUNCTION OF AUTOPHAGY ON CILIOGENESIS

Autophagy mutants such as *Atg5-/-* or *Atg7-/-* are used to study the role of autophagy over ciliogenesis. In addition to the genetic mutants, autophagy can be impaired by blocking the formation of autophagosomes with the PI3K inhibitor 3-methyladenine [86].

Initially, ciliary length and percentage of cells that contain a cilium (percentage of ciliated cells) are quantified in basal conditions. For that purpose, using classical immunocytochemistry methods, the ciliary axoneme and basal body are visualized with antibodies against acetylated tubulin and against gamma-tubulin, respectively. After image acquisition, the percentage of ciliated cells and the ciliary length are quantified by Image J or other image-processing softwares. These results will help understanding the role of autophagy over ciliary structure. A good and recommended complementary method is to use scanning electron microscopy, which allows high-resolution imaging and visualization of ultrastructural changes in the cilium.

Finally, ciliogenesis can be induced in autophagy-impaired models by starvation or other methods that lead to cell cycle arrest. Cells can also be treated with Cytochalasin D, a potent inhibitor of actin depolymerization, for 3 hours to maximally induce ciliogenesis. After these treatments and immunostaining, length of the cilium and percentage of cells containing a cilium will serve to determine if autophagy is required for inducible autophagy.

6.8 CONCLUSIONS AND FUTURE DIRECTIONS

The bidirectional interplay between autophagy and ciliogenesis has added a new regulation level to autophagy. Besides Hh signaling, other ciliary pathways can potentially modulate autophagy, which opens the window to new treatment options for ciliopathies and neurodegenerative diseases that course with autophagy malfunction.

As many tissues and noncycling cells rely on the PC for many of their physiological functions, it is necessary to determine the molecular players that govern the interaction between autophagy and ciliogenesis in nonnutritional conditions. For example, it would be worthy deciphering if other signaling pathways are also increasing autophagy through the PC in kidney tubules, or if ependymal flow sensed by the cilium can modify autophagic function. In the same way, it would be worthy clarifying if defects in autophagy function, such as that described for HD, do have the ability to modify cilia structure and function in neurons, and do contribute to the deleterious effects in neuronal death.

REFERENCES

1. Efeyan, A., W.C. Comb, and D.M. Sabatini. Nutrient-sensing mechanisms and pathways. *Nature*, 2015. **517**(7534): 302–10.
2. Louvi, A. and E.A. Grove. Cilia in the CNS: The quiet organelle claims center stage. *Neuron*, 2011. **69**(6): 1046–60.
3. Rabinowitz, J.D. and E. White. Autophagy and metabolism. *Science*, 2010. **330**(6009): 1344–8.
4. Sancak, Y., et al. Ragulator-Rag complex targets mTORC1 to the lysosomal surface and is necessary for its activation by amino acids. *Cell*, 2010. **141**(2): 290–303.
5. Arias, E., et al. Lysosomal mTORC2/PHLPP1/Akt regulate chaperone-mediated autophagy. *Mol Cell*, 2015. **59**(2): 270–84.

6. Arias, E. Lysosomal mTORC2/PHLPP1/Akt axis: A new point of control of chaperone-mediated autophagy. *Oncotarget*, 2015. **6**(34): 35147–8.

7. Bettencourt-Dias, M. and D.M. Glover. Centrosome biogenesis and function: Centrosomics brings new understanding. *Nat Rev Mol Cell Biol*, 2007. **8**(6): 451–63.

8. Dawe, H.R., H. Farr, and K. Gull. Centriole/basal body morphogenesis and migration during ciliogenesis in animal cells. *J Cell Sci*, 2007. **120**(1): 7–15.

9. Reiter, J.F., O.E. Blacque, and M.R. Leroux. The base of the cilium: Roles for transition fibres and the transition zone in ciliary formation, maintenance and compartmentalization. *EMBO Rep*, 2012. **13**(7): 608–18.

10. Orhon, I., et al. Autophagy and regulation of cilia function and assembly. *Cell Death Differ*, 2015. **22**(3): 389–97.

11. Ishikawa, H. and W.F. Marshall. Ciliogenesis: Building the cell's antenna. *Nat Rev Mol Cell Biol*, 2011. **12**(4): 222–34.

12. Davis, E.E. and N. Katsanis. The ciliopathies: A transitional model into systems biology of human genetic disease. *Curr Opin Genet Dev*, 2012. **22**(3): 290–303.

13. Cole, D.G. and W.J. Snell. SnapShot: Intraflagellar transport. *Cell*, 2009. **137**(4): 784–784.e1.

14. Finetti, F., et al. Intraflagellar transport: A new player at the immune synapse. *Trends Immunol*, 2011. **32**(4): 139–45.

15. Pampliega, O., et al. Functional interaction between autophagy and ciliogenesis. *Nature*, 2013. **502**(7470): 194–200.

16. Finetti, F., et al. Specific recycling receptors are targeted to the immune synapse by the intraflagellar transport system. *J Cell Sci*, 2014. 1;129 (Pt 9): 1924–37.

17. Huangfu, D., et al. Hedgehog signalling in the mouse requires intraflagellar transport proteins. *Nature*, 2003. **426**(6962): 83–7.

18. Goetz, S.C. and K.V. Anderson. The primary cilium: A signalling centre during vertebrate development. *Nat Rev Genet*, 2010. **11**(5): 331.

19. Liem, K.F., et al. The IFT-A complex regulates Shh signaling through cilia structure and membrane protein trafficking. *J Cell Biol*, 2012. **197**(6): 789–800.

20. Pazour, G.J. and R.A. Bloodgood. Targeting proteins to the ciliary membrane, in *Current Topics in Developmental Biology*, ed., Bradley K. Yoder, 2008, Elsevier, The Netherlands, Volume 85, 115–49.

21. Valente, E.M., et al. Primary cilia in neurodevelopmental disorders. *Nat Rev Neurol*, 2014. **10**(1): 27–36.

22. Kee, H.L., et al. A size-exclusion permeability barrier and nucleoporins characterize a ciliary pore complex that regulates transport into cilia. *Nat Cell Biol*, 2012. **14**(4): 431–437.

23. Benmerah, A. The ciliary pocket. *Curr Opin Cell Biol*, 2013. **25**(1): 78–84.

24. Ghossoub, R., et al., The ciliary pocket: A once-forgotten membrane domain at the base of cilia. *Biol Cell*, 2011. **103**(3): 131–44.

25. Yano, J., et al. Proteomic analysis of the cilia membrane of *Paramecium tetraurelia*. *J Proteomics*, 2013. **78**: 113–22.

26. Kuhlmann, K., et al. The membrane proteome of sensory cilia to the depth of olfactory receptors. *Mol Cell Proteomics: MCP*, 2014. **13**(7): 1828–43.

27. DeCaen, P.G., et al. Direct recording and molecular identification of the calcium channel of primary cilia. *Nature*, 2013. **504**(7479): 315–18.

28. Delling, M., et al. Primary cilia are specialized calcium signalling organelles. *Nature*, 2013. **504**(7479): 311–14.

29. Bloodgood, R.A. The future of ciliary and flagellar membrane research. *Mol Biol Cell*, 2012. **23**(13): 2407–11.

30. Wheway, G., D.A. Parry, and C.A. Johnson. The role of primary cilia in the development and disease of the retina. *Organogenesis*, 2014. **10**(1): 69–85.

31. Sorokin, S.P. Reconstructions of centriole formation and ciliogenesis in mammalian lungs. *J Cell Sci*, 1968. **3**(2): 207–30.
32. Zaiming, T., et al. Autophagy promotes primary ciliogenesis by removing OFD1 from centriolar satellites. *Nature*, 2013. **502**(7470): 254–7.
33. Rodenfels, J., et al. Production of systemically circulating Hedgehog by the intestine couples nutrition to growth and development. *Gene Dev*, 2014. **28**(23): 2636–51.
34. Jimenez-Sanchez, M., et al. The Hedgehog signalling pathway regulates autophagy. *Nat Commun*, 2012. **3**: 1200.
35. Li, H., et al. Sonic hedgehog promotes autophagy of vascular smooth muscle cells. *Am J Physiol Heart Circ Physiol*, 2012. **303**(11): H1319–31.
36. Petralia, R.S., et al. Sonic hedgehog promotes autophagy in hippocampal neurons. *Biol Open*, 2013. **2**(5): 499–504.
37. Schneider, L., et al. PDGFRalphaalpha signaling is regulated through the primary cilium in fibroblasts. *Curr Biol*, 2005. **15**(20): 1861–6.
38. Schneider, L., et al. Directional cell migration and chemotaxis in wound healing response to PDGF-AA are coordinated by the primary cilium in fibroblasts. *Cell Physiol Biochem*, 2010. **25**(2–3): 279–92.
39. Umberger, N.L. and T. Caspary. Ciliary transport regulates PDGF-AA/αα signaling via elevated mammalian target of rapamycin signaling and diminished PP2A activity. *Mol Biol Cell*, 2015. **26**(2): 350–8.
40. Takeuchi, H., et al. Inhibition of platelet-derived growth factor signalling induces autophagy in malignant glioma cells. *Br J Cancer*, 2004. **90**(5): 1069–75.
41. Salabei, J.K., et al. PDGF-mediated autophagy regulates vascular smooth muscle cell phenotype and resistance to oxidative stress. *Biochem J*, 2013. **451**(3): 375–88.
42. Li, J., et al., c-Ski inhibits autophagy of vascular smooth muscle cells induced by oxLDL and PDGF. *PLoS One*, 2014. **9**(6): e98902.
43. Zhu, D., et al. Growth arrest induces primary-cilium formation and sensitizes IGF-1-receptor signaling during differentiation induction of 3T3-L1 preadipocytes. *J Cell Sci*, 2009. **122**(15): 2760–8.
44. Yeh, C., et al. IGF-1 Activates a cilium-localized noncanonical Gβγ signaling pathway that regulates cell-cycle progression. *Dev Cell*, 2013. **26**(4): 358–68.
45. Wang, H., et al. Hsp90α forms a stable complex at the cilium neck for the interaction of signalling molecules in IGF-1 receptor signalling. *J Cell Sci*, 2015. **128**(1): 100–8.
46. Troncoso, R., et al. Energy-preserving effects of IGF-1 antagonize starvation-induced cardiac autophagy. *Cardiovasc Res*, 2012. **93**(2): 320–9.
47. Baly, C., et al. Leptin and its receptors are present in the rat olfactory mucosa and modulated by the nutritional status. *Brain Res*, 2007. **1129**: 130–41.
48. Davenport, J.R., et al. Disruption of intraflagellar transport in adult mice leads to obesity and slow-onset cystic kidney disease. *Curr Biol*, 2007. **17**(18): 1586–94.
49. Seo, S., et al. Requirement of bardet-biedl syndrome proteins for leptin receptor signaling. *Hum Mol Genet*, 2009. **18**(7): 1323–31.
50. Rahmouni, K., et al. Leptin resistance contributes to obesity and hypertension in mouse models of Bardet-Biedl syndrome. *J Clin Invest*, 2008. **118**(4): 1458–67.
51. Han, Y.M., et al. Leptin-promoted cilia assembly is critical for normal energy balance. *J Clin Invest*, 2014. **124**(5): 2193–7.
52. Kang, G.M., et al. Leptin elongates hypothalamic neuronal cilia via transcriptional regulation and actin destabilization. *J Biol Chem*, 2015. **290**(29): 18146–55.
53. Berbari, N.F., et al. Leptin resistance is a secondary consequence of the obesity in ciliopathy mutant mice. *Proc Natl Acad Sci U S Am*, 2013. **110**(19): 7796–801.
54. Malik, S.A., et al. Neuroendocrine regulation of autophagy by leptin. *Cell Cycle*, 2011. **10**(17): 2917–23.

55. Nepal, S., et al. Autophagy induction by leptin contributes to suppression of apoptosis in cancer cells and xenograft model: Involvement of p53/FoxO3A axis. *Oncotarget*, 2015. **6**(9): 7166–81.

56. Zhang, Y., et al. Adipose-specific deletion of autophagy-related gene 7 (atg7) in mice reveals a role in adipogenesis. *Proc Natl Acad Sci U S A*, 2009. **106**(47): 19860–5.

57. Malhotra, R., et al. Loss of Atg12, but not Atg5, in pro-opiomelanocortin neurons exacerbates diet-induced obesity. *Autophagy*, 2015. **11**(1): 145–54.

58. Hartman, T.R., et al. The tuberous sclerosis proteins regulate formation of the primary cilium via a rapamycin-insensitive and polycystin 1-independent pathway. *Hum Mol Genet*, 2009. **18**(1): 151–63.

59. Yuan, S., et al. Target-of-rapamycin complex 1 (Torc1) signaling modulates cilia size and function through protein synthesis regulation. *Proc Natl Acad Sci U S A*, 2012. **109**(6): 2021–6.

60. Jin, G., et al. Skp2-mediated RagA ubiquitination elicits a negative feedback to prevent amino-acid-dependent mTORC1 hyperactivation by recruiting GATOR1. *Mol Cell*, 2015. **58**(6): 989–1000.

61. Kim, S. and B.D. Dynlacht. Assembling a primary cilium. *Curr Opin Cell Biol*, 2013. **25**(4): 506–11.

62. Shillingford, J.M., et al. The mTOR pathway is regulated by polycystin-1, and its inhibition reverses renal cystogenesis in polycystic kidney disease. *Proc Natl Acad Sci U S A*, 2006. **103**(14): 5466–71.

63. Wahl, P.R., et al. Inhibition of mTOR with sirolimus slows disease progression in Han:SPRD rats with autosomal dominant polycystic kidney disease (ADPKD). *Nephrol Dial Transplant*, 2006. **21**(3): 598–604.

64. Boehlke, C., et al. Primary cilia regulate mTORC1 activity and cell size through Lkb1. *Nat Cell Biol*, 2010. **12**(11): 1115–22.

65. Dunlop, E.A. and A.R. Tee. mTOR and autophagy: A dynamic relationship governed by nutrients and energy. *Semin Cell Dev Biol*, 2014. **36**: 121–9.

66. Kim, Y.C. and K.L. Guan. mTOR: A pharmacologic target for autophagy regulation. *J Clin Invest*, 2015. **125**(1): 25–32.

67. Gerdes, J.M. and N. Katsanis. Ciliary function and Wnt signal modulation. *Curr Top Dev Biol*, 2008. **85**: 175–95.

68. Gerdes, J.M., et al. Disruption of the basal body compromises proteasomal function and perturbs intracellular Wnt response. *Nat Genet*, 2007. **39**(11): 1350–60.

69. Lancaster, M.A., J. Schroth, and J.G. Gleeson. Subcellular spatial regulation of canonical Wnt signalling at the primary cilium. *Nat Cell Biol*, 2011. **13**(6): 700–7.

70. Ohata, S., et al. Loss of dishevelleds disrupts planar polarity in ependymal motile cilia and results in hydrocephalus. *Neuron*, 2014. **83**(3): 558–71.

71. Shnitsar, I., et al. PTEN regulates cilia through Dishevelled. *Nat Commun*, 2015. **6**: 8388.

72. Gao, C., et al. Autophagy negatively regulates Wnt signalling by promoting dishevelled degradation. *Nat Cell Biol*, 2010. **12**(8): 781–90.

73. Ma, B., et al. The Wnt signaling antagonist dapper1 accelerates dishevelled2 degradation via promoting its ubiquitination and aggregate-induced autophagy. *J Biol Chem*, 2015. **290**(19): 12346–54.

74. Guemez-Gamboa, A., G.C. Nicole, and G.G. Joseph. Primary cilia in the developing and mature brain. *Neuron*, 2014. **82**(3): 511–21.

75. Lam, H.C., et al. Histone deacetylase 6–mediated selective autophagy regulates COPD-associated cilia dysfunction. *J Clin Invest*, 2013. **123**(12): 5212–30.

76. Keryer, G., et al. Ciliogenesis is regulated by a huntingtin-HAP1-PCM1 pathway and is altered in Huntington disease. *J Clin Invest*, 2011. **121**(11): 4372–82.

77. Rui, Y.N., et al. Huntingtin functions as a scaffold for selective macroautophagy. *Nat Cell Biol*, 2015. **17**(3): 262–75.
78. Martinez-Vicente, M., et al. Cargo recognition failure is responsible for inefficient autophagy in Huntington's disease. *Nat Neurosci*, 2010. **13**(5): 567–76.
79. Chapin, H.C. and M.J. Caplan. The cell biology of polycystic kidney disease. *J Cell Biol*, 2010. **191**(4): 701–10.
80. Moyer, J.H., et al. Candidate gene associated with a mutation causing recessive polycystic kidney disease in mice. *Science*, 1994. **264**(5163): 1329–33.
81. Pazour, G.J., et al. Chlamydomonas IFT88 and its mouse homologue, polycystic kidney disease gene tg737, are required for assembly of cilia and flagella. *J Cell Biol*, 2000. **151**(3): 709–18.
82. De Rechter, S.,et al. Autophagy in renal diseases. *Pediatr Nephrol*, 2016. **31**(5): 737–52.
83. Salinas, P.C., Wnt signaling in the vertebrate central nervous system: From axon guidance to synaptic function. *Cold Spring Harb Perspect Biol*, 2012. **4**(2). pii: a008003.
84. Logan, C.V., Z. Abdel-Hamed, and C.A. Johnson. Molecular genetics and pathogenic mechanisms for the severe ciliopathies: Insights into neurodevelopment and pathogenesis of neural tube defects. *Mol Neurobiol*, 2011. **43**(1): 12–26.
85. Lee, K.M., S.K. Hwang, and J.A. Lee. Neuronal autophagy and neurodevelopmental disorders. *Exp Neurobiol*, 2013. **22**(3): 133–42.
86. Klionsky, D.J., et al. Guidelines for the use and interpretation of assays for monitoring autophagy. *Autophagy*, 2012. **8**(4): 445–544.

7 Regulation of Lipophagy

Nuria Martinez-Lopez
Albert Einstein College of Medicine
Bronx, NY

CONTENTS

7.1 AN INTRODUCTION TO AUTOPHAGY

7.1.1 DEFINITION AND FUNCTIONS OF AUTOPHAGY

Autophagy is derived from the Greek *auto*, meaning self, and *phagy*, meaning *eating*, and is a cellular homeostatic mechanism that degrades cytoplasmic components such as long-lived proteins, macromolecular aggregates, and dysfunctional and/or aged organelles, among others, in acidic compartments known as lysosomes. Autophagy is evolutionarily conserved from small organisms to humans and functions in a constitutive manner to perform its essential roles in cellular quality control. On the contrary, autophagy is stimulated under cellular stress such as during nutrient deprivation in order to provide an alternate source of macromolecular building blocks. In addition to quality control and provision of energy, additional functions have been attributed to autophagy including its roles in unconventional secretion (Duran et al. 2010; Manjithaya et al. 2010), cellular signaling (Martinez-Lopez et al. 2013a), tissue differentiation (Singh et al. 2009a; Martinez-Lopez et al. 2013b), cellular immunity (reviewed by Levine et al. 2011), as an antiaging pathway (reviewed by Rubinsztein et al. 2011; Cuervo and Wong 2014), and in lipid metabolism (Singh et al. 2009b; Kaushik et al. 2011, 2012). Thus, it is conceivable that perturbations in autophagy will severely impact human health. Over the past decades, numerous studies have revealed the contribution of altered autophagy to a variety of diseases such as cancer (Liang et al. 1999; Shimizu et al. 2012), neurodegeneration (Komatsu et al. 2006), muscular disorders (Masiero et al. 2009; Yamada et al. 2012; Martinez-Lopez et al. 2013b), and infectious diseases (Yuk et al. 2012), to mention a few examples, highlighting the relevance of intact autophagy in maintenance of human health. Clearly, deeper understanding in the molecular mechanism of autophagy and its regulatory signaling cascades would help in the development of therapeutic strategies to prevent/treat impaired autophagy-associated diseases.

7.1.2 DIFFERENT TYPES OF AUTOPHAGY

In mammals, three primary forms of autophagy have been described depending on the type of cargo and the mechanism by which the cargo is identified and delivered to the lysosomes: macroautophagy, microautophagy, and chaperone-mediated autophagy (CMA). Macroautophagy is the most well-studied form of autophagy. It consists of an *in-bulk* degradation mechanism by which whole cytoplasmic portions are engulfed by *de novo* formation of double-membrane structures called autophagosomes. Autophagosomes fuse with the lysosomes forming autolysosomes wherein lysosomal-resident acid hydrolases degrade the engulfed cargo. The products derived from the breakdown, such as amino acids, fatty acids, and nucleic acids, are released back into the cytosol through lysosomal permeases and reused by the cell (Figure 7.1). While macroautophagy was initially considered a nonselective degradation pathway, selective targets have been described including the specific engulfment of mitochondria via *mitophagy* (Kim et al. 2007), peroxisomes via *pexophagy* (Kim and Klionsky 2000), ribosomes by *ribophagy* (Kraft et al. 2008), endoplasmic reticulum through *reticulophagy* (Tasdemir et al. 2007), and lipid droplets by a process termed *lipophagy* (Singh et al. 2009b). Interestingly, this selectivity is not restricted to organelles, since macroautophagy has been shown to degrade microbes via xenophagy (Levine 2005). The degradation of lipid droplets via lipophagy will remain the primary focus of this chapter. Microautophagy, the least-characterized form of autophagy, involves the direct sequestration of cytoplasmic content by invaginations of the lysosomal and late endosomal membranes (endosomal microautophagy) (Sahu et al. 2011) with the consequent vesicle scission and cargo degradation (Marzella et al. 1981; Li et al. 2012) (Figure 7.1). The third form of autophagy, CMA, is a more specialized form of autophagy wherein cytosolic proteins containing a specific pentapeptide motif (KFERQ) in its primary sequence are recognized by cytosolic chaperone heat-shock cognate of 70 kDa (Hsc70) and then unfolded at the lysosomal surface and internalized inside lysosomes by the lysosomal-associated membrane protein 2A (LAMP-2A) receptor

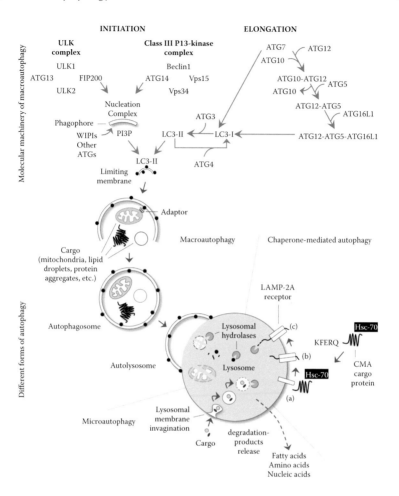

FIGURE 7.1 The molecular machinery of autophagy. Stress, such as starvation, induces three forms of autophagy. In macroautophagy, the activation and mobilization of the ULK complex and the Class III phosphotidylinositide 3-kinase (PI3K) complex to the nascent limiting membrane result in the generation of phosphoinositol 3-phosphate (PI3P) lipid species, recruiting additional ATG proteins. In parallel, ATG7 activates the conjugation cascades ATG12-ATG5 and LC3 that results in the conversion of cytosolic LC3-I into the limiting membrane-bound LC3-II. LC3-II-decorated autophagic membranes expand, engulf cargo, seal upon itself, and fuse with the lysosomes where the resident acid hydrolases degrade the cargo. Microautophagy degrades cytosolic cargo by invagination of the lysosomal membrane. Chaperone-mediated autophagy is a selective form of autophagy by which proteins containing the KFERQ motif are recognized by the heat-shock cognate of 70 kDa (Hsc-70) chaperone and recruited to the lysosomes. There the protein (a) binds to the lysosomal-associated membrane protein 2A (LAMP-2A) receptor that in turn (b) oligomerizes, unfolds the protein, and (c) translocates the protein into the lysosomes where it is degraded. The products from the degradation are released back to the cytosol by lysosomal permeases.

with the assistance of the Hsc70 (Figure 7.1) (Dice 1990; reviewed by Kaushik and Cuervo 2012). These three forms of autophagy coexist in almost all cell types. Furthermore, functional interplay between these pathways has been reported, since impairment in macroautophagy is compensated by constitutive activation of CMA (Kaushik et al. 2008). In this chapter, we will primarily focus on macroautophagy (henceforth autophagy), since it is the form of autophagy initially described to participate in cellular turnover of lipid droplets (Singh et al. 2009b), and thereafter we will comment on the recently described role of CMA in cellular lipid hydrolysis (Kaushik and Cuervo 2015).

7.2 APPROACHES TO ASSESS LIPOPHAGY IN CULTURED CELLS

In essence, analyses of lipophagy include the following groups of assays: (1) the assessment of the degree to which regulatory autophagy-related genes (ATG) proteins and the autophagosome marker LC3 are recruited to lipid droplets in a given condition, and (2) the assessment of the extent of turnover of lipid droplets in lysosomes by using lysosomal inhibitors. Given that autophagosomes directly sequester lipid droplets during lipophagy, the first step in lipophagy analyses relies on image-based studies to determine the colocalization of LC3 with typical lipid droplet markers, BODIPY 493/503 or 4', 6-diamidino-2-phenylindole (DAPI)-range dye 1,6-diphenyl-1,3,5-hexatriene (DPH; excitation/emission λmax = 350 nm/420 nm). Increased colocalization of LC3 with BODIPY 493/503 in a given condition indicates abundance of LC3-I and LC3-II on lipid droplets, which points toward autophagic sequestration of lipid droplets. Quantitative analyses of the colocalization of LC3 with BODIPY 493/503 (using NIH ImageJ-supported plugins) reveal the relative percentage of the total cellular LC3 pool associated with lipid droplets, and the number of lipid droplets that associates with LC3. These studies are particularly valuable since one can reliably distinguish whether changes in lipid droplet size or proteome (via co-labeling for lipid droplet coat proteins) play a role in lipid droplet sequestration by LC3 in the condition being tested. However, it should be noted that increased lipid droplet sequestration does not necessarily correlate positively with lysosomal lipolysis, since autophagosomes *per se* do not contain lipases for initiation of lipolysis. Consequently, assays detailed in the next category should be performed in conjunction with these LC3-BODIPY 493/503 colocalization experiments, and in totality, these approaches will reliably indicate the true lipophagy status. Since LC3-I and LC3-II are each enriched in lipid droplets during active lipophagy, fractionation studies are helpful at this stage to assess the relative enrichment of LC3-I and LC3-II in the isolated lipid droplets. Quantifying for LC3-II levels in lipid droplets is helpful to assess the amount of autophagosomes in lipid droplets in the given experimental condition. While presence of LC3-II on lipid droplet fractions points toward increased autophagic sequestration, it is advisable to employ LC3 immuno-gold electron microscopy for qualitative evidence of the engulfment of lipid droplets by LC3-II-positive membranes.

Caution should be exercised so as not to overinterpret the enrichment of LC3-II in lipid droplets as activation of lipophagy. It is, in our opinion, a good practice to determine that conditions appearing to activate lipophagy should also associate with active LC3-II flux in whole homogenates. Since increased abundance of LC3-II in lipid droplets could also ensue from the failure of lipid-containing autophagosomes to fuse with lysosomes, increased LC3-II flux data in this condition will exclude the possibility of decreased autophagosome–lysosome fusion. Therefore, in our view, increased enrichment of LC3-II in lipid droplets *per se* is not sufficient to conclude that lipophagy is active under the conditions being evaluated.

To conclusively demonstrate the precise status of lipophagy in a given condition, the aforementioned approach must be supplemented with procedures testing lysosomal lipid delivery and consumption. Colocalizing BODIPY 493/503 with lysosomal/endosomal marker LAMP1 in presence or absence of lysosomal inhibitors will reveal information pertaining to (1) lysosomal delivery of lipid droplets and (2) the extent of turnover of lipid droplets in lysosomes. Any lipid delivered to lysosomes will be revealed by increased colocalization of BODIPY 493/503 with LAMP1. In addition, since any lipid delivered to lysosomes is likely to be rapidly degraded, including a lysosomal inhibitor to cells will allow accumulation of lipid delivered to lysosomes. Subtracting the lipid signal colocalizing with LAMP1 in lysosomal inhibitor–untreated cells from corresponding values obtained from lysosomal inhibitor–treated cells will reveal the net lipid content turned over by lysosomes. This readout will reliably demonstrate the extent of lipid consumed via lipophagy in a given experimental condition. Thus, these assays in conjunction with those in (1) are more likely to reveal rates of lipid consumption via lipophagy under the condition being tested. Additional experiments required at this stage are evaluation of whether knocking down the factor under investigation leads to increased lipid droplet content, and whether lipid accumulation associates with decreased LC3-BODIPY 493/503 and BODIPY 493/503-LAMP1 colocalization in cells cultured in presence of lysosomal inhibitors. Although determination of net lipid droplet turnover by use of lysosomal inhibitors conclusively demonstrates turnover of lipid droplets in lysosomes, it is prudent to demonstrate that cells with decreased lipophagy also display reduction in mitochondrial fatty acid oxidation as a functional readout of decreased free fatty acid availability, and that changes in lipid droplets accumulation do not occur secondarily to differences in lipogenesis. In totality, these strategies are likely to yield the true status of lipophagy and autophagy-mediated lipid turnover in cultured cells.

7.3 THE MOLECULAR INTERPLAYERS OF AUTOPHAGY

7.3.1 The Biogenesis of an Autophagosome

It was in 1950s when lysosomes were first observed by De Duve and coworkers in mammalian cells using electron microscopy (Novikoff et al. 1956) and years later in 1965, the former coined the term *autophagy* (reviewed by Klionsky 2007). Subsequently, molecular genetic approaches in yeast allowed the identification of ATG proteins (Tsukad and Ohsumi 1993; Harding et al. 1996; Schlumpberger et al. 1997) and further studies identified their corresponding orthologs in mammals (Mizushima et al. 1998). The formation of an autophagosome is a well-coordinated multistep process orchestrated by more than 30 ATGs. This complex network of modulators participate in the initiation, nucleation, and elongation of a limiting membrane, which expands to form a double-membrane autophagic structure around targeted cytosolic cargo for their lysosomal degradation (Figure 7.1) (He and Klionsky 2009). A crucial question regarding autophagosome formation, which is still under investigation, is the source of the phagophore from where the limiting membrane originates. Several studies have identified the plasma membrane (Ravikumar et al. 2010), endoplasmic reticulum (Roberts and Ktistakis 2013), mitochondria (Hailey et al. 2010), endoplasmic reticulum–mitochondrial contact sites (Hamasaki et al. 2013), and Golgi (Ge et al. 2013) as membrane donors for the autophagosome biogenesis. Moreover, Shpilka et al. (2015) have recently observed that in yeast, the lipid droplets can also contribute to the formation of autophagosomes during conditions of nitrogen starvation.

The site of the phagophore assembly constitutes the initiation of the autophagosome biogenesis via two essential ATG protein complexes: (1) the kinase-containing Unc-51-like kinase (ULK) (the yeast Atg1) and (2) class III phosphatidylinositide 3-kinase (PI3K) (recently reviewed by Carlsson and Simonsen 2015). The ULK complex is composed by ULK1, ULK2, ATG13, and the focal adhesion kinase family–interacting protein of 200 kD (FIP200), and ULK1 and ULK2 display different affinities for the components of this complex (Jung et al. 2009). The activation of the ULK complex drives starvation-induced autophagy (Egan et al. 2011). By contrast, when nutrients are available, the nutrient sensor of the cell, mechanistic target of rapamycin (mTOR) phosphorylates ULK1 and ATG13, thus inhibiting ULK kinase activity (Egan et al. 2011). Conversely, upon starvation, mTOR-mediated inhibition of the ULK complex is reversed, and AMPK-mediated ULK phosphorylation ensues (Egan et al. 2011). Subsequently, ULK1-mediated autophosphorylation and phosphorylation of ATG13 and FIP200 enhance ULK-FIP200 interaction and ULK activity. As a consequence, ULK1 mobilizes to the autophagosomal structures to participate in the formation of the autophagosome (Hara et al. 2008; Jung et al. 2009). The next step in the formation of the nucleation complex involves the mammalian ortholog of ATG6, Beclin-1. Upon introducing cellular stress, Beclin-1 dissociates from its inhibitory partner Bcl-2 (Pattingre et al. 2005; Wei et al. 2008) and associates with ATG14, vacuolar protein sorting (VPS)15, and VPS34 that give rise to the functional PI3K complex (Itakura et al. 2008). The PI3K complex generates phosphoinositol 3-phosphate (PI3P) lipid species that recruit additional ATGs, for instance, the WD-repeat protein interacting with phosphoinositide (WIPI) family proteins, involved in the maturation of the limiting membrane (Polson et al. 2010; Tooze and Yoshimori 2010). The elongation of the limiting membrane occurs through the action of ATG7, a ubiquitin E1-like ligase that activates in parallel two different ubiquitin-like conjugation cascades: the ATG12-ATG5 and microtubule-associated protein 1 light chain 3 (LC3) systems (Tanida et al. 2001). For this purpose, ATG7 activates ATG12, transfers ATG12 to ATG10, and links ATG12 covalently to ATG5 to form the ATG12-ATG5 conjugate. ATG12-ATG5 interacts with ATG16L1 giving rise to the ATG12-ATG5-ATG16L1 complex (Kuma et al. 2002) that is then recruited to the nucleation complex to participate in expansion of the limiting membrane (Hanada et al. 2007). This recruitment is also mediated by the interaction of WIPI2 with ATG16L1 (Dooley et al. 2014). In parallel, ATG7 catalyzes the binding of the soluble form of LC3 (LC3-I), through its carboxyl terminus–exposed glycine, to the phosphatidylethanolamine of the growing autophagic membranes generating lipidated LC3 (LC3-II) (Hanada et al. 2007; Walczak and Martens 2013). Both the E2-like enzyme ATG3 and the ATG5-ATG12-ATG16L1 conjugate facilitate this process (Hanada et al. 2007; Walczak and Martens 2013). The relevance of LC3-II in the formation and cargo-recognition function of autophagosomes makes this lipidated protein the molecular identity of the autophagosomes, and it is used by the researches to reliably monitor autophagy.

7.3.2 CARGO RECOGNITION AND AUTOPHAGOSOME–LYSOSOME FUSION

The mechanism by which autophagosomes recognize cargo is via presence of cargo-recognition receptors that label autophagic substrates. In mammals, different cargo receptors have been identified including SQSTM1/p62, NBR1, and NDP52 (Johansen and Lamark 2011), with p62 being the best characterized cargo receptor (Pankiv et al. 2007). p62 recognizes tags in the cargo, for instance, polyubiquitin chains, through the ubiquitin-binding domain. In addition, these adaptors contain an LC3-interacting region (LIR) motif that allows them to bind to LC3-II (Noda et al. 2008), thus serving as a bridge between the cargo and LC3. Contrary to this notion, recent studies performed in mouse fibroblasts suggest that the localization of p62 to the autophagosome formation site is dependent on its self-oligomerization, and not on the interaction with LC3-II (Itakura and Mizushima 2011). Nevertheless, after sequestering the cargo, the limiting membrane seals upon itself to form double-membrane structures called autophagosomes, wherein both the inner and outer membranes are decorated by LC3-II. The fusion of the autophagosome with the lysosome generates autolysosomes. In this step, the outer membrane of the autophagosome fuses with the lysosome, and the inner membrane together with the engulfed cargo is degraded in the late endosome and/or lysosome (Figure 7.1) (He and Klionsky 2009). The LC3-II molecules on the outer membrane are cleaved off and recycled back to the cytosol by the cysteine protease ATG4 (Tanida et al. 2004). While much remains to be known about the mechanism of autophagosome–lysosome fusion, recent studies indicate the crucial role of N-ethylmaleimide-sensitive factor–activating protein (SNAREs) in mediating fusion events. In fact, the hairpin-type tail-anchored SNARE syntaxin 17 has been shown to regulate the fusion of autophagosomes with endosomes and lysosomes (Itakura et al. 2012). Indeed, coimmunoprecipitation and mass spectrometry analyses revealed that components of the homotypic fusion and protein sorting (HOPS)-tethering complex, vacuolar protein sorting 33A (VPS33A) and VPS16, interact with syntaxin 17, and that silencing HOPS complex members blocks autophagy flux (Jiang et al. 2014). Thus, the HOPS complex and its interaction with syntaxin 17 determines autophagy activity. More recently, O-GlcNAc (β-N-acetylglucosamine)-modification of the snare protein SNAP-29 has also been shown to regulate autophagosome–lysosome fusion (Guo et al. 2014). Thus, a complex network of SNARE proteins and post-translational modifications "fine-tune" the degree to which autophagosomes fuse with lysosomes, which in turn determines autophagy flux.

7.4 NUTRIENT-SENSING NETWORKS REGULATING AUTOPHAGY

7.4.1 NUTRIENT-SENSING mTOR AND AMPK PATHWAYS

Two major pathways integrate nutrient-related signals in the cell: the mTOR pathway and the 5′ adenosine monophosphate (AMP)-activated protein kinase (AMPK) pathway. Both signaling cascades converge in a diverging manner on effector molecules that regulate autophagy (Kim et al. 2011). The mTOR pathway is a central amino acid sensor in the cell that gets activated in presence of nutrients, in particular, branched-chain amino acids (Hara et al. 1998). The mTOR serine/threonine kinase can exist in two different multimeric complexes: mTOR complex 1 (mTORC1) and mTORC2, both of which exhibit wide-ranging and distinct functions (reviewed by Laplante and Sabatini 2009). mTORC1 has emerged as one of the key negative regulators of autophagy (Ravikumar et al. 2004) and is exquisitely sensitive to the antibiotic rapamycin. In fact, the use of rapamycin has been a useful tool to stimulate autophagy (Sarkar et al. 2009). By contrast, mTORC2 has been identified as a rapamycin-insensitive complex, although recent studies show that chronic exposure to rapamycin inhibits the formation of new mTORC2 complexes (Sarbassov et al. 2006). As described earlier, nutrient-driven activation of mTORC1 prevents ULK1 activation by its phosphorylation at serine 757 (Ser757), thus inhibiting autophagy (Kim et al. 2011). On the contrary, the serine/threonine kinase AMPK is the energy sensor of the cell whose activity is linked to the activation of autophagy (Meley et al. 2006). Conditions resulting in increased AMP/ATP ratio, such as nutrient starvation, trigger AMPK activity. Once active, AMPK directly phosphorylates ULK1 at residues distinct from those phosphorylated by mTOR, for instance, Ser317, Ser555, and Ser777, which in turn leads to recruitment of ULK1 to the site of autophagosome formation and activation of autophagy (Egan et al. 2011; Kim et al. 2011). In addition, AMPK can indirectly activate autophagy by inhibiting mTORC1 at multiple levels by (1) phosphorylation and activation of the negative regulator of mTORC1, tuberous sclerosis complex 2 (TCS2) (Inoki et al. 2003), (2) by direct inhibitory phosphorylation of mTOR kinase (Cheng et al. 2004), and (3) of raptor, an essential component of the mTORC1 complex (Gwinn et al. 2008). Interestingly, recent work from Munson et al. describes that formation of a cellular pool of PI3P at the lysosomes by the VPS34 kinase activity prevents lysosomal tubulation in a mechanism dependent on mTOR-mediated phosphorylation of UVRAG (Munson et al. 2015), a component of the Class III PI3K complex 2 involved in endosomal trafficking (Itakura et al. 2008). While the specific consequences of mTOR-regulated lysosomal tubulation on lysosomal function remain unclear, it is tempting to speculate that induction of VPS34 activity via mTOR is mechanistically linked to maintenance of lysosomal integrity. (4) Finally, mTORC1 has been shown to phosphorylate and inactivate transcription factor EB (TFEB), a recently elucidated master regulator of autophagy gene expression (Settembre et al. 2011). mTORC1-driven TFEB phosphorylation restricts it in the cytosol and thus interferes with autophagy gene expression.

7.4.2 TRANSCRIPTION MACHINERY REGULATING AUTOPHAGY

As briefly mentioned in the previous section, transcriptional regulation is emerging as a key mechanism to modulate autophagy. In particular, transcription factor

EB (TFEB), a member of the bHLH leucine-zipper family of transcription factors, has emerged as a master regulator of autophagy by coordinating the expression of genes belonging to the coordinated lysosomal expression and regulation (CLEAR) network (Sardiello and Ballabio 2009; Settembre et al. 2011). Signaling pathways that respond to nutrients and growth factors, for instance, mTOR (described above) and ERK2, control TFEB activity (Peña-Llopis et al. 2011; Settembre et al. 2011, 2012). During cell growth, ERK2-mediated phosphorylation of TFEB at Ser142 retains the transcription factor in the cytoplasm. By contrast, starvation blunts ERK2-driven inhibitory phosphorylation and rapidly induces TFEB shuttling to the nucleus, where TFEB promotes expression of network of autophagy and lysosomal genes which results in increased autophagy flux (Settembre et al. 2011). Although, mTORC1 has been identified as a TFEB kinase, additional regulatory mechanisms have also been proposed. For instance, studies by Peña-Llopis et al. have identified mTORC1-mediated phosphorylation of TFEB at a C-terminal serine-rich motif, which is distinct from the region described by Settembre and coworkers for ERK2, which induces TFEB nuclear localization and expression of vacuolar ATPase genes necessary for maintaining endosomal/lysosomal acidification (Peña-Llopis et al. 2011). By contrast, several groups have postulated mTORC1 as a negative regulator of TFEB through phosphorylation at Ser142 and a recently identified residue, Ser211 (Martina et al. 2012; Settembre et al. 2012). This regulatory mechanism involves the interaction between TFEB and mTORC1 at the lysosomal surface where mTORC1-mediated inhibitory phosphorylation of TFEB occurs, with the consequent downregulation of autophagy and lysosomal genes. The transcriptional control of autophagy via TFEB is relevant in physiological terms since overexpressing the TFEB ortholog in *Caenorhabditis elegans,* HLH-30, as described by the Hansen group, has been shown to activate autophagy and extend life span (Lapierre et al. 2013). This evidence supports the general notion that autophagy declines with age (Kaushik et al. 2012; Cuervo and Wong 2014; Grimmel et al. 2015; Schneider et al. 2015) and suggests that transcriptional regulation of the autophagy machinery could be a potential mechanism to restore autophagy during aging and to combat age-related diseases. In this context, mention must also be made of ZKSCAN3, which has been shown to be a master transcriptional repressor of autophagy (Chauhan et al. 2013). Inhibiting ZKSCAN3 led to repression of a large set of genes that are direct autophagy regulators or integral components of the autophagic machinery, for instance, *Map1lC3b* and *Wipi2*. Furthermore, ZKSCAN3 and TFEB were shown to be differentially regulated by nutrient deprivation, and thus ZKSCAN3 and TFEB play opposing roles in lysosome biogenesis and autophagy. Altogether, there is sufficient evidence to suggest that nutrient-related signaling networks regulate autophagy at multiple levels, and that cross talk between these signaling modulators points to the complexity of the autophagy process *per se.*

7.5 AUTOPHAGY REGULATES LIPID METABOLISM

In humans, alterations in lipid metabolism is intimately linked with the risk of developing metabolic disorders that are typically observed during obesity and in type 2 diabetes (reviewed by Greenberg et al. 2011). This strong association underscores the need for tight regulation of cell- and tissue-intrinsic lipid metabolism, which in turn will translate into maintained overall energy balance. Therefore, deep insight into the biology of mobilization of lipid droplets is necessary to understand how cells and tissues handle excessive fat stores. This information is required to fully understand the molecular mechanisms of development of obesity and type 2 diabetes, and how age-associated conditions, which frequently associate with insulin resistance, can impact human health. Given the role of autophagy in removal of cell-intrinsic lipid stores via lipophagy, the subsequent sections will discuss how autophagy contributes to cellular lipid turnover.

7.5.1   Lipid Droplets: More than Just Storing Lipids

Cells store fat in cytosolic lipid deposits known as lipid droplets. These lipid stores serve as reservoirs that the cell utilizes as fuel according to the energetic demands. Structurally, lipid droplets are formed by a core of neutral lipids (primarily triacylglycerols and cholesterol esters) surrounded by a monolayer of phospholipids, with phosphatidylcholine, phosphatidylethanolamine, and phosphatidylinositol being the principal lipid species in mammalian cells (Bartz et al. 2007). This lipid phase is separated from the cytosolic aqueous environment by a coat of structural proteins known as perilipins (PLINs) (Kimmel et al. 2010). PLIN1/perilipin 1 is the most characteristic feature of lipid droplets in adipocytes (Greenberg et al. 1991), while PLIN2/adipophilin and PLIN3/TIP47 are ubiquitously expressed (Brasaemle et al. 1997; Heid et al. 1998). Recently, studies have identified a number of additional lipid droplet coat proteins, for instance, OXPAT (Wolins et al. 2006), S3-12 (Wolins et al. 2003), and HILPDA (Gimm et al. 2010) to mention a few. Lipid droplets are not static entities. Instead, these undergo dynamic remodeling, for instance, by lipid transfer between lipid droplets (Paar et al. 2012). Dynamic interactions between lipid droplets and autophagic markers LC3 and LAMP1 have been reported, which are considered to participate in the "pinching off" of portions of lipid droplets through lipophagy (Singh et al. 2009b). In the sense of this dynamic remodeling of lipid droplets, small nascent droplets are primarily decorated by PLIN3 and S3-12 (Wolins et al. 2001, 2003), which during growth and maturation of the lipid droplet are replaced by PLIN2. In adipocytes, this organelle eventually grows to form a large and central lipid droplet enriched with PLIN1 (Brasaemle et al. 1997). All cells contain lipid droplets; however, the size and the number of lipid droplets vary between different cell populations from 100 nm to 10 µm in diameter. In particular, cells specially dedicated to lipid storage, like white adipocytes, contain a large, unilocular lipid droplet of a diameter of upto 100 µm that fills the entire cytoplasm (reviewed by Suzuki et al. 2011). Lipid droplets have been traditionally considered as mere energy depots; however, recently, they are being considered to play a role in intracellular lipid trafficking by providing with an important source of lipids to other cellular compartments (Bartz et al. 2007).

Interestingly, lipid droplets have been reclassified as organelles due to the growing number of nonconventional functions attributed to them in recent years. In this sense, evidence from the Welte Laboratory demonstrates that the lipid droplet surface serves as a temporary platform to store proteins such as histones during embryogenesis in *Drosophila melanogaster* (Cermelli et al. 2006; Li et al. 2012). The presence of viral proteins has also been reported during hepatitis C virus (HCV) infection, where the virus utilizes the lipid droplet surface for its own assembly during the replication process (Miyanari et al. 2007). In addition, lipid droplets can also sequester proteins that otherwise would be toxic for the cell, that is, apolipoprotein B-100 or α-synuclein (reviewed in Fujimoto and Parton 2011), thus revealing the availability of the lipid droplets as highly dynamic organelles that assist in diverse cellular functions.

7.5.2 THE ORIGIN OF LIPID DROPLETS

Several hypotheses have been postulated to explain the biogenesis of the lipid droplets. The majority of the findings support that the lipid droplets arise from the endoplasmic reticulum where the enzymes that synthetize the neutral lipids reside. While the initial steps of this process are still unclear, the most prevailing model proposes that neutral lipids accumulate in the bilayer of the endoplasmic reticulum that buds off to form a lipid droplet (reviewed by Fujimoto and Parton 2011; Thiam et al. 2013). After its scission from the endoplasmic reticulum, this newly formed lipid droplet grows as a result of several possible mechanisms such as (1) lipid synthesis by local enzymes such as GPAT4 (Wilfling et al. 2013), (2) lipid transfer from the small lipid droplets to larger ones at the lipid droplet contact sites (Gong et al. 2011) or even without physical interaction (Paar et al. 2012), and/or (3) fusion of smaller lipid droplets through the action of microtubules (Boström et al. 2005), SNARE proteins (Boström et al. 2007), and Rab GTPases (Wu et al. 2014).

7.5.3 LIPID DROPLET MOBILIZATION BY CYTOSOLIC LIPASES

In the cell, the mobilization and breakdown of lipid droplets occur via lipolysis. During starvation, triglycerides in the neutral lipid core are catabolized by the sequential action of cytosolic neutral lipases: adipose triglyceride lipase (ATGL), hormone-sensitive lipase (HSL), and monoacylglycerol lipase (MGL) (reviewed by Zechner et al. 2012). While lipolysis via cytosolic lipases occurs with varying degrees in all cell types, this biochemical process has been extensively studied in adipocytes. Following lipolytic stimulation, for example, upon treatment with catecholamines, the activation of the cAMP-dependent protein kinase A (PKA) directly triggers PLIN1 phosphorylation at the lipid droplet surface (Souza et al. 2002), which leads to the dissociation of PLIN1 from the PLIN1/CGI-58 complex (Granneman et al. 2007). The resulting displacement of PLIN1 allows CGI-58 to activate ATGL that, in turn, provides access to lipases for the lipid substrates (Lass et al. 2006; Granneman et al. 2007). As a result, ATGL hydrolyzes the triacylglycerols to diacylglycerols and fatty acids. In addition, PKA phosphorylates HSL, facilitating the mobilization of the lipase to PLIN1-positive lipid droplets (Brasaemle et al. 2000; Moore et al. 2005; Granneman et al. 2007). This mechanism is dependent on PLIN1 phosphorylation and the physical interaction between PLIN1 and HSL (Granneman et al. 2007; Shen et al. 2009). Once at the lipid droplet surface, HSL hydrolyzes diacylglycerols to monoacylglycerols and fatty acids. Ultimately, monoacylglycerols are converted to glycerol and fatty acids by MGL in the cytosol. The resulting free fatty acids and glycerol are released into circulation from where the fatty acids are taken up by other tissues to undergo mitochondrial β-oxidation to generate energy in the form of adenosine triphosphate (ATP).

7.5.4 Lipophagy: Autophagy-Mediated Lipid Droplet Turnover

As previously mentioned, during periods of nutrient scarcity, cells utilize lipid stores to meet the bioenergetic demands. During nutrient deprivation, autophagy is robustly activated. Given that starvation triggers autophagy to provide nutrients to the cell, it was proposed that lysosomal-mediated degradation of lipid droplets could provide an alternate source of energy during starvation (Singh et al. 2009b). This new function of selective lipid mobilization via autophagy was termed macrolipophagy or *lipophagy*. The first evidence for the existence of lipophagy was provided from studies in mouse hepatocytes (Singh et al. 2009b). In these studies, the authors used pharmacological inhibition of autophagy with 3-methyladenine (inhibitor of the class III PI3K activity) and genetic knockdown of autophagy by silencing *Atg5* gene using small interference RNA technology and observed the accumulation of lipid droplets under both conditions. These findings were validated *in vivo* using liver-specific mice knocked out for the critical autophagy gene *Atg7* (Albumin$^{Atg7\text{-}KO}$ mice), which revealed that autophagy blockage remarkably increased triglyceride and lipid droplet accumulation (Singh et al. 2009b). Consequently, cells knocked down for *Atg5* displayed a significant decrease in cellular β-oxidation rates when challenged with oleate (fatty acid) or cultured in methionine- and choline-deficient medium (MCDM). These observations indicate that the accumulation of lipid droplets is not secondary to increased triglyceride synthesis but is a consequence of impairment in autophagy. In support of this notion, the induction of autophagy activity by the presence of autophagy activators such as rapamycin and lithium chloride decreased the number of lipid droplets, clearly indicating that autophagy is required for lipid droplet breakdown. While the molecular mechanism of lipophagy still remains elusive, it has demonstrated the association of the autophagic machinery components LC3 and the lysosomal-associated membrane protein 1 (LAMP1) with lipid droplets. Upon lipid stimulation with MCDM, the colocalization of LAMP1 with the intracellular neutral lipid dye BODIPY 493/503 was shown; however, in unstimulated cells, this association was only detected in the presence of lysosomal inhibitors indicating autophagy-mediated rapid lipid turnover even during basal conditions. These observations correlate with the increase in LC3-II flux when cells were provided with a lipid stimulus. Direct evidence for lipophagy in hepatocytes came from electron microscopic studies, wherein double-membraned structures engulfing or containing lipid droplets were identified due possibly by the sequestration of an entire small lipid droplet or by the "chewing off" of portions of large lipid droplets (Figure 7.2). Indeed, this effect was increased when hepatocytes were exposed to lipid stimulus, reinforcing the concept of selective lipid degradation by autophagy. Further, support for a direct role of autophagosomes in sequestration of lipid droplets came from immunogold labeling strategies to tag LC3 molecules on the lipid droplets fractions isolated from starved mice when compared with fed cohorts. In addition, isolation of autophagic structures revealed enrichment of the lipid droplet coat proteins PLIN2 and PLIN3 in autophagosomes and lysosomes from starved livers when compared with fed samples.

Interestingly, mice exposed to high-fat diet (HFD) for 16 weeks resulted in lipophagy failure, and severely compromised the ability of autophagy to mobilize

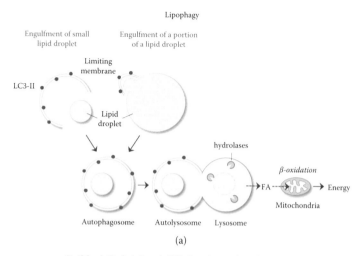

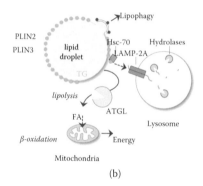

FIGURE 7.2 Autophagy-mediated lipid droplets turnover. (a) Nutrient deprivation activates autophagy. The newly formed LC3-II-positive double-membranes engulf small lipid droplets or portions of large lipid droplets. Once the limiting membrane seals upon itself, the autophagosome delivers the lipid cargo within the lysosomes where the luminal acid hydrolases degrade the lipids. The resulting fatty acids (FA) are released back into the cytosol and translocated to the mitochondria where they undergo β-oxidation to generate energy. (b) Activation of chaperone-mediated autophagy degrades the lipid droplets coat proteins perilipin 2 (PLIN2) and 3 (PLIN3). As a result, the lipid droplet surface becomes accessible to the action of (1) cytosolic lipase ATGL that hydrolyzes the lipid droplet resident triglycerides (TG) generating FA; (2) autophagic machinery proteins that form autophagosomes in order to sequester lipid droplet portions as described in (a). The resulting FA from both processes are oxidized in the mitochondria to generate energy.

lipid stores as indicated by increased lipid droplet content and decreased enrichment of LC3-II on lipid droplets during starvation when compared with regular chow-fed mice. There results indicate the inability of the autophagic process to accommodate to the increased lipid load such as those observed in mice models of diet-induced obesity. Altogether, these findings demonstrated a selective role of autophagy in lipid removal, and identified lipophagy as a completely new mechanism that regulates lipohomeostasis.

Since the discovery of lipophagy in hepatocytes, this mechanism has also been shown to be active in other cell types, for instance, in distinct neuronal populations such as hypothalamic neurons (Kaushik et al. 2011) and striatal neurons (Martinez-Vicente et al. 2010), macrophages (Ouimet et al. 2011; Xu et al. 2013), immune cells (Hubbard et al. 2010), hepatic stellate cells (Hernández-Gea et al. 2012), intestinal cells (Khaldoun et al. 2014), and cancer cells (Kaini et al. 2012) to mention a few. Hence, these findings point to the existence of lipophagy as a ubiquitous mechanism in cellular lipid homeostasis.

In contrast, several groups have reported recently that lipid droplets are required for efficient autophagy activation (Dupont et al. 2014; Shpilka et al. 2015). Dupont et al. (2014) described that lipid droplets, through the activity of the triglyceride lipase patatin-like phospholipase domain–containing protein 5 (PNPLA5), contribute lipid intermediates facilitating autophagosome membrane biogenesis. In addition, a study by Shpilka et al. (2015) has elucidated that in yeast under conditions of nitrogen starvation, lipid droplets *per se* contribute to *de novo* formation of autophagosomes by providing the triacylglycerols and steryl esters necessary for autophagosome biogenesis at the contact sites where endoplasmic reticulum and the lipid droplets converge. In this process, the enzymes required for biosynthesis of triacylglycerol (Dga1 and Lro1) and steryl esters (Are1 and Are2), and the corresponding enzymes driving their lipolysis, Ayr1 and Yeh1, respectively, together with the lipase/hydrolase Ldh1 are critical for formation of autophagosomes. It is not surprising that lipid droplets participate in autophagosome biogenesis since it has been previously shown that the very organelles that are degraded by autophagy, for example, endoplasmic reticulum (Khaminets et al. 2015) and portions of the plasma membrane (Bejarano et al. 2012), are also involved in autophagosome biogenesis (Ravikumar et al. 2010).

The molecular mechanism by which autophagy selectively targets lipid droplets as cargo is not known. Interestingly, abnormal lipid accumulation has been described in cells harboring mutations in huntingtin, a protein critically involved in the development of Huntington's disease. Mutant huntingtin results in failure to recognize cargo that, in turn, leads to inefficient cellular clearance and subsequent cellular toxicity (Martinez-Vicente et al. 2010). This would suggest a possible role of huntingtin as an adaptor molecule in lipid droplet turnover via lipophagy; in fact, a recent work has shown that huntingtin serves as a scaffold protein for cargo degradation through macroautophagy (Rui et al. 2015). Ancient ubiquitous protein 1 (AUP1), a quality control protein that modulates the degradation of misfolded proteins in the endoplasmic reticulum, has been identified at the lipid droplet surface (Klemm et al. 2011; Spandl et al. 2011). AUP1 has the ability to bind to the E2 ubiquitin conjugase Ube2g2 by its C-terminal region, which recruits Ube2g2 to the lipid droplet (Spandl et al. 2011). Moreover, a role of AUP1 in lipid droplet accumulation/formation has been described by Klemm et al. (2011), allowing us to hypothesize that proteins involved in the ubiquitination machinery may serve as scaffold/adaptors for targeted lipid droplet degradation via lipophagy. In addition, several groups have reported the role of Rab GTPases in regulation of lipophagy by mediating the interaction between lipid droplets and endosomal/lysosomal compartments (Liu et al. 2007; Lizaso et al. 2013; Schroeder et al. 2015). A role of the GTPase dynamin-2 has also been shown to contribute to lipid droplet turnover in hepatocytes. Through its role in the generation of lysosomal pools from autolysosomal tubule scission, dynamin-2 induces autophagy-mediated lipid droplet utilization (Schulze et al. 2013).

7.5.5 Additional Forms of Autophagy and Regulation of Lipid Catabolism

This chapter focuses on autophagy of lipids, that is, lipophagy; however, a brief mention must be made of the recently reported work showing roles for other forms of autophagy in the modulation of lipid droplet mobilization. More specifically in yeast, lipid droplets are degraded in vacuoles in a process morphologically resembling microautophagy (van Zutphen et al. 2014). Furthermore, a recent work from the Cuervo Laboratory has described the new role of CMA in lipid droplet degradation by selective degradation of PLINs (Kaushik and Cuervo 2015) (Figure 7.2). Indeed, previous studies from the same group have shown that CMA is an important regulator of lipid metabolism and whole-body energy balance (Schneider et al. 2014). More specifically, the group observed an increase in the number of lipid droplets and development of hepatic steatosis in liver-specific LAMP-2A knockout mice (Schneider et al. 2014). Recent work by the Cuervo Laboratory describes now a new molecular mechanism of CMA-dependent lipid droplet turnover by selective degradation of PLIN2 and PLIN3, both of which contain CMA-selective KFERQ motifs (Kaushik and Cuervo 2015). The CMA-dependent degradation of PLIN proteins allows both cytosolic lipases and ATG proteins to access the lipid droplet surface and induce lipolysis during starvation (Kaushik and Cuervo 2015). In consistency with this notion, deletion of LAMP-2A led to decreased association of ATG proteins and cytosolic lipases to the lipid droplets both *in vivo* and *in vitro*. Moreover, targeted mutation of the KFERQ motif in PLIN2 recapitulated the effects observed in the LAMP-2A knockout systems, clearly demonstrating that CMA is required for the mobilization of lipid droplets via lipophagy and lipases (Kaushik and Cuervo 2015).

7.5.6 Transcriptional Regulators of Lipophagy

In recent years, the study of the regulation of lipid metabolism has focused on the nutrient-sensing networks that modulate lipid mobilization. The identification of TFEB as a master regulator of autophagy gene expression has strengthened the connection between transcriptional control of autophagy and lipid metabolism. To this end, the *C. elegans* TFEB ortholog, HLH-30, and MXL-3 have been shown to link the nutrient status to lipophagy through the modulation of the lysosomal lipases LIPL1- and LIPL3-mediated lipid breakdown (O'Rourke and Ruvkun 2013). In the presence of nutrients, MXL-3 represses *lipl1* and *lipl3* genes, whereas in response to fasting HLH-30 induces the expression of these lipases (O'Rourke and Ruvkun 2013). In addition to worms, mammalian TFEB has also been reported to activate lipid metabolism (Settembre et al. 2013). Concretely, TFEB induces PPARα-mediated lipid catabolism by direct upregulation of PGC1α expression in a starvation-dependent manner. Interestingly, autophagy *per se* is necessary for TFEB-mediated lipid degradation as TFEB overexpression failed to decrease lipid droplet number in liver-specific *Atg7* knockout mice (Settembre et al. 2013). Moreover, in primary neurons, the treatment with the lipid-lowering drug, gemfibrozil, enhances TFEB-driven lysosomal biogenesis in a mechanism controlled by PPARα (Ghosh et al. 2015). In line with the notion that the nutrient-activated pathways inversely regulate lipophagy, the cAMP response element–binding protein (CREB) has been shown to directly increase transcription of TFEB during starvation. Intriguingly, the opposite effects are observed when nutrients are available, which lead to nuclear receptor farnesoid X receptor (FXR)-mediated suppression of lipophagy (Seok et al. 2014). In a similar mechanism observed by Lee et al., starvation-induced PPARα activity leads to blockage of FXR-mediated inhibition of autophagy under fed conditions (Lee et al. 2014). Finally, in addition to TFEB, additional transcriptional mechanisms involving forkhead box protein O1 (FoxO1) have also been demonstrated. In a study performed by Lettieri Barbato et al., starvation-induced FoxO1 activity promotes lipid catabolism through lysosomal acid lipase activity (Lettieri Barbato et al. 2013). Thus, transcriptional mechanisms including those influenced by TFEB and FoxO1 have been shown to control lipophagy.

7.5.7 LIPOPHAGY IN AGRP NEURONS IN CONTROL OF FEEDING

As briefly discussed previously, lipophagy has been shown to function in distinct neuronal populations including the nutrient-sensing agouti-related peptide (AgRP) neurons. In this subsection, our objective will be to expand upon the role of lipophagy in hypothalamic nutrient sensing. The mediobasal hypothalamus (MBH) consists of the orexigenic AgRP and the anorexigenic proopiomelanocortin (POMC) neurons that contribute to a neural circuitry essential for integrating nutritional, hormonal, and neural information to determine feeding and energy metabolism (Belgardt and Bruning 2010). In particular, AgRP neurons secrete AgRP and provide inhibitory GABAergic projections on POMC neurons that drive feeding by suppressing melanocortinergic signaling originating from POMC neurons (Flier 2006). By contrast, POMC neurons release α-melanocyte-stimulating hormone (MSH) that suppresses feeding and increases energy expenditure (Mountjoy 2010).

Extensive studies have revealed crucial roles for hypothalamic mTOR (Cota et al. 2006) and AMPK (Claret et al. 2007; Lopez et al. 2008) in control of feeding and energy balance, and intriguingly, both mTOR and AMPK regulate autophagy (Jung et al. 2009; Lee et al. 2010; Neufeld 2010; Egan et al. 2011). Furthermore, neuronal free fatty acid availability has been shown to control food intake (Lam et al. 2005), and in fact, a study suggests that neuronal free fatty acid oxidation fulfills the energetic needs for neuronal activation (Andrews et al. 2008). Despite these developments, the mechanisms regulating neuronal free fatty acid availability remained unclear. Since hypothalamic mTOR suppresses feeding (Cota et al. 2006), and given that hypothalamic free fatty acid availability increases feeding, we hypothesized that lipophagy induction in AgRP neurons is mechanistically coupled to feeding by providing free fatty acids necessary to activate orexigenic neurons.

Indeed, our studies reveal that starvation-associated signals stimulate autophagy in AgRP-positive hypothalamic GT1-7 cells and in MBH tissue (Kaushik et al. 2011) (Figure 11.3). In parallel with our findings of lipophagy in cultured hepatocytes, exposing cultured hypothalamic cells to an acute lipid stimulus increased autophagic activity, and increased protein turnover by lysosomes (Kaushik et al. 2011). Interestingly, free fatty acid–driven autophagy activity occurred in parallel with increases in hypothalamic levels of phosphorylated AMPK and ULK1 (Kaushik et al. 2011) that are upstream activators of autophagy (Jung et al. 2009; Lee et al. 2010; Egan et al. 2011). Activation of autophagy in the hypothalamus occurred in unison with increased triglyceride content in MBH tissue as determined by thin layer chromatography. Since starvation in intact animals is typically associated with raised free fatty acids in circulation, we reasoned that the source of hypothalamic triglycerides during starvation is the periphery, as suggested previously (Andrews et al. 2008). In fact, cultured hypothalamic cells when exposed to serum from starved rodents displayed an increase in their neuronal lipid droplets when contrasted against cells treated with serum from fed rodents. Intriguingly, the fate of free fatty acids taken up by starved hypothalamic cells was synthesis of triglycerides and formation of lipid droplets, since triacsin C (inhibitor of triglycerides synthesis)–treated starved cells displayed decreased triglycerides levels during starvation (Kaushik et al. 2011). This suggested that a system was in place to rapidly degrade any of the newly formed lipid droplets as a means to increase availability of cell-intrinsic free fatty acids during starvation—enter lipophagy! Indeed, treating cultured hypothalamic cells and primary hypothalamic neurons with fatty acids enhanced the colocalization of LC3 and LAMP1 with BODIPY 493/503, which demonstrated that as detected in liver, lipophagy in hypothalamic neurons degrades lipid droplets in lysosomes (Kaushik et al. 2011).

The physiological relevance of lipophagy in starved hypothalamic cells is for controlled generation of neuronal free fatty acids, since interfering with lysosomal function or knocking down *Atg5* in hypothalamic cells decreased free fatty acid levels as detected by thin layer chromatography (Kaushik et al. 2011). It was shown that activation of lipophagy is linked to the generation of AgRP, the orexigenic signal for feeding, since blocking lysosomal function decreased fatty acid–induced increases in AgRP levels (Kaushik et al. 2011) (Figure 7.3). In addition, knocking out AgRP neuron–selective autophagy in mice decreased MBH neuronal AgRP levels and suppressed fasting-induced increases in food intake (Kaushik et al. 2011).

Consistently, knockout mice displayed reduced adiposity and lower body weights. Deleting *Atg7* in AgRP neurons associated with increased hypothalamic POMC levels and increased anorexigenic α-MSH levels that, in turn, likely led to reduced body weight and adiposity by driving locomotor activity and raising adipose triglyceride lipase levels in fat pads of knockout mice (Kaushik et al. 2011). These findings suggest that activation of lipophagy in a subset of hypothalamic neurons during starvation is sufficient to trigger changes in feeding patterns and energy expenditure in mice—highlighting important roles for lipophagy in these neurons.

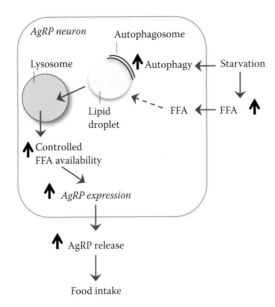

FIGURE 7.3 Hypothalamic autophagy regulates *AgRP* expression in response to starvation. Starvation induces increased circulating free fatty acids (FFA), which drives fatty acid uptake and triglyceride synthesis in the hypothalamic AgRP neurons. In turn, fatty acid uptake induces neuronal lipophagy generating endogenous free fatty acids that increase orexigenic *AgRP* gene expression. As a consequence, increased AgRP secretion stimulates food intake during fasting.

7.6 CONCLUDING REMARKS

Lipophagy impacts on lipid metabolism and thereby on whole-body energy balance by degrading lipid stores. Since lipophagy was first described in hepatocytes, several works have reported the existence of this process in multiple cell types, pointing to lipophagy as a ubiquitous mechanism in cellular lipid homeostasis. However, several questions still need to be elucidated, among others, the detailed molecular mechanism for the recognition of the lipid droplets by the autophagic machinery. A better understanding of the mechanism of lipophagy and its regulatory signaling networks would be helpful to develop new strategies for the treatment of metabolic diseases associated with perturbed autophagic function.

ACKNOWLEDGMENTS

Work in the laboratory of Dr. Rajat Singh is supported by the National Institutes of Health AG043517 and AG031782, an Ellison Medical Foundation New Scholar Award (R.S.), and institutional funds from the Albert Einstein College of Medicine.

REFERENCES

Andrews, Z.B., Liu, Z.W., Walllingford, N., et al. 2008. UCP2 mediates ghrelin's action on NPY/AgRP neurons by lowering free radicals. *Nature*. 454:846–51.

Bartz, R., Li, W.H., Venables, B., et al. 2007. Lipidomics reveals that adiposomes store ether lipids and mediate phospholipid traffic. *J Lipid Res.* 48(4):837–47.

Bejarano, E., Girao, H., Yuste, A., et al. 2012. Autophagy modulates dynamics of connexins at the plasma membrane in a ubiquitin-dependent manner. *Mol Biol Cell.* 11:2156–69.

Belgardt, B.F. and Bruning, J.C. 2010. CNS leptin and insulin action in the control of energy homeostasis. *Ann N Y Acad Sci.* 1212:97–113.

Boström, P., Andersson, L., Rutberg, M., et al. 2007. SNARE proteins mediate fusion between cytosolic lipid droplets and are implicated in insulin sensitivity. *Nat Cell Biol.* 9:1286–93.

Boström, P., Rutberg, M., Ericsson, J., et al. 2005. Cytosolic lipid droplets increase in size by microtubule-dependent complex formation. *Arterioscler Thromb Vasc Biol.* 25(9):1945–51.

Brasaemle, D.L., Barber, T., Wolins, N.E., et al. 1997. Adipose differentiation-related protein is an ubiquitously expressed lipid storage droplet-associated protein. *J Lipid Res.* 38(11):2249–63.

Brasaemle, D.L., Levin, D.M., Adler-Wailes, D.C., et al. 2000. The lipolytic stimulation of 3T3-L1 adipocytes promotes the translocation of hormone-sensitive lipase to the surfaces of lipid storage droplets. *Biochim Biophys Acta.* 1483(2):251–62.

Carlsson, S.R., and Simonsen, A. 2015. Membrane dynamics in autophagosome biogenesis. *J Cell Sci.* 128(2):193–205.

Cermelli, S., Guo, Y., Gross, SP., et al. 2006. The lipid-droplet proteome reveals that droplets are a protein-storage depot. *Curr Biol.* 16:1783–95.

Chauhan, S., Goodwin, J.G., Chauhan, S., et al. 2013. ZKSCAN3 is a master transcriptional repressor of autophagy. *Mol Cell.* 50(1):16–28.

Cheng, S.W., Fryer, L.G., Carling, D., et al. 2004. Thr2446 is a novel mammalian target of rapamycin (mTOR) phosphorylation site regulated by nutrient status. *J Biol Chem.* 279(16):15719–22.

Claret, M., Smith, M.A., Batterham, R.L., et al. 2007. AMPK is essential for energy homeostasis regulation and glucose sensing by POMC and AgRP neurons. *J Clin Invest.* 117:2325–36.

Cota, D., Proulx, K., Smith, K.A., et al. 2006. Hypothalamic mTOR signaling regulates food intake. *Science.* 312:927–30.

Cuervo, A.M., and Wong, E. 2014. Chaperone-mediated autophagy: Roles in disease and aging. *Cell Res.* 24(1):92–104.

Dice, F.J. 1990. Peptide sequences that target cytosolic proteins for lysosomal proteolysis. *Trends Biochem Sci* 15(8): 305–9.

Dooley, H.C., Razi, M., Polson, H.E., Girardin, S.E., Wilson, M.I., and Tooze, S.A. 2014. WIPI2 links LC3 conjugation with PI3P, autophagosome formation, and pathogen clearance by recruiting Atg12-5-16L1. *Mol Cell.* 55:238–52.

Dupont, N., Chauhan, S., Arko-Mensah, J., et al 2014. Neutral lipid stores and lipase PNPLA5 contribute to autophagosome biogenesis. *Curr Biol.* 24(6):609–20.

Duran, J.M., Anjard, C., Stefan, C., et al. 2010. Unconventional secretion of Acb1 is mediated by autophagosomes. *J Cell Biol.* 188(4):527–36.

Egan, D.F., Shackelford, D.B., Mihaylova, M.M., et al. 2011. Phosphorylation of ULK1 (hATG1) by AMP-activated protein kinase connects energy sensing to mitophagy. *Science.* 331:456–61.

Flier, J.S. 2006. AgRP in energy balance: Will the real AgRP please stand up? *Cell Metab.* 3:83–85.

Fujimoto, T., and Parton, R.G. 2011. Not just fat: The structure and function of the lipid droplet. *Cold Spring Harb Perspect Biol.* 3:a004838.

Ge, L., Melville, D., Zhang, M., et al. 2013. The ER-Golgi intermediate compartment is a key membrane source for the LC3 lipidation step of autophagosome biogenesis. *eLife.* 2:e00947.

Ghosh, A., Jana, M., Modi, K., et al. 2015. Activation of peroxisome proliferator-activated receptor α induces lysosomal biogenesis in brain cells: Implications for lysosomal storage disorders. *J Biol Chem.* 290(16):10309–24.

Gimm, T., Wiese, M., Teschemacher, B., et al. 2010. Hypoxia-inducible protein 2 is a novel lipid droplet protein and a specific target gene of hypoxia-inducible factor-1. *FASEB J.* 24(11):4443–58.

Gong, J., Sun, Z., Wu, L., et al. 2011. Fsp27 promotes lipid droplet growth by lipid exchange and transfer at lipid droplet contact sites. *J Cell Biol.* 195(6):953–63.

Granneman, J.G., Moore, H.P., Granneman, R.L., et al. 2007. Analysis of lipolytic protein trafficking and interactions in adipocytes. *J Biol Chem.* 282(8):5726–35.

Greenberg, A.S., Coleman, R.A., Kraemer, F.B., et al. 2011. The role of lipid droplets in metabolic disease in rodents and humans. *J Clin Invest.* 121(6): 2102–10.

Greenberg, A.S., Egan, J.J., Wek, S.A., et al. 1991. Perilipin, a major hormonally regulated adipocyte-specific phosphoprotein associated with the periphery of lipid storage droplets. *J Biol Chem.* 266(17):11341–6.

Grimmel, M., Backhaus, C., Proikas-Cezanne, T. 2015. WIPI-Mediated autophagy and longevity. *Cells.* 4(2):202–17.

Guo, B., Liang, Q., Li, L., et al. 2014. O-GlcNAc-modification of SNAP-29 regulates autophagosome maturation. *Nat Cell Biol.* 16(12):1215–26.

Gwinn, D.M., Shackelford, D.B., Egan, D.F., et al. 2008. AMPK phosphorylation of raptor mediates a metabolic checkpoint. *Mol Cell.* 30(2):214–26.

Hailey, D.W., Rambold, A.S., Satpute-Krishnan, P., et al. 2010. Mitochondria supply membranes for autophagosome biogenesis during starvation. *Cell.* 141(4):656–67.

Hamasaki, M., Furuta, N., Matsuda, A., et al. 2013. Autophagosomes form at ER-mitochondria contact sites. *Nature.* 495(7441):389–93.

Hanada, T., Noda, N.N., Satomi, Y., et al. 2007. The Atg12-Atg5 conjugate has a novel E3-like activity for protein lipidation in autophagy. *J Biol Chem.* 282(52):37298–302.

Hara, K., Yonezawa, K., Weng, Q.P., et al. 1998. Amino acid sufficiency and mTOR regulate p70 S6 kinase and eIF-4E BP1 through a common effector mechanism. *J Biol Chem.* 273(23):14484–94.

Harding, T.M., Hefner-Gravink, A., Thumm, M., et al. 1996. Genetic and phenotypic overlap between autophagy and the cytoplasm to vacuole protein targeting pathway. *J Biol Chem.* 271(30):17621–4.

He, C., Klionsky, D.J. 2009. Regulation mechanisms and signaling pathways of autophagy. *Annu Rev Genet.* 43:67–93.

Heid, H.W., Moll, R., Schwetlick, I., et al. 1998. Adipophilin is a specific marker of lipid accumulation in diverse cell types and diseases. *Cell Tissue Res.* 294:309–21.

Hernández-Gea, V., Ghiassi-Nejad, Z., Rozenfeld, R., et al. 2012. Autophagy releases lipid that promotes fibrogenesis by activated hepatic stellate cells in mice and in human tissues. *Gastroenterology.* 142(4):938–46.

Hubbard, V.M., Valdor, R., Patel, B., et al. 2010. Macroautophagy regulates energy metabolism during effector T cell activation. *J Immunol.* 185(12):7349–57.

Inoki, K., Zhu, T., Guan, K.L. 2003. TSC2 mediates cellular energy response to control cell growth and survival. *Cell.* 115(5):577–90.

Itakura, E., Kishi, C., Inoue, K., et al. 2008. Beclin 1 forms two distinct phosphatidylinositol 3-kinase complexes with mammalian Atg14 and UVRAG. *Mol Biol Cell.* 19(12):5360–72.

Itakura, E., Kishi-Itakura, C., Mizushima, N., et al. 2012. The hairpin-type tail-anchored SNARE syntaxin 17 targets to autophagosomes for fusion with endosomes/lysosomes. *Cell.* 151(6):1256–69.

Itakura, E., and Mizushima, N. 2011. p62 Targeting to the autophagosome formation site requires self-oligomerization but not LC3 binding. *J Cell Biol.* 192:17–27.

Jiang, P., Nishimura, T., Sakamaki, Y., et al. 2014. The HOPS complex mediates autophagosome-lysosome fusion through interaction with syntaxin 17. *Mol Biol Cell.* 25(8):1327–37.

Johansen, T., and Lamark, T. 2011. Selective autophagy mediated by autophagic adapter proteins. *Autophagy.* 7(3):279–96.

Jung, C.H., Jun, C.B., Ro, SH., et al. 2009. ULK-Atg13-FIP200 complexes mediate mTOR signaling to the autophagy machinery. *Mol Biol Cell.* 20:1992–2003.

Kaini, R.R., Sillerud, L.O., Zhaorigetu, S., et al. 2012. Autophagy regulates lipolysis and cell survival through lipid droplet degradation in androgen-sensitive prostate cancer cells. *Prostate* 72:1412–22.

Kaushik, S., Arias, E., Kwon, H., et al. 2012. Loss of autophagy in hypothalamic POMC neurons impairs lipolysis. *EMBO Rep.* 13(3):258–65.

Kaushik, S., and Cuervo, A.M. 2012. Chaperone-mediated autophagy: A unique way to enter the lysosome world. *Trends Cell Biol.* 22(8): 407–17.

Kaushik, S., Cuervo, A.M. 2015. Degradation of lipid droplet-associated proteins by chaperone-mediated autophagy facilitates lipolysis. *Nat Cell Biol.* 17(6):759–70.

Kaushik, S., Massey, A.C., Mizushima, N., et al. 2008. Constitutive activation of chaperone-mediated autophagy in cells with impaired macroautophagy. *Mol Biol Cell.* 19(5):2179–92.

Kaushik, S., Rodriguez-Navarro, J.A., Arias, E., et al. 2011. Autophagy in hypothalamic AgRP neurons regulates food intake and energy balance. *Cell Metab.* 14(2):173–83.

Khaldoun, S.A., Emond-Boisjoly, M.A., Chateau, D., et al. 2014. Autophagosomes contribute to intracellular lipid distribution in enterocytes. *Mol Biol Cell.* 25(1):118–32.

Khaminets, A., Heinrich, T., Mari, M., et al. 2015. Regulation of endoplasmic reticulum turnover by selective autophagy. *Nature.* 522(7556):354–8.

Kim, I., Rodriguez-Enriquez, S., Lemasters, J.J. 2007. Selective degradation of mitochondria by mitophagy. *Arch Biochem Biophys.* 462:245–53.

Kim, J., and Klionsky, D.J. 2000. Autophagy, cytoplasm-to-vacuole targeting pathway, and pexophagy in yeast and mammalian cells. *Annu Rev Biochem.* 69:303–42.

Kim, J., Kundu, M., Viollet, B., et al. 2011. AMPK and mTOR regulate autophagy through direct phosphorylation of Ulk1. *Nat Cell Biol.* 13(2):132–41.

Kimmel, A.R., Brasaemle, D.L., McAndrews-Hill, M., et al. 2010. Adoption of PERILIPIN as a unifying nomenclature for the mammalian PAT-family of intracellular lipid storage droplet proteins. *J Lipid Res.* 51(3):468–71.

Klemm, E.J., Spooner, E., and Ploegh, H.L. 2011. Dual role of ancient ubiquitous protein 1 (AUP1) in lipid droplet accumulation and endoplasmic reticulum (ER) protein quality control. *J Biol Chem.* 286(43):37602–14.

Klionsky, D.J. 2007. Autophagy: From phenomenology to molecular understanding in less than a decade. *Nat Rev Mol Cell Biol.* 8:931–37.

Komatsu, M., Waguri, S., Chiba, T., et al. 2006. Loss of autophagy in the central nervous system causes neurodegeneration. *Nature.* 441:880–84.

Kraft, C., Deplazes, A., Sohrmann, M., et al. 2008. Mature ribosomes are selectively degraded upon starvation by an autophagy pathway requiring the Ubp3p/Bre5p ubiquitin protease. *Nat Cell Biol.* 10:602–10.

Kuma, A., Mizushima, N., Ishihara, N., et al. 2002. Formation of the approximately 350-kDa Apg12-Apg5.Apg16 multimeric complex, mediated by Apg16 oligomerization, is essential for autophagy in yeast. *J Biol Chem.* 277(21):18619–25.

Lam, T.K., Schwartz, G.J., and Rossetti, L. 2005. Hypothalamic sensing of fatty acids. *Nat Neurosci.* 8:579–84.

Lapierre, L.R., De Magalhaes Filho, C.D., McQuary, P.R., et al. 2013. The TFEB orthologue HLH-30 regulates autophagy and modulates longevity in *Caenorhabditis elegans. Nat Commun.* 4:2267.

Laplante, M., and Sabatini, D.M. 2009. mTOR signaling at a glance. *J Cell Sci.* 122(Pt 20):3589–94.

Lass, A., Zimmermann, R., Haemmerle, G., et al. 2006. Adipose triglyceride lipase-mediated lipolysis of cellular fat stores is activated by CGI-58 and defective in Chanarin-Dorfman syndrome. *Cell Metab.* 3:309–19.

Lee, J.M., Wagner, M., Xiao, R., et al. 2014. Nutrient-sensing nuclear receptors coordinate autophagy. *Nature.* 516:112–15.

Lee, J.W., Park, S., Takahashi, Y., et al. 2010. The association of AMPK with ULK1 regulates autophagy. *PLoS One.* 5:e15394.

Lettieri Barbato, D., Tatulli, G., Aquilano, K., et al. 2013. FoxO1 controls lysosomal acid lipase in adipocytes: Implication of lipophagy during nutrient restriction and metformin treatment. *Cell Death Dis.* 4:e861.

Levine, B., 2005. Eating oneself and uninvited guests: Autophagy-related pathways in cellular defense. *Cell.* 120(2):159–62.

Levine, B., Mizushima, N., and Virgin, H.W., 2011. Autophagy in immunity and inflammation. *Nature.* 469(7330):323–35.

Li, W.W., Li, J., and Bao, J.K., 2012. Microautophagy: Lesser-known self-eating. *Cell Mol Life Sci.* 69:1125–1136.

Li, Z., Thiel, K., Thul, P.J., et al. 2012. Lipid droplets control the maternal histone supply of Drosophila embryos. *Curr Biol.* 22(22):2104–13.

Liang, X.H., Jackson, S., Seaman, M., et al. 1999. Induction of autophagy and inhibition of tumorigenesis by beclin 1. *Nature.* 402(6762):672–6.

Liu, P., Bartz, R., Zehmer, J.K., et al. 2007. Rab-regulated interaction of early endosomes with lipid droplets. *Biochim Biophys Acta.* 1773(6):784–93.

Lizaso, A., Tan, K.T., and Lee, Y.H., 2013. β-adrenergic receptor-stimulated lipolysis requires the RAB7-mediated autolysosomal lipid degradation. *Autophagy.* 9(8):1228–43.

Lopez, M., Lage, R., Saha, A.K., et al. 2008. Hypothalamic fatty acid metabolism mediates the orexigenic action of ghrelin. *Cell Metab.* 7:389–99.

Manjithaya, R., Anjard, C., Loomis, W.F., et al. 2010. Unconventional secretion of Pichia pastoris Acb1 is dependent on GRASP protein, peroxisomal functions, and autophagosome formation. *J Cell Biol.* 188(4):537–46.

Martina, J.A., Chen, Y., Gucek, M., et al. 2012. MTORC1 functions as a transcriptional regulator of autophagy by preventing nuclear transport of TFEB. *Autophagy.* 8(6):903–14.

Martinez-Lopez, N., Athonvarangkul, D., Mishall, P., et al. 2013a. Autophagy proteins regulate ERK phosphorylation. *Nat Commun.* 4:2799.

Martinez-Lopez, N., Athonvarangkul, D., Sahu, S., et al. 2013b. Autophagy in Myf5+ progenitors regulates energy and glucose homeostasis through control of brown fat and skeletal muscle development. *EMBO Rep.* 14(9):795–803.

Martinez-Vicente, M., Talloczy, Z., Wong, E., et al. 2010. Cargo recognition failure is responsible for inefficient autophagy in Huntington's disease. *Nat Neurosci.* 13:567–76.

Marzella, L., Ahlberg, J., and Glaumann, H., 1981. Autophagy, heterophagy, microautophagy and crinophagy as the means for intracellular degradation. *Virchows Arch B Cell Pathol Incl Mol Pathol.* 36:219–34.

Masiero, E., Agatea, L., Mammucari, C., et al. 2009. Autophagy is required to maintain muscle mass. *Cell Metab.* 10(6):507–15.

Meley, D., Bauvy, C., Houben-Weerts, J.H., et al. 2006. AMP-activated protein kinase and the regulation of autophagic proteolysis. *J Biol Chem.* 281(46):34870–9.

Miyanari, Y., Atsuzawa, K., Usuda, N., et al. 2007. The lipid droplet is an important organelle for hepatitis c virus production. *Nat Cell Biol.* 9:1089–1097.

Mizushima, N., Sugita, H., Yoshimori, T., et al. 1998. A new protein conjugation system in human. The counterpart of the yeast Apg12p conjugation system essential for autophagy. *J Biol Chem.* 18;273(51):33889–92.

Moore, H.P., Silver, R.B., Mottillo, E.P., et al. 2005. Perilipin targets a novel pool of lipid droplets for lipolytic attack by hormone-sensitive lipase. *J Biol Chem.* 280(52):43109–20.

Mountjoy, K.G., 2010. Functions for pro-opiomelanocortin-derived peptides in obesity and diabetes. *Biochem J*. 428:305–24.

Munson, M.J., Allen, G.F., Toth, R., et al. 2015. mTOR activates the VPS34-UVRAG complex to regulate autolysosomal tubulation and cell survival. *EMBO J*. 34(17):2272–90.

Neufeld, T.P. 2010. TOR-dependent control of autophagy: Biting the hand that feeds. *Curr Opin Cell Biol*. 22:157–68.

Noda, N.N., Kumeta, H., Nakatogawa, H., et al. 2008. Structural basis of target recognition by Atg8/LC3 during selective autophagy. *Genes Cells*. 13(12):1211–8.

Novikoff, A.B., Beaufay, H., and de Duve, C., 1956. Electron microscopy of lysosome-rich fractions from rat liver. *J Biophys Biochem Cytol*. 2(4):179–184.

O'Rourke, E.J., and Ruvkun, G., 2013. MXL-3 and HLH-30 transcriptionally link lipolysis and autophagy to nutrient availability. *Nat Cell Biol*. 15:668–76.

Ouimet, M., Franklin, V., Mak, E., et al. 2011. Autophagy regulates cholesterol efflux from macrophage foam cells via lysosomal acid lipase. *Cell Metab*. 13(6):655–67.

Paar, M., Jüngst, C., Steiner, N.A., et al. 2012. Remodeling of lipid droplets during lipolysis and growth in adipocytes. *J Biol Chem*. 287(14):11164–73.

Pankiv, S., Clausen, T.H., Lamark, T., et al. 2007. p62/SQSTM1 binds directly to Atg8/LC3 to facilitate degradation of ubiquitinated protein aggregates by autophagy. *J Biol Chem*. 282(33):24131–45.

Pattingre, S., Tassa, A., Qu, X., et al. 2005. Bcl-2 antiapoptotic proteins inhibit Beclin 1-dependent autophagy. *Cell*. 122(6):927–39.

Peña-Llopis, S., Vega-Rubin-de-Celis, S., Schwartz, J.C., et al. 2011. Regulation of TFEB and V-ATPases by mTORC1. *EMBO J*. 30(16):3242–58.

Polson, H.E., de Lartigue, J., Rigden, D.J., et al. 2010. Mammalian Atg18 (WIPI2) localizes to omegasome-anchored phagophores and positively regulates LC3 lipidation. *Autophagy*. 6(4):506–22.

Ravikumar, B., Moreau, K., Jahreiss, L., et al. 2010. Plasma membrane contributes to the formation of pre-autophagosomal structures. *Nat Cell Biol*. 12:747–757.

Ravikumar, B., Vacher, C., Berger, Z., et al. 2004. Inhibition of mTOR induces autophagy and reduces toxicity of polyglutamine expansions in fly and mouse models of Huntington disease. *Nat Genet*. 36:585–95.

Roberts, R., Ktistakis, N.T. 2013. Omegasomes: PI3P platforms that manufacture autophagosomes. *Essays Biochem*. 55:17–27.

Rubinsztein, D.C., Mariño, G., and Kroemer, G., 2011. Autophagy and aging. *Cell*. 146(5):682–95.

Rui, Y.N., Xu, Z., Patel, B., et al. 2015. Huntingtin functions as a scaffold for selective macroautophagy. *Nat Cell Biol*. 3:262–75.

Sahu, R., Kaushik, S., Clement, C.C., et al. 2011. Microautophagy of cytosolic proteins by late endosomes. *Dev Cell*. 20(1):131–9.

Sarbassov, D.D., Ali, S.M., Sengupta, S., et al. 2006. Prolonged rapamycin treatment inhibits mTORC2 assembly and Akt/PKB. *Mol Cell*. 22(2):159–68.

Sardiello, M., and Ballabio, A., 2009. Lysosomal enhancement: A CLEAR answer to cellular degradative needs. *Cell Cycle*. 8(24):4021–2.

Sarkar, S., Ravikumar, B., Floto, R.A., et al. 2009. Rapamycin and mTOR independent autophagy inducers ameliorate toxicity of polyglutamine-expanded huntingtin and related proteinopathies. *Cell Death Differ*. 16(1):46–56.

Schlumpberger, M., Schaeffeler, E., Straub, M., et al. 1997. AUT1, a gene essential for autophagocytosis in the yeast *Saccharomyces cerevisiae*. *J Bacteriol*. 179:1068–76.

Schneider, J.L., Suh, Y., Cuervo, A.M. 2014. Deficient chaperone-mediated autophagy in liver leads to metabolic dysregulation. *Cell Metab*. 20(3):417–32.

Schneider, J.L., Villarroya, J., Diaz-Carretero, A., et al. 2015. Loss of hepatic chaperone-mediated autophagy accelerates proteostasis failure in aging. *Aging Cell*. 14(2):249–64.

Schroeder, B., Schulze, R.J., Weller, S.G., et al. 2015. The small GTPase Rab7 as a central regulator of hepatocellular lipophagy. *Hepatology*. 61(6):1896–907.

Schulze, R.J., Weller, S.G., Schroeder, B., et al. 2013. Lipid droplet breakdown requires dynamin 2 for vesiculation of autolysosomal tubules in hepatocytes. *J Cell Biol.* 203(2):315–26.

Seok, S., Fu, T., Choi, S.E., et al. 2014. Transcriptional regulation of autophagy by an FXR-CREB axis. *Nature*. 516:108–11.

Settembre, C., De Cegli, R., Mansueto, G., et al. 2013. TFEB controls cellular lipid metabolism through a starvation-induced autoregulatory loop. *Nat Cell Biol.* 15:647–58.

Settembre, C., Di Malta, C., Polito, V.A., et al. 2011. TFEB links autophagy to lysosomal biogenesis. *Science*. 332(6036):1429–33.

Settembre, C., Zoncu, R., Medina, D.L., et al. 2012. A lysosome-to-nucleus signaling mechanism senses and regulates the lysosome via mTOR and TFEB. *EMBO J.* 31(5):1095–108.

Shen, W.J., Patel, S., Miyoshi, H., et al. 2009. Functional interaction of hormone-sensitive lipase and perilipin in lipolysis. *J Lipid Res.* 50(11):2306–13.

Shimizu, S., Takehara, T., Hikita, H., et al. 2012. Inhibition of autophagy potentiates the antitumor effect of the multikinase inhibitor sorafenib in hepatocellular carcinoma. *Int J Cancer.* 131(3):548–57.

Shpilka, T., Welter, E., Borovsky, N., et al. 2015. Lipid droplets and their component triglycerides and steryl esters regulate autophagosome biogenesis. *EMBO J.* 34(16):2117–31.

Singh, R., Kaushik, S., Wang, Y., et al. 2009b. Autophagy regulates lipid metabolism. *Nature* 458, 1131–5.

Singh, R., Xiang, Y., Wang, Y., et al. 2009a. Autophagy regulates adipose mass and differentiation in mice. *J Clin Invest.* 119(11):3329–39.

Souza, S.C., Muliro, K.V., Liscum, L., et al. 2002. Modulation of hormone-sensitive lipase and protein kinase A-mediated lipolysis by perilipin A in an adenoviral reconstituted system. *J Biol Chem.* 277:8267–72.

Spandl, J., Lohmann, D., Kuerschner, L., et al. 2011. Ancient ubiquitous protein 1 (AUP1) localizes to lipid droplets and binds the E2 ubiquitin conjugase G2 (Ube2g2) via its G2 binding region. *J Biol Chem.* 286(7):5599–606.

Suzuki, M., Shinohara, Y., Ohsaki, Y, et al. 2011. Lipid droplets: Size matters. *J Electron Microsc (Tokyo).* 60 Suppl 1:S101–16.

Tanida, I., Tanida-Miyake, E., Ueno, T., et al. 2001. The human homolog of Saccharomyces cerevisiae Apg7p is a Protein-activating enzyme for multiple substrates including human Apg12p, GATE-16, GABARAP, and MAP-LC3. *J Biol Chem.* 276(3):1701–6.

Tanida, I., Sou, Y.S., Ezaki, J., et al. 2004. HsAtg4B/HsApg4B/autophagin-1 cleaves the carboxyl termini of three human Atg8 homologues and delipidates microtubule-associated protein light chain 3- and GABAA receptor-associated protein-phospholipid conjugates. *J Biol Chem.* 279(35):36268–76.

Tasdemir, E., Maiuri, M.C., Tajeddine, N., et al. 2007. Cell cycle-dependent induction of autophagy, mitophagy and reticulophagy. *Cell Cycle* 6:2263–7.

Thiam, A.R., Farese, R.V. Jr., and Walther, T.C., 2013. The biophysics and cell biology of lipid droplets. *Nat Rev Mol Cell Biol.* 14(12):775–86.

Tooze, S.A., and Yoshimori, T. 2010. The origin of the autophagosomal membrane. *Nat Cell Biol.* 12:831–35.

Tsukada, M., and Ohsumi, Y. 1993. Isolation and characterization of autophagy-defective mutants of Saccharomyces cerevisiae. *FEBS Lett.* 333:169–74.

Van Zutphen, T., Todde, V., de Boer, R., et al. 2014. Lipid droplet autophagy in the yeast Saccharomyces cerevisiae. *Mol Biol Cell.* 25(2):290–301.

Walczak, M., and Martens, S. 2013. Dissecting the role of the Atg12-Atg5-Atg16 complex during autophagosome formation. *Autophagy.* 9(3):424–5.

Wei, Y., Pattingre, S., Sinha, S., et al. 2008. JNK1-mediated phosphorylation of Bcl-2 regulates starvation-induced autophagy. *Mol Cell.* 30(6):678–88.

Wilfling, F., Wang, H., Haas, J.T., et al. 2013. Triacylglycerol synthesis enzymes mediate lipid droplet growth by relocalizing from the ER to lipid droplets. *Dev Cell.* 24(4):384–99.

Wolins, N.E., Rubin, B., and Brasaemle, D.L., 2001. TIP47 associates with lipid droplets. *J Biol Chem.* 276(7):5101–8.

Wolins, N.E., Skinner, J.R., Schoenfish, M.J., et al. 2003. Adipocyte protein S3-12 coats nascent lipid droplets. *J Biol Chem.* 278(39):37713–21.

Wolins, N.E., Quaynor, B.K., Skinner, J.R., et al. 2006. OXPAT/PAT-1 is a PPAR-induced lipid droplet protein that promotes fatty acid utilization. *Diabetes.* 55(12):3418–28.

Wu, L., Xu, D., Zhou L., et al. 2014. Rab8a-AS160-MSS4 regulatory circuit controls lipid droplet fusion and growth. *Dev Cell.* 30(4):378–93.

Xu, X., Grijalva, A., Skowronski, A., et al. 2013. Obesity activates a program of lysosomal-dependent lipid metabolism in adipose tissue macrophages independently of classic activation. *Cell Metab.* 18:816–30.

Yamada, E., Bastie, CC., Koga, H., et al. 2012. Mouse skeletal muscle fiber-type-specific macroautophagy and muscle wasting are regulated by a Fyn/STAT3/Vps34 signaling pathway. *Cell Rep.* 1(5):557–69.

Yuk, J.M., Yoshimori, T., Jo, E.K. 2012. Autophagy and bacterial infectious diseases. *Exp Mol Med.* 44(2):99–108.

Zechner, R., Zimmermann, R., Eichmann, T.O., et al. 2012. FAT SIGNALS - Lipases and lipolysis in lipid metabolism and signaling. *Cell Metab.* 15(3):279–91.

Section IV

Autophagy in Neural
Homeostasis and
Neurodegeneration

8 Mitophagy and Neurodegeneration

Kah-Leong Lim
National Neuroscience Institute, Singapore
National University of Singapore, Singapore

Hui-Ying Chan and Grace G.Y. Lim
National Neuroscience Institute, Singapore

Tso-Pang Yao
Duke University School of Medicine, Durham, NC

CONTENTS

8.1   INTRODUCTION

Mitochondria, known as the "powerhouse of the cell," are the principal sites of adenosine triphosphate (ATP) production in aerobic, nonphotosynthetic eukaryotic cells. Most classical textbooks depict these double membrane–bound organelles as solitary and static structures. However, we now know that mitochondria are complex, dynamic, and mobile organelles that constantly undergo membrane remodeling through repeated cycles of fusion and fission, as well as regulated turnover. Collectively, these varied processes help maintain the quality, and thereby the optimal function of mitochondria, as well as allow the organelle to respond rapidly to changes in cellular energy status. The dynamic nature of mitochondria is particularly important for neuronal function, whose unique demands for energy require a highly adaptable mitochondrial network to support. The high energy demand of neurons is critical for several bioenergetically expensive neuronal processes that include axonal transport of macromolecules and organelles (including mitochondria) toward distally located synaptic terminals, maintenance of membrane potential, neurotransmitters uptake and release, and buffering of cytosolic calcium (Schwarz 2013). Among these, perhaps the need for active transportation of components over large distances is one that best distinguishes neurons from their nonneuronal counterparts and arguably also the most fascinating feature about these polarized cells. Although the dimensions of the majority of cells in our body are in the micrometers range, neurons can extend their processes (especially the axon) for much longer distances. For example, the axonal length of a motor neuron in humans is about 1 m. In blue whales, spinal tracts can reach an unimaginable 30 m length (Durcan et al. 2014)! Even when confined to the human brain, the axonal length of projection neurons such as the substantia nigra dopaminergic neurons, including its arborization, can be as long as 0.5 m, and each axon in turn can support nearly 400,000 synapses (Matsuda et al. 2009). The maintenance of an active transport system to supply energy to distally located synapses presents an exquisite challenge to neurons. Furthermore, synapses are themselves metabolically extremely demanding. With every synaptic vesicle release, tens of millions of ions will enter the postsynaptic side as a result of the opening of ion channels. To return the postsynaptic activated neuron to the basal state, one could imagine the large number of ATP that needs to be hydrolyzed to transport the influxed ions out to the extracellular space. The same scenario happens with every action potential fired as the neuronal membrane restores itself back to the resting potential. Notwithstanding this, it is perhaps still amazing to note from a recent imaging study that a single cortical neuron consumes nearly 5 billion ATP per second (Zhu et al. 2012)! One could therefore readily appreciate the urgency for neurons to maintain a constant pool of bioenergetically competent mitochondria that are appropriately distributed to all regions of the cell and to organize these organelles into a dynamic network that could respond rapidly to the changing landscape of neuronal ATP needs. As mentioned above, the remodeling of mitochondrial network also includes its turnover. This is particularly important for postmitotic neurons that need to survive through the entire lifespan of an organism. Comparatively, the lifespan of injured mitochondria is much shortened. Hence, timely removal of these damaged mitochondria is of utmost importance to maintain healthy mitochondrial network to support neuronal survival. The constant turnover of old and/or dysfunctional mitochondria is achieved by a regulated process known as "mitophagy." Significant insights have been obtained in the last decade or so regarding the process of intracellular mitochondrial clearance. However, much of what we know about the mechanisms underlying mitophagy is gleaned from studies in nonneuronal cells that are usually conducted in the presence of chemical uncouplers such as carbonyl cyanide m-chlorophenylhydrazone (CCCP). These chemicals serve to collapse the mitochondrial membrane potential ($\Delta\Psi m$), typically at a concentration that is generally regarded as nonphysiological. Nonetheless, the information obtained has been useful in guiding researchers toward the elucidation of a similar mechanism underlying mitophagy in neurons. In the following sections, we will provide a brief overview of the mechanisms underlying mitophagy and an update on the current understandings of the process in neurons along with its potential role in neurodegeneration.

8.2 MITOPHAGY: PAST AND PRESENT

Mitophagy is a process whereby mitochondria are selectively targeted to the lysosomes for degradation via autophagy. This phenomenon usually occurs in the presence of mitochondrial damage as a form of intracellular quality control mechanism to prevent the accumulation of unwanted materials that could compromise cellular functions. Although mitophagy has been a highly researched and studied field in the recent years, the degradation of mitochondria by lysosomes, especially in the presence of cellular starvation, has already been appreciated for more than half a century. In 1962, Ashford and Porter examined liver cells perfused with glucagon by means of electron microscopy and found that the number of mitochondria-enriched lysosomes was dramatically increased in these cells relative to control preparations. In addition, these mitochondria exhibit "varying degree of structural decay" (Ashford and Porter 1962). The term "mitophagy" was however coined more recently by Lemasters (2005), who suggested the presence of selectivity in the process, particularly in reference to a report by Kissova et al. (2004) who demonstrated that Uth1p, a specific outer membrane protein in yeast, is required for efficient mitochondrial autophagy. A series of papers by other groups of researchers around the same time further showed that depolarized mitochondria are selectively targeted for elimination (Elmore et al. 2001; Priault et al. 2005) and that organelle fission is required to segregate dysfunctional mitochondria from the healthy population to permit their specific removal by mitophagy (Twig et al. 2008). However, little was known about the proteins regulating the mitophagy process in mammalian cells although the Bcl-2 homology 3 (BH3)-only family member BCL2/adenovirus E1B 19 kDa protein-interacting protein 3-like (BNIP3L)/NIX was found to promote programmed mitochondrial clearance during reticulocyte maturation (Schweers et al. 2007).

A major breakthrough in the understanding of the molecular mechanisms underlying mitophagy came from the seminal discovery by Richard Youle's group who found that Parkin, a Parkinson's disease (PD)-linked ubiquitin ligase, is a key mammalian regulator of the process. Subsequent studies by his group and several others revealed that Parkin collaborates with another PD-linked gene product known as PINK1 (encoding a mitochondrial targeted serine/threonine kinase) to mediate mitophagy (Geisler et al. 2010; Matsuda et al. 2010; Narendra et al. 2010; Vives-Bauza et al. 2010). Collectively, these initial reports triggered an explosion of interest among the global mitochondrial research community in delineating the pathways involved in Parkin/PINK1-mediated mitophagy, with the excitement ensuing to this date. A model that has emerged from these studies is shown in Figure 8.1. Until recently, the model describes a linear sequence of events occurring in response to mitochondrial depolarization that culminates in their removal. According to the proposed model (Youle and Narendra 2011), a key initial event that occurs upon mitochondrial depolarization is the selective accumulation of PINK1 on the outer membrane of the damaged organelle. This accumulation allows PINK1 to recruit Parkin (Okatsu et al. 2012), whose latent ubiquitin ligase activity becomes unmasked along the way in part due to its phosphorylation by PINK1 (Kondapalli et al. 2012; Matsuda et al. 2010). PINK1 also phosphorylates ubiquitin, which binds and activates Parkin (Kane et al. 2014; Koyano et al. 2014).

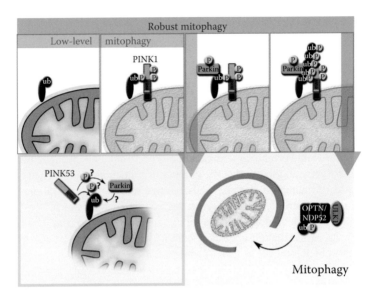

FIGURE 8.1 Model of PINK1/Parkin-mediated mitophagy. Upon mitochondrial depolarization, full-length PINK1 accumulates on the OMM, leading to the phosphorylation of ubiquitin on the surface of the mitochondria. This results in the recruitment of the autophagy receptors OPTN and NDP52 and the consequent activation of the mitophagy process, albeit at a low level. PINK1-mediated ubiquitin phosphorylation also leads to the translocation of Parkin to the OMM, which promotes the polyubiquitination of mitochondrial substrates that, in turn, provides more ubiquitin substrates for PINK1 to phosphorylate. This amplifies the signal for the recruitment of autophagy receptors and results in robust mitophagy. Like full-length PINK1, PINK1-53 upon its stabilization is also capable of promoting mitophagy, although the mechanism underlying PINK1-53-mediated mitophagy remains to be clarified.

Activated Parkin then promotes the ubiquitination and subsequent degradation of many outer mitochondrial membrane (OMM) proteins (Chan et al. 2011; Yoshii et al. 2011). During the process, Parkin-decorated mitochondria progressively cluster toward the perinuclear region to form mito-aggresomes (Lee et al. 2010), which by virtue of their association with lysosomal components are removed with time in an autophagy-dependent manner. Although this linear model explains the selectivity of the mitophagy process rather elegantly, recent discoveries again led by the Youle's group revealed that a more complicated network of molecular interactions is involved in the disposal of unwanted mitochondria (Figure 8.1) (Lazarou et al. 2015). In essence, the group found that mitochondrial-localized PINK1 is able to recruit the autophagy receptors optineurin (OPTN) and nuclear dot protein 52 kDa (NDP52) in the absence of Parkin and that this phenomenon alone is sufficient to trigger mitophagy, albeit at a low level. The recruitment of OPTN and NDP52 is dependent on PINK1-mediated phosphorylation of ubiquitin, which serves as an autophagy signal on the mitochondria. Parkin promotes the process by amplifying the phospho-ubiquitin level generated by PINK1 that otherwise occurs at a low level due to the limited basal ubiquitin availability on mitochondria. In so doing, Parkin mediates a positive feed-forward cycle that generates more ubiquitin substrates on the OMM for PINK1 to phosphorylate that, in turn, facilitates the recruitment of autophagy receptors and thereby enhances the mitophagy process. Corroborating with this study, Heo et al. similarly found that PINK1/Parkin pathway recruits OPTN/NDP52 to damage mitochondria. More importantly, they further found that mitochondrial depolarization leads to PINK1 and Parkin-dependent phosphorylation of TBK1 kinase (a known interactor of OPTN), which activates the kinase to phosphorylate autophagy adaptors to facilitate mitochondrial capture by the autophagosomes (Heo et al. 2015). Thus, what used to be a model that is dominated by Parkin-mediated ubiquitination of mitochondrial proteins has now evolved into one where PINK1 takes the center stage as an active initiator of mitophagy. Adding to this complexity is the recent elucidation by several groups that deubiquitinating enzymes (DUBs) such as USP8, USP15, USP30, and USP35 can counteract Parkin-mediated mitochondrial ubiquitination and consequently mitophagy, although they generally do not affect the recruitment of Parkin to damaged mitochondria (Bingol et al. 2014; Cornelissen et al. 2014; Durcan et al. 2014; Wang et al. 2015). In the case of USP8, the antagonistic effect on mitophagy is due to the selective removal of Parkin-mediated K6-linked ubiquitin chains by this DUB (Durcan et al. 2014). Moreover, given that ubiquitin phosphorylation is key to the mitophagy process and that this posttranslational modification is also a reversible reaction, we anticipate that the tapestry of molecular events surrounding mitophagy will be further complicated by the participation of phosphatases. Finally, we have recently found that the cytosolic form of PINK1 (PINK1-53), which is normally rapidly degraded by the proteasome, could also participate in mitophagy. Our study demonstrated that PINK1-53 may be stabilized by NFκB signaling via TRAF6-mediated K63-linked ubiquitination, whereupon it initiates the mitophagy cascade, albeit in apparently healthy mitochondria (Lim et al. 2015). We speculate that this "non-selective" form of mitophagy may potentially help counteract the buildup of reactive oxygen species (ROS) in cells undergoing oxidative stress and as such represent a cytoprotective response. Taken together, it appears that PINK1/Parkin-mediated mitophagy may be regulated at multiple levels depending on the prevailing cellular conditions. Before we discuss the implications of this pathway in neuronal function in the next section, we wish to highlight to our readers that PINK1/Parkin is not the only mitophagy pathway in mammals. Other than BNIP3L/NIX described above, the OMM protein FUNDC1 has recently been elucidated as an essential factor for hypoxia-dependent mitophagy (Liu et al. 2012). Furthermore, Allen et al. (2013) showed that loss of iron triggers PINK1/Parkin-independent mitophagy. However, the link between these regulators and neurodegeneration is less clear.

8.3 MITOPHAGY IN NEURONS

Notwithstanding the logic and the elegance of the Parkin/PINK1-mediated mitophagy model described in the previous section, whether mitophagy observed in large part in dividing cells is relevant at all to postmitotic neurons has been a controversial topic. As mentioned above, the bioenergetic demands of neurons are rather different from other cell types in our body. Because of their unique dependence on mitochondrial respiration, neurons are highly unlikely to embark on a degradation scheme that serves to rid them of a large complement of their mitochondria. In contrast, non-neuronal cells, especially those in culture, could better tolerate mitochondria loss as they could fulfill their energy requirements by generating ATP via glycolysis from the high amount of glucose typically present in the medium (e.g., DMEM contains between 1 and 4.5g/mL glucose depending on the formulation). Indeed, a study by Van Laar et al. demonstrated that Parkin's translocation to depolarized mitochondria is attenuated in cultured cells that are forced into dependence on mitochondrial respiration (i.e., cultured in glucose-free medium) (Van Laar et al. 2011). Furthermore, the group also found that CCCP treatment of primary neurons neither promotes the translocation of Parkin (both exogenous and endogenous) to the mitochondria nor triggers mitophagy. In a related study, Rakovic et al. (2013) examined primary human fibroblasts and induced pluripotent stem cell (iPS)–derived neurons from controls and PINK1 mutation carriers and found that endogenous Parkin is insufficient to initiate mitophagy in these models regardless of the functional status of PINK1. They further found that although Parkin overexpression can rescue the defective mitophagy, the restoration happened only in fibroblasts but not in iPS-derived neurons. In contrast to these reports, several groups have documented mitophagy taking place in primary neurons, although the majority of these studies relied on Parkin recruitment to depolarized mitochondria as readout, that is, they stopped short of showing mitochondrial clearance (Table 8.1). For example, Seibler et al. (2011) reported that the impairment of Parkin recruitment to depolarized mitochondria in iPS-derived dopaminergic neurons from PINK1-related PD patients can be rescued by reintroduction of wild-type PINK1 into PINK1-deficient neurons, which is consistent with previous reports that demonstrated the accumulation of Parkin on mitochondria of primary neurons treated with CCCP (Narendra et al. 2008; Vives-Bauza et al. 2010). In a related study, Cai et al. (2012) observed that Parkin-targeted mitochondria in cortical neurons dissipated of mitochondrial potential are mostly accumulated in the somatodendritic region, a site where mature lysosomes are also localized. Importantly, this group was able to observe via time-lapse imaging the dynamic formation and elimination of Parkin- and LC3-ring-like structures surrounding depolarized mitochondria through the autophagy–lysosomal pathway, suggesting that mitophagy indeed takes place in these neurons. This compartmentally restricted neuronal mitophagy process is apparently accompanied by a reduction in anterograde mitochondrial transport, which suggests a mechanism whereby damaged mitochondria are transported back to the soma in a retrograde fashion for efficient removal by mature lysosomes. Corroborating with this, Miller and Sheetz (2004) have previously documented a correlation between axonal mitochondrial transport and potential, that is, mitochondria with high potential tend to be transported toward the growth cone, whereas those with low potential are usually transported toward the cell body. However, this correlation is not unequivocally observed (Verburg and Hollenbeck 2008).

In another interesting development, which contradicts the proposal that bioenergetically incompetent mitochondria are retrogradely transported to the soma for clearance, Wang et al. (2011) found that mitochondrial damage in distal axons actually arrested their motility through degradation of the mitochondrial motor adaptor protein Miro (that normally anchors kinesin to the surface of mitochondria).

TABLE 8.1

Mitophagy in Neurons

System	Uncoupling Agent	Parkin-mito Translocation	Mitochondria Clearance	References
Rat cortical neurons	5 h, 10 μM CCCP	Yes	NA	Narendra et al. (2010)
Mouse cortical neurons	1 h, 100 nM CCCP	Yes	NA	Vives-Bauza et al. (2010)
Rat cortical neurons, striatal/midbrain neurons (cell body and neurites)	6 h, 10 μM, 100 nM CCCP	No	No	Van Laar et al. (2011)
Mouse cortical neurons (cell body and neurites)	24 h, 10 μM CCCP with lysosomal inhibitors	Yes (somatodendritic) No (axons and distal dendritic)	Enhanced LC3, LAMP1 mitochondrial colocalization in somatodendritic neurons	Cai et al. (2012)
Mouse primary neurons	3 h, 30 μM CCCP	Yes	NA	Koyano et al. (2013)
Mouse cortical neurons	1–3h, 5 μM CCCP	Yes (only in the absence of B27)	NA	Joselin et al. (2012)
Mouse and rat hippocampal neurons (distal axons)	Mito-KR activation, 20 min, 40 μM Antimycin A	Yes (distal axons)	Significant increase in LC3 and LAMP1 mitochondrial colocalization; significant decrease in mitochondrial size	Ashrafi et al. (2014)
Mouse cortical neurons	Short mitochondria isoform of ARF	Yes	Significant increase in LC3 mitochondrial colocalization	Grenier et al. (2014)
Rat hippocampal neurons	None	NA	Presence of lysosomal mitochondria stained via mt-Keima at baseline condition (in the absence of mitochondrial uncoupler), which was attenuated with Parkin and PINK1 shRNA	Bingol et al. (2014)
Rat cortical neurons	Glutamate	Yes	Yes; in the presence of N-acetyl cysteine	Van Laar et al. (2015)

(Continued)

TABLE 8.1 (Continued)
Mitophagy in Neurons

System	Uncoupling Agent	Parkin-mito Translocation	Mitochondria Clearance	References
WT or hAPP cortical neurons	24 h, 10 µM CCCP, hAPP	Yes	NA; only looked at mitophagosomes in the absence of uncoupler	Ye et al. (2015)
Mouse cortical neurons	18 h, 20 nM CCCP, or 2 nM rotenone	NA—moderate stress leads to mitochondrial hyperfusion	NA	Norris et al. (2015)
iPS-derived dopaminergic neurons (controls and PD patient with PINK1 V170G missense mutation)	16 h, 1 µM valinomycin	Yes	No	Rakovic et al. (2013)
iPS-derived dopaminergic neurons (2 iPSC line from PD patient with mutant PINK1 c.1366C>T and healthy family member)	12 h, 1 µM valinomycin	Yes	Reduced mtDNA	Seibler et al. (2011)

This pathway is also dependent on PINK1 and Parkin collaboration. Upon mitochondrial damage, PINK1 phosphorylates Miro that activates its degradation in a Parkindependent manner. Removal of Miro helps quarantine damaged mitochondria to facilitate their localized clearance, thereby limiting the spread of oxidative damage. In a follow-up study providing support to this proposal, the same group demonstrated that localized damage sustained by axonal mitochondria that mimics physiological levels of mitochondrial dysfunction resulted in the rapid recruitment of Parkin, LC3-positive autophagosomes, and LAMP1-positive lysosomes to damaged mitochondria (Ashrafi et al. 2014). Notably, the activation of this pathway from the recruitment of the mitophagy machinery to the apparent clearance of damaged axonal mitochondria can occur within less than an hour, which highlights the need of examining this process in a fairly acute manner. Undoubtedly, the proposed mechanism for axonal mitophagy is attractive and logical. However, a pertinent question that remains to be addressed is: "how does PINK1 participate in this process?" Full-length PINK1 is widely recognized as an extremely labile protein that accumulates only in the presence of mitochondrial potential dissipation. Assuming that PINK1 turnover is similarly regulated in neurons, it would mean that significant amounts of PINK1 need to be transported to axonal mitochondria. Alternatively, local *de novo* synthesis of PINK1 needs to occur at distal sites of neurons. At present, the question remains unresolved. Finally, using *Drosophila* as a model, a recent study provided evidence that Parkin-mediated mitophagy does occur *in vivo* (Vincow et al. 2013). The investigators found that the brains of Parkin-mutant flies exhibit a significantly decreased rate of mitochondrial protein turnover, which is similar to that produced by general autophagy blockade induced by genetic ablation of *atg7*. Their results suggest that Parkin indeed promotes mitochondrial turnover through autophagy and that the process is physiologically relevant, albeit in the fly. Taken together, considering the different lines of evidence supporting neuronal mitophagy, it is probably safe to say that mitophagy in neurons does occur in these cells but perhaps in a limited way such that the overall aerobic respiration linked to their function is not overtly compromised.

8.4 MITOPHAGY AND NEURODEGENERATION

Given the high energy demands of neurons, it is intuitive to accept that impairments in mitochondrial quality control (QC) will contribute to neuronal dysfunction and, consequently, neurodegeneration. However, we would like to remind the readers that mitochondrial QC is not restricted to mitophagy but involved a continuum of mitochondrial processes including biogenesis, fusion, fission, and anterograde/retrograde movements that are all interwoven in an integrated and coordinated network to ensure that the cell is populated with an optimal level of bioenergetically competent mitochondria. Accordingly, disruption in any of these pathways, and not just mitophagy, could result in neuronal dysfunction. For example, mutations in *Mfn2* (a mitochondrial fusion regulator) cause Charcot–Marie–Tooth type 2A, a classic axonal peripheral sensorimotor neuropathy characterized by degeneration of long peripheral nerves (Zuchner et al. 2004). Similarly, mutations in the pro-fusion *OPA1* promote neuronal loss of retinal ganglia cells in a condition known as autosomal dominant optic atrophy, which is the most common form of inherited childhood blindness (Alexander et al. 2000; Delettre et al. 2000). Notably, *OPA1* missense mutations have recently been linked to syndromic Parkinsonism and dementia that are comorbid with chronic progressive external ophthalmoplegia (Carelli et al. 2015). Furthermore, a dominant negative mutation in *Drp1* (a mitochondrial pro-fission regulator) results in neonatal lethality that is characterized by severe neurological defects in the newborn including abnormal brain development, microcephaly, and optic atrophy, among other phenotypic manifestations (Waterham et al. 2007). Abnormal *Drp1* expression is also implicated in several age-related neurodegenerative diseases (Cornelissen et al. 2014). Thus, disruptions in mitochondrial homeostasis at any level can result in neurodegeneration. While keeping this in mind, for the purpose of this chapter, we will focus on the relationship between mitophagy and neurodegeneration, in particular on PD, given that mutations in Parkin and PINK1 are both causative links of familial Parkinsonism.

8.4.1 MITOPHAGY IN PARKINSON'S DISEASE

PD is a prevalent neurodegenerative disease that is characterized clinically by a constellation of motor deficits arising principally from the loss of midbrain dopaminergic neurons in the substantia nigra pars compacta (SNpc). The specific pattern of neurodegeneration in PD is often accompanied by the presence of eosinophilic intracytoplasmic inclusions known as Lewy bodies (LBs) in surviving neurons in the SN and other affected brain areas (Braak et al. 2003). Notably, LBs are enriched with ubiquitin and α-synuclein (a presynaptic protein whose mutations are causative of autosomal dominant PD). Although the etiology of PD remains unclear, a broad range of studies conducted over the past few decades have consistently implicated a few pathogenic culprits, including mitochondrial dysfunction, as key contributors to the pathogenesis of the disease (Lim and Zhang 2013). The idea that mitochondrial dysfunction could contribute to the development of PD is actually not new. It originally surfaced in the early 1980s after Langston et al. (1983) noticed that drug abusers exposed to 1-methyl-4-phenyl-1,2,3,4-tertahydropyridine (MPTP), an inhibitor of mitochondrial complex I function, display motor behaviors that bear uncanny resemblance to those exhibited by sporadic PD patients. Thereafter, several groups conducted postmortem analysis of PD brains and recorded a significant reduction in the activity of mitochondrial complex I as well as ubiquinone (coenzyme Q10) in the SN of diseased samples (Schapira et al. 1989; Shults et al. 1997). Moreover, mitochondrial poisoning recapitulates PD features in humans and represents a popular strategy to model the disease in animals (Dauer and Przedborski 2003). Comparatively, the concept of mitophagy defects potentially underlying mitochondrial abnormalities as seen in the PD brain is more recent, which stemmed from the discoveries in the past 5–8 years regarding the role of Parkin and PINK1 in the clearance of damaged mitochondria. Supporting this, we and others have demonstrated that disease-associated mutations in Parkin compromise mitophagy due to differential defects at recognition, transportation, or ubiquitination of impaired mitochondria, albeit in nonneuronal cells (Lee et al. 2010; Matsuda et al. 2010). Several reports, as mentioned in the previous section, also documented mitophagy defects in iPS-derived human neurons generated from somatic cells of Parkin- or PINK1-related PD patients (Rakovic et al. 2013; Seibler et al. 2011) (Table 8.1), which lend further support to the proposal that PD neurodegeneration (at least in cases that involve Parkin or PINK1 mutations) may arise from defective mitophagy. However, these evidences are arguably largely indirect. Moreover, virtually, all these studies relied on acute chemically induced mitochondrial damage to trigger mitophagy. Interestingly, we have recently found that chronic, low-dose CCCP treatment promotes Parkin-mediated mitochondrial fusion instead of mitophagy (Norris et al. 2015). In principle, active fusion could protect mitochondrial integrity either by complementing damaged mitochondria with their healthy counterparts or by limiting the production of ROS. This adaptive mitochondrial fusion strategy that we have found in the presence of low but chronic mitochondrial stress is likely more physiologically relevant to the disease than high dose of mitochondrial toxins commonly used to induce acute neuronal dysfunction. More importantly, we further showed that Parkin, PINK1, and α-synuclein form a regulatory circuit to regulate mitochondrial stress response, thus functionally connecting three key PD-linked genes to the pathogenesis of PD. To clarify, our results do not exclude the participation of mitophagy defects in PD *per se*, especially in Parkin/PINK1-related patients or in advanced stages of the disease. Rather, it once again highlights the need to pay close attention to the conditions used in mitophagy-related experiments and that mitochondrial QC involves not just its turnover.

Perhaps one of the most challenging tasks at hand is to demonstrate unequivocally that mitophagy impairment, instead of a generalized impairment in the autophagy process (that consequently impairs mitophagy), contributes directly to neurodegeneration *in vivo*. This would require the genetic differentiation of targeted components that are exclusively involved in mitophagy. Currently, key components of mitophagy and autophagy tend to overlap, which makes interpretation difficult. However, a recent study (Fiesel et al. 2015) using phosphorylated ubiquitin as a surrogate marker for mitophagy has provided some important insights regarding its pathophysiological relevance. As revealed by several groups, the phosphorylation of ubiquitin at Serine 65 (S65) by PINK1 is a key and essential step in the initiation of the mitophagy process (Kane et al. 2014; Koyano et al. 2014). By means of phospho-specific ubiquitin immunostaining that only recognizes the presence of S65 phosphorylated ubiquitin (pS65-Ub), Fiesel et al. (2015) recently found that the level of pS65-Ub (that is barely detectable under basal condition) is elevated in cells in response to mitochondrial stress. Importantly, they further found that pS65-Ub accumulates in human brain during aging and in postmortem PD brain samples in the form of cytoplasmic granules that partially overlap with mitochondrial and lysosomal markers. In contrast, the pS65-Ub signal is absent in brain samples from a patient with compound heterozygous PINK1 mutation. Taking these findings into perspective, it is perhaps logical to propose that mitophagy is compromised at the initiation stage in PINK1-deficient PD brains (i.e., lack of pS65-Ub signal) but perhaps at a later stage in the aging or sporadic PD brains, because PINK1-mediated pS65-Ub, and hence mitophagy initiation, does not appear to be affected (although the phosphorylated ubiquitin species should not have exhibited significant accumulation either). Of course, it is entirely possible that a (hitherto unknown) phosphatase returning pS65-Ub to its basal state is deficient in aged or sporadic PD brains that had resulted in the observed increase in pS65-Ub level. Alternatively, Parkin function may be affected in aged and diseased brains that concomitantly impair the efficiency of mitochondrial clearance. Relevant to this is the finding by several groups including ours, which demonstrated that Parkin dysfunction could arise in the PD brain in the absence of apparent mutations, that is, through modifications of the wild-type protein leading to its functional impairments. This could be a result of stress-induced biochemical alterations including oxidation and nitrosylation, posttranslational modifications, or aberrant protein–protein interaction that can alter the catalytic function of the E3 ligase either directly or indirectly through promoting its aggregation or degradation (Tan et al. 2009). Interestingly, normal Parkin in the brain also becomes progressively more detergent-insoluble (and therefore nonfunctional) with aging (Pawlyk et al. 2003), which may provide an explanation to why age represents a risk factor for PD. In all these cases, the loss of Parkin function is expected to compromise the efficiency of PINK1/Parkin-mediated mitophagy. Thus, deficient mitochondrial QC may not necessarily be restricted to cases where Parkin (or PINK1) is overtly mutated. As it stands now, the evidence supporting the role of mitophagy defects in PD pathogenesis is intuitive and attractive but remains indirect. However, this does not exclude it from being an important player in the whole orchestral of neuronal adaptive mitochondrial stress response that has gone awry in PD. Moreover, in a rare form of PD linked to mutations in *ATP13A2* that result in lysosomal dysfunction (Usenovic et al. 2012), the mitophagic machinery by virtue of its dependence on the lysosomal system is expected to be functioning suboptimally, and this impairment would certainly be one of the contributors in the pathogenesis of *ATP13A2*-related PD.

8.4.2 MITOPHAGY IN OTHER NEURODEGENERATIVE DISEASES

Mitochondrial structural alterations are a prominent feature in the brains of individuals afflicted with Alzheimer's disease (AD) (Baloyannis 2006), the presence of which would suggest a failure in their clearance. Consistent with this, the maturation of autophagolysosomes and their retrograde transport appear to be significantly impeded in AD neurons, which collectively results in a massive accumulation of unproductive autophagic vacuoles within large swellings along dystrophic and degenerating neuritis (Boland et al. 2008; Nixon 2007). Because mitophagy requires a competent autophagy apparatus to take place, impairments in autophagy are expected to compromise mitochondria QC in the AD brain, although mitophagy defects in this case are probably a consequence rather than an initiator of AD pathogenesis. Notably, mutations in both the mitophagy regulators (i.e., Parkin and PINK1) are associated with frank Parkinsonism rather than with dementia, notwithstanding the findings of a recent study conducted in a mouse model of AD which demonstrated that Parkin mitigates AD progression by stimulating beclin-dependent molecular cascade of autophagy and thereby facilitates clearance of vesicles containing defective mitochondria (Khandelwal et al. 2011). Similar to the situation in AD brains, the pathology of Huntington's disease (HD) is also associated with impairments in autophagic cargo recognition that leads to the accumulation of damaged mitochondria (Martinez-Vicente et al. 2010). Moreover, the expression of disease-linked mutant huntingtin can trigger mitochondrial dysfunction (Bossy-Wetzel et al. 2008). In amyotrophic lateral sclerosis (ALS), the aggregation of mutant SOD1 is known to result in the reduction of OPA1 level and upregulation of Drp1 level. This imbalance in OPA1/Drp1 ratio leads to excessive mitochondrial fragmentation that is expected to activate the PINK1/Parkin-mediated mitophagy pathway. However, the clearance process will be impeded as autophagy is also dysregulated in ALS (Chen et al. 2012). In the case of FUS or TDP-43-related ALS, their functional deficiency could apparently result in the reduction of Parkin expression (Lagier-Tourenne et al. 2012). Being a key mitophagy regulator, the depletion of Parkin expression would obviously exert profound impact on neuronal mitophagy, although whether and how this contributes to the progression of ALS is still to be established.

Taken together, defective mitophagy, which can arise directly from overt dysfunctions in its regulators (e.g., Parkin/PINK1 mutations) or indirectly as a consequence of a generalized impairment in the autophagy system, has emerged as an important player in the pathogenesis of neurodegenerative diseases.

8.5 CONCLUDING REMARKS

Given the exquisitely high energy demands of neurons, a reduction in the bioenergetic efficacy due to defective mitochondrial QC will have serious implications for the survival of neurons. Although such defects could arise at multiple levels in the life cycle of mitochondria, impairments in mitophagy would deny the neuron the opportunity to clear the unwanted damaged mitochondria. The resulting reduction in the net bioenergetic efficacy will have an impact on neuronal processes, especially axonal transport and electrical properties that are highly dependent on the spatial and temporal availability of sufficient ATP supply. Thus, it is easy to appreciate the role of mitophagy in neurodegeneration, whether as a direct or indirect contributor. By the same token, it is also easy to accept that deficiency in Parkin and PINK1 functions, which are important for mitophagy, can trigger neuronal loss. However, why mutations in Parkin/PINK1 are specifically causing PD is less clear, since mitophagy should occur in a pan-neuronal fashion in these cases. Nonetheless, the elucidation of Parkin/PINK1-mediated mitophagy pathway has certainly significantly improved our understanding of the relationship between mitochondrial dysfunction and neurodegeneration. Undoubtedly, the excitement surrounding the mechanisms underlying Parkin/PINK1 pathway and precisely how mitophagy defects contribute to neurodegenerative diseases will persist.

ACKNOWLEDGMENTS

We apologize to our colleagues whose works are not cited in this chapter due to space limitations. We thank Liting Hang for her help with illustrations. This work was supported by grants from the National Medical Research Council—Collaborative Research Grant, Open-Fund Individual Research Grant & Translational Clinical Research Grant in Parkinson's disease. (K.-L.L) and National Institutes of Health (NIH) 2R01-NS054022 (T.-P.Y.). C.-H.Y. is supported by a graduate scholarship from the National University of Singapore Graduate School for Integrative Sciences and Engineering.

REFERENCES

Alexander, C., M. Votruba, U. E. Pesch, D. L. Thiselton, S. Mayer, A. Moore, M. Rodriguez, U. Kellner, B. Leo-Kottler, G. Auburger, et al. 2000. OPA1, encoding a dynamin-related GTPase, is mutated in autosomal dominant optic atrophy linked to chromosome 3q28. *Nat Genet* 26 (2):211–5. doi: 10.1038/79944.

Allen, G. F., R. Toth, J. James, and I. G. Ganley. 2013. Loss of iron triggers PINK1/Parkin-independent mitophagy. *EMBO Rep* 14 (12):1127–35. doi: 10.1038/embor.2013.168.

Ashford, T. P., and K. R. Porter. 1962. Cytoplasmic components in hepatic cell lysosomes. *J Cell Biol* 12:198–202.

Ashrafi, G., J. S. Schlehe, M. J. LaVoie, and T. L. Schwarz. 2014. Mitophagy of damaged mitochondria occurs locally in distal neuronal axons and requires PINK1 and Parkin. *J Cell Biol* 206 (5):655–70. doi: 10.1083/jcb.201401070.

Baloyannis, S. J. 2006. Mitochondrial alterations in Alzheimer's disease. *J Alzheimers Dis* 9 (2):119–26.

Bingol, B., J. S. Tea, L. Phu, M. Reichelt, C. E. Bakalarski, Q. Song, O. Foreman, D. S. Kirkpatrick, and M. Sheng. 2014. The mitochondrial deubiquitinase USP30 opposes parkin-mediated mitophagy. *Nature* 510 (7505):370–5. doi: 10.1038/nature13418.

Boland, B., A. Kumar, S. Lee, F. M. Platt, J. Wegiel, W. H. Yu, and R. A. Nixon. 2008. Autophagy induction and autophagosome clearance in neurons: Relationship to autophagic pathology in Alzheimer's disease. *J Neurosci* 28 (27):6926–37. doi: 10.1523/JNEUROSCI.0800-08.2008.

Bossy-Wetzel, E., A. Petrilli, and A. B. Knott. 2008. Mutant huntingtin and mitochondrial dysfunction. *Trends Neurosci* 31 (12):609–16. doi: 10.1016/j.tins.2008.09.004.

Braak, H., K. Del Tredici, U. Rub, R. A. de Vos, E. N. Jansen Steur, and E. Braak. 2003. Staging of brain pathology related to sporadic Parkinson' disease. *Neurobiol Aging* 24 (2):197–211.

Cai, Q., H. M. Zakaria, A. Simone, and Z. H. Sheng. 2012. Spatial parkin translocation and degradation of damaged mitochondria via mitophagy in live cortical neurons. *Curr Biol* 22 (6):545–52. doi: 10.1016/j.cub.2012.02.005.

Carelli, V., O. Musumeci, L. Caporali, C. Zanna, C. La Morgia, V. Del Dotto, A. M. Porcelli, M. Rugolo, M. L. Valentino, L. Iommarini, et al. 2015. Syndromic parkinsonism and dementia associated with OPA1 missense mutations. *Ann Neurol* 78 (1):21–38. doi: 10.1002/ana.24410.

Chan, N. C., A. M. Salazar, A. H. Pham, M. J. Sweredoski, N. J. Kolawa, R. L. Graham, S. Hess, and D. C. Chan. 2011. Broad activation of the ubiquitin-proteasome system by Parkin is critical for mitophagy. *Hum Mol Genet* 20 (9):1726–37. doi: 10.1093/hmg/ddr048.

Chen, S., X. Zhang, L. Song, and W. Le. 2012. Autophagy dysregulation in amyotrophic lateral sclerosis. *Brain Pathol* 22 (1):110–6. doi: 10.1111/j.1750-3639.2011.00546.x.

Cornelissen, T., D. Haddad, F. Wauters, C. Van Humbeeck, W. Mandemakers, B. Koentjoro, C. Sue, K. Gevaert, B. De Strooper, P. Verstreken, et al. 2014. The deubiquitinase USP15 antagonizes Parkin-mediated mitochondrial ubiquitination and mitophagy. *Hum Mol Genet* 23 (19):5227–42. doi: 10.1093/hmg/ddu244.

Dauer, W., and S. Przedborski. 2003. Parkinson' disease: Mechanisms and models. *Neuron* 39 (6):889–909. doi: 10.1016/S0896-6273(03)00568-3.

Delettre, C., G. Lenaers, J. M. Griffoin, N. Gigarel, C. Lorenzo, P. Belenguer, L. Pelloquin, J. Grosgeorge, C. Turc-Carel, E. Perret, et al. 2000. Nuclear gene OPA1, encoding a mitochondrial dynamin-related protein, is mutated in dominant optic atrophy. *Nat Genet* 26 (2):207–10. doi: 10.1038/79936.

Durcan, T. M., M. Y. Tang, J. R. Perusse, E. A. Dashti, M. A. Aguileta, G. L. McLelland, P. Gros, T. A. Shaler, D. Faubert, B. Coulombe, et al. 2014. USP8 regulates mitophagy by removing K6-linked ubiquitin conjugates from parkin. *EMBO J* 33 (21):2473–91. doi: 10.15252/embj.201489729.

Elmore, S. P., T. Qian, S. F. Grissom, and J. J. Lemasters. 2001. The mitochondrial permeability transition initiates autophagy in rat hepatocytes. *FASEB J* 15 (12):2286–7. doi: 10.1096/fj.01-0206fje.

Fiesel, F. C., M. Ando, R. Hudec, A. R. Hill, M. Castanedes-Casey, T. R. Caulfield, E. L. Moussaud-Lamodiere, J. N. Stankowski, P. O. Bauer, O. Lorenzo-Betancor, et al. 2015. (Patho-)physiological relevance of PINK1-dependent ubiquitin phosphorylation. *EMBO Rep* 16 (9):1114–30. doi: 10.15252/embr.201540514.

Geisler, S., K. M. Holmstrom, D. Skujat, F. C. Fiesel, O. C. Rothfuss, P. J. Kahle, and W. Springer. 2010. PINK1/Parkin-mediated mitophagy is dependent on VDAC1 and p62/SQSTM1. *Nat Cell Biol* 12 (2):119–31.

Grenier, K., M. Kontogiannea, and E. A. Fon. 2014. Short mitochondrial ARF triggers Parkin/PINK1-dependent mitophagy. *J Biol Chem* 289 (43):29519–30. doi: 10.1074/jbc. M114.607150.

Heo, J. M., A. Ordureau, J. A. Paulo, J. Rinehart, and J. W. Harper. 2015. The PINK1-PARKIN mitochondrial ubiquitylation pathway drives a program of OPTN/NDP52 recruitment and TBK1 activation to promote mitophagy. *Mol Cell* 60:7–20. doi: 10.1016/j.molcel.2015.08.016.

Joselin, A. P., S. J. Hewitt, S. M. Callaghan, R. H. Kim, Y. H. Chung, T. W. Mak, J. Shen, R. S. Slack, and D. S. Park. 2012. ROS-dependent regulation of Parkin and DJ-1 localization during oxidative stress in neurons. *Hum Mol Genet* 21 (22):4888–903. doi: 10.1093/hmg/dds325.

Kane, L. A., M. Lazarou, A. I. Fogel, Y. Li, K. Yamano, S. A. Sarraf, S. Banerjee, and R. J. Youle. 2014. PINK1 phosphorylates ubiquitin to activate Parkin E3 ubiquitin ligase activity. *J Cell Biol* 205 (2):143–53. doi: 10.1083/jcb.201402104.

Khandelwal, P. J., A. M. Herman, H. S. Hoe, G. W. Rebeck, and C. E. Moussa. 2011. Parkin mediates beclin-dependent autophagic clearance of defective mitochondria and ubiquitinated Abeta in AD models. *Hum Mol Genet* 20 (11):2091–102. doi: 10.1093/hmg/ddr091.

Kissova, I., M. Deffieu, S. Manon, and N. Camougrand. 2004. Uth1p is involved in the autophagic degradation of mitochondria. *J Biol Chem* 279 (37):39068–74. doi: 10.1074/jbc.M406960200.

Kondapalli, C., A. Kazlauskaite, N. Zhang, H. I. Woodroof, D. G. Campbell, R. Gourlay, L. Burchell, H. Walden, T. J. Macartney, M. Deak, et al. 2012. PINK1 is activated by mitochondrial membrane potential depolarization and stimulates Parkin E3 ligase activity by phosphorylating Serine 65. *Open Biol* 2 (5):120080.

Koyano, F., K. Okatsu, S. Ishigaki, Y. Fujioka, M. Kimura, G. Sobue, K. Tanaka, and N. Matsuda. 2013. The principal PINK1 and Parkin cellular events triggered in response to dissipation of mitochondrial membrane potential occur in primary neurons. *Genes Cells* 18 (8):672–81. doi: 10.1111/gtc.12066.

Koyano, F., K. Okatsu, H. Kosako, Y. Tamura, E. Go, M. Kimura, Y. Kimura, H. Tsuchiya, H. Yoshihara, T. Hirokawa, et al. 2014. Ubiquitin is phosphorylated by PINK1 to activate parkin. *Nature* 510 (7503):162–6. doi: 10.1038/nature13392.

Lagier-Tourenne, C., M. Polymenidou, K. R. Hutt, A. Q. Vu, M. Baughn, S. C. Huelga, K. M. Clutario, S. C. Ling, T. Y. Liang, C. Mazur, et al. 2012. Divergent roles of ALS-linked proteins FUS/TLS and TDP-43 intersect in processing long pre-mRNAs. *Nat Neurosci* 15 (11):1488–97. doi: 10.1038/nn.3230.

Langston, J. W., P. Ballard, J. W. Tetrud, and I. Irwin. 1983. Chronic Parkinsonism in humans due to a product of meperidine-analog synthesis. *Science* 219 (4587):979–80.

Lazarou, M., D. A. Sliter, L. A. Kane, S. A. Sarraf, C. Wang, J. L. Burman, D. P. Sideris, A. I. Fogel, and R. J. Youle. 2015. The ubiquitin kinase PINK1 recruits autophagy receptors to induce mitophagy. *Nature* 524 (7565):309–14. doi: 10.1038/nature14893.

Lee, J. Y., Y. Nagano, J. P. Taylor, K. L. Lim, and T. P. Yao. 2010. Disease-causing mutations in parkin impair mitochondrial ubiquitination, aggregation, and HDAC6-dependent mitophagy. *J Cell Biol* 189 (4):671–9. doi: 10.1083/jcb.201001039.

Lemasters, J. J. 2005. Selective mitochondrial autophagy, or mitophagy, as a targeted defense against oxidative stress, mitochondrial dysfunction, and aging. *Rejuvenation Res* 8 (1):3–5. doi: 10.1089/rej.2005.8.3.

Lim, G. G., D. S. Chua, A. H. Basil, H. Y. Chan, C. Chai, T. Arumugam, and K. L. Lim. 2015. Cytosolic PTEN-induced putative kinase 1 is stabilized by the NF-kappaB pathway and promotes non-selective mitophagy. *J Biol Chem* 290 (27):16882–93. doi: 10.1074/jbc.M114.622399.

Lim, K. L., and C. W. Zhang. 2013. Molecular events underlying Parkinson's disease—An interwoven tapestry. *Front Neurol* 4:33. doi: 10.3389/fneur.2013.00033.

Liu, L., D. Feng, G. Chen, M. Chen, Q. Zheng, P. Song, Q. Ma, C. Zhu, R. Wang, W. Qi, et al. 2012. Mitochondrial outer-membrane protein FUNDC1 mediates hypoxia-induced mitophagy in mammalian cells. *Nat Cell Biol* 14 (2):177–85. doi: 10.1038/ncb2422.

Martinez-Vicente, M., Z. Talloczy, E. Wong, G. Tang, H. Koga, S. Kaushik, R. de Vries, E. Arias, S. Harris, D. Sulzer, et al. 2010. Cargo recognition failure is responsible for inefficient autophagy in Huntington's disease. *Nat Neurosci* 13 (5):567–76. doi: 10.1038/nn.2528.

Matsuda, N., S. Sato, K. Shiba, K. Okatsu, K. Saisho, C. A. Gautier, Y. S. Sou, S. Saiki, S. Kawajiri, F. Sato, et al. 2010. PINK1 stabilized by mitochondrial depolarization recruits Parkin to damaged mitochondria and activates latent Parkin for mitophagy. *J Cell Biol* 189 (2):211–21. doi: 10.1083/jcb.200910140.

Matsuda, W., T. Furuta, K. C. Nakamura, H. Hioki, F. Fujiyama, R. Arai, and T. Kaneko. 2009. Single nigrostriatal dopaminergic neurons form widely spread and highly dense axonal arborizations in the neostriatum. *J Neurosci* 29 (2):444–53. doi: 10.1523/JNEUROSCI.4029-08.2009.

Miller, K. E., and M. P. Sheetz. 2004. Axonal mitochondrial transport and potential are correlated. *J Cell Sci* 117 (Pt 13):2791–804. doi: 10.1242/jcs.01130.

Narendra, D., A. Tanaka, D. F. Suen, and R. J. Youle. 2008. Parkin is recruited selectively to impaired mitochondria and promotes their autophagy. *J Cell Biol* 183 (5):795–803.

Narendra, D. P., S. M. Jin, A. Tanaka, D. F. Suen, C. A. Gautier, J. Shen, M. R. Cookson, and R. J. Youle. 2010. PINK1 is selectively stabilized on impaired mitochondria to activate Parkin. *PLoS Biol* 8 (1):e1000298. doi: 10.1371/journal.pbio.1000298.

Nixon, R. A. 2007. Autophagy, amyloidogenesis and Alzheimer disease. *J Cell Sci* 120 (Pt 23):4081–91. doi: 10.1242/jcs.019265.

Norris, K. L., R. Hao, L. F. Chen, C. H. Lai, M. Kapur, P. J. Shaughnessy, D. Chou, J. Yan, J. P. Taylor, S. Engelender, et al. 2015. Convergence of Parkin, PINK1, and alpha-synuclein on stress-induced mitochondrial morphological remodeling. *J Biol Chem* 290 (22):13862–74. doi: 10.1074/jbc.M114.634063.

Okatsu, K., T. Oka, M. Iguchi, K. Imamura, H. Kosako, N. Tani, M. Kimura, E. Go, F. Koyano, M. Funayama, et al. 2012. PINK1 autophosphorylation upon membrane potential dissipation is essential for Parkin recruitment to damaged mitochondria. *Nat Commun* 3:1016. doi: 10.1038/ncomms2016.

Pawlyk, A. C., B. I. Giasson, D. M. Sampathu, F. A. Perez, K. L. Lim, V. L. Dawson, T. M. Dawson, R. D. Palmiter, J. Q. Trojanowski, and V. M. Lee. 2003. Novel monoclonal antibodies demonstrate biochemical variation of brain parkin with age. *J Biol Chem* 278 (48):48120–8.

Priault, M., B. Salin, J. Schaeffer, F. M. Vallette, J. P. di Rago, and J. C. Martinou. 2005. Impairing the bioenergetic status and the biogenesis of mitochondria triggers mitophagy in yeast. *Cell Death Differ* 12 (12):1613–21. doi: 10.1038/sj.cdd.4401697.

Rakovic, A., K. Shurkewitsch, P. Seibler, A. Grunewald, A. Zanon, J. Hagenah, D. Krainc, and C. Klein. 2013. Phosphatase and tensin homolog (PTEN)-induced putative kinase 1 (PINK1)-dependent ubiquitination of endogenous Parkin attenuates mitophagy: Study in human primary fibroblasts and induced pluripotent stem cell-derived neurons. *J Biol Chem* 288 (4):2223–37. doi: 10.1074/jbc.M112.391680.

Schapira, A. H., J. M. Cooper, D. Dexter, P. Jenner, J. B. Clark, and C. D. Marsden. 1989. Mitochondrial complex I deficiency in Parkinson's disease. *Lancet* 1 (8649):1269.

Schwarz, T. L. 2013. Mitochondrial trafficking in neurons. *Cold Spring Harb Perspect Biol.* 2013. Jun 1;5(6). pii: a011304. doi: 10.1101/cshperspect.a011304. Review.

Schweers, R. L., J. Zhang, M. S. Randall, M. R. Loyd, W. Li, F. C. Dorsey, M. Kundu, J. T. Opferman, J. L. Cleveland, J. L. Miller, et al. 2007. NIX is required for programmed mitochondrial clearance during reticulocyte maturation. *Proc Natl Acad Sci U S A* 104 (49):19500–5. doi: 10.1073/pnas.0708818104.

Seibler, P., J. Graziotto, H. Jeong, F. Simunovic, C. Klein, and D. Krainc. 2011. Mitochondrial Parkin recruitment is impaired in neurons derived from mutant PINK1-induced pluripotent stem cells. *J Neurosci* 31 (16):5970–6. doi: 10.1523/JNEUROSCI.4441-10.2011.

Shults, C. W., R. H. Haas, D. Passov, and M. F. Beal. 1997. Coenzyme Q10 levels correlate with the activities of complexes I and II/III in mitochondria from parkinsonian and nonparkinsonian subjects. *Ann Neurol* 42 (2):261–4.

Tan, J. M., E. S. Wong, and K. L. Lim. 2009. Protein misfolding and aggregation in Parkinson's disease. *Antioxid Redox Signal* 11 (9):2119–34.

Twig, G., A. Elorza, A. J. Molina, H. Mohamed, J. D. Wikstrom, G. Walzer, L. Stiles, S. E. Haigh, S. Katz, G. Las, et al. 2008. Fission and selective fusion govern mitochondrial segregation and elimination by autophagy. *EMBO J* 27 (2):433–46. doi: 10.1038/sj.emboj.7601963.

Usenovic, M., E. Tresse, J. R. Mazzulli, J. P. Taylor, and D. Krainc. 2012. Deficiency of ATP13A2 leads to lysosomal dysfunction, alpha-synuclein accumulation, and neurotoxicity. *The Journal of neuroscience : the official journal of the Society for Neuroscience* 32 (12):4240–6. doi: 10.1523/JNEUROSCI.5575-11.2012.

Van Laar, V. S., B. Arnold, S. J. Cassady, C. T. Chu, E. A. Burton, and S. B. Berman. 2011. Bioenergetics of neurons inhibit the translocation response of Parkin following rapid mitochondrial depolarization. *Hum Mol Genet* 20 (5):927–40. doi: 10.1093/hmg/ddq531.

Van Laar, V. S., N. Roy, A. Liu, S. Rajprohat, B. Arnold, A. A. Dukes, C. D. Holbein, and S. B. Berman. 2015. Glutamate excitotoxicity in neurons triggers mitochondrial and endoplasmic reticulum accumulation of Parkin, and, in the presence of N-acetyl cysteine, mitophagy. *Neurobiol Dis* 74:180–93. doi: 10.1016/j.nbd.2014.11.015.

Verburg, J., and P. J. Hollenbeck. 2008. Mitochondrial membrane potential in axons increases with local nerve growth factor or semaphorin signaling. *J Neurosci* 28 (33):8306–15. doi: 10.1523/JNEUROSCI.2614-08.2008.

Vincow, E. S., G. Merrihew, R. E. Thomas, N. J. Shulman, R. P. Beyer, M. J. Maccoss, and L. J. Pallanck. 2013. The PINK1-Parkin pathway promotes both mitophagy and selective respiratory chain turnover in vivo. *Proc Natl Acad Sci U S A* 110 (6):6400–5. doi: 10.1073/pnas.1221132110.

Vives-Bauza, C., C. Zhou, Y. Huang, M. Cui, R. L. de Vries, J. Kim, J. May, M. A. Tocilescu, W. Liu, H. S. Ko, et al. 2010. PINK1-dependent recruitment of Parkin to mitochondria in mitophagy. *Proc Natl Acad Sci U S A* 107 (1):378–83. doi: 10.1073/pnas.0911187107.

Wang, X., D. Winter, G. Ashrafi, J. Schlehe, Y. L. Wong, D. Selkoe, S. Rice, J. Steen, M. J. LaVoie, and T. L. Schwarz. 2011. PINK1 and Parkin target Miro for phosphorylation and degradation to arrest mitochondrial motility. *Cell* 147 (4):893–906. doi: 10.1016/j.cell.2011.10.018.

Wang, Y., M. Serricchio, M. Jauregui, R. Shanbhag, T. Stoltz, C. T. Di Paolo, P. K. Kim, and G. A. McQuibban. 2015. Deubiquitinating enzymes regulate PARK2-mediated mitophagy. *Autophagy* 11 (4):595–606. doi: 10.1080/15548627.2015.1034408.

Waterham, H. R., J. Koster, C. W. van Roermund, P. A. Mooyer, R. J. Wanders, and J. V. Leonard. 2007. A lethal defect of mitochondrial and peroxisomal fission. *N Engl J Med* 356 (17):1736–41. doi: 10.1056/NEJMoa064436.

Ye, X., X. Sun, V. Starovoytov, and Q. Cai. 2015. Parkin-mediated mitophagy in mutant hAPP neurons and Alzheimer's disease patient brains. *Hum Mol Genet* 24 (10):2938–51. doi: 10.1093/hmg/ddv056.

Yoshii, S. R., C. Kishi, N. Ishihara, and N. Mizushima. 2011. Parkin mediates proteasome-dependent protein degradation and rupture of the outer mitochondrial membrane. *J Biol Chem* 286 (22):19630–40. doi: 10.1074/jbc.M110.209338.

Youle, R. J., and D. P. Narendra. 2011. Mechanisms of mitophagy. *Nat Rev Mol Cell Biol* 12 (1):9–14. doi: 10.1038/nrm3028.

Zhu, X. H., H. Qiao, F. Du, Q. Xiong, X. Liu, X. Zhang, K. Ugurbil, and W. Chen. 2012. Quantitative imaging of energy expenditure in human brain. *Neuroimage* 60 (4):2107–17. doi: 10.1016/j.neuroimage.2012.02.013.

Zuchner, S., I. V. Mersiyanova, M. Muglia, N. Bissar-Tadmouri, J. Rochelle, E. L. Dadali, M. Zappia, E. Nelis, A. Patitucci, J. Senderek, et al. 2004. Mutations in the mitochondrial GTPase mitofusin 2 cause Charcot-Marie-Tooth neuropathy type 2A. *Nat Genet* 36 (5):449–51. doi: 10.1038/ng1341.

9 Aggrephagy

Sijie Tan and Esther Wong
Nanyang Technological University
Singapore

CONTENTS

9.1   INTRODUCTION: PROTEIN AGGREGATION AND NEURODEGENERATION

Formation of pathological protein inclusions involves aggregation of mutant or misfolded proteins. These deleterious proteins generally show an increase in their surface hydrophobicity and propensity to aggregate due to a variety of factors, which broadly include (1) mutations or abnormal post-translational modifications (PTMs), (2) stressors that compromise proper protein folding dynamics, and/or (3) impairments in protein quality control systems such as chaperones, proteasome, and autophagy [1–6]. Accumulation of protein lesions is often associated with development of proteinopathies. For example, pathogenic buildup of protein aggregates in the form of hyaline and Mallory–Denk bodies in hepatocytes is observed in hepatocellular carcinoma (liver cancer) and nonalcoholic steatohepatitis (fatty liver disease) [7,8]. In the pancreas, accumulation of toxic protein aggregates to form islet amyloids in β-cells leads to the development of type 2 diabetes [9]. In general, aggregation-induced toxicity can affect different tissues and organs and is implicated in numerous human diseases [10]. However, of all the different tissues and cell types, the accumulation of protein aggregates results in greater cytotoxicity in post-mitotic cells such as neurons in the brain. Unlike dividing cells, post-mitotic neurons cannot rely on mitosis as a mechanism to prevent toxic accumulation of harmful damaged proteins through cellular dilution during cell division. Hence, protein quality control systems play a crucial role in protecting neurons against toxic protein species. Many lines of evidence have shown that loss of proper protein quality control can lead to neuronal cell death and the development of neurodegenerative diseases [11].

Different types of aggregation-prone proteins are associated with different forms of neurodegenerative diseases. The deleterious effects of protein aggregates on neuronal survival and cognitive functions have been extensively studied in Alzheimer's disease (AD), Parkinson's disease (PD), and Huntington's disease (HD). AD is characterized by the occurrence of amyloid plaques and neurofibrillary tangles due to aggregation of natively unfolded amyloid-beta (Aβ) peptides and hyperphosphorylated tau proteins, respectively. Accumulation of these aggregates is found in the neocortex region, including the temporal and parietal lobes as well as the frontal cortex [12]. α-Synucleinopathies, including PD and dementia with Lewy bodies (LBs), are characterized by the presence of LBs enriched with aggregated α-synuclein proteins [13]. Polyglutamine (PolyQ) diseases are a group of neurodegenerative disorders caused by the abnormal expansion of the polyglutamine (CAG) repeats, which affects protein folding and stability. Prominent PolyQ diseases include HD and spinocerebellar ataxia (SCA), which are characterized with protein inclusions formed by abnormal CAG repeat expansion in the mutant huntingtin (HTT) and ataxin proteins, respectively [14].

Formation of neuropathological lesions is a dynamic process. In a simple model, misfolded protein species can assemble into either rope or ribbon-like fibrillary structures containing extensive β-sheet conformations (protofibrils) or amorphous protein oligomers. These intermediates are transient and soluble, and act as a focal point for further polymerization and aggregation of misfolded proteins to form larger insoluble aggregates known as amyloid fibrils or amorphous protein inclusion bodies [10]. An important progress in the field of protein aggregation is the discovery of spatial propagation of the aggregation process within or between cells. Studies have shown that matured amyloids can fragment into multiple prefibrillary structures that can spread within or between cells to associate with the pre-existing pool of prefibrils. This will accelerate the pathogenicity spread of aggregated proteins in a mechanism similar to the propagation of prion proteins [15–17]. Currently, α-synuclein, tau protein, and Aβ peptides have been reported to have self-propagating properties similar to the prion proteins. Hence, its oligomers are capable to multiply through protein seeding and spreading [18,19].

The appearance of neuronal protein deposits in neurodegenerative diseases may suggest a connection between protein inclusions and neurotoxicity. This view remains highly debatable as both protective and toxic roles have been accorded to the occurrence of protein inclusions. Till date, what is widely observed among different neurodegenerative disorders is that the soluble protofibrillar or oligomeric intermediates are the major pathogenic species responsible for the neuronal insults in various disease states [20–26]. The mechanism underlying the neurotoxicity of these intermediate species is not fully understood, but structural analysis of the different array of α-synuclein oligomers revealed that these intermediates have larger hydrophobic surface and higher toxicity when compared with the larger insoluble α-synuclein protein inclusions. The high hydrophobic surface area to volume ratio of these intermediates could encourage aberrant interactions with proteinaceous and lipid cellular components to disrupt cellular processes [24–26]. Recent evidence seems to support a protective role of protein inclusions, where cells were found to utilize active sequestration of misfolded proteins into distinct protein deposition sites. This serves as a strategy to differentially target misfolded proteins to various proteolytic systems for degradation during proteotoxic stress (see Section 3). It is likely that these protein inclusions are protective only when they are efficiently removed from the cells by the degradative pathways.

Ubiquitin proteasome system (UPS) and macroautophagy (will be termed autophagy henceforth) are intracellular protein quality control pathways that remove unwanted proteins. UPS plays a critical role in degrading soluble proteins, while the autophagy pathway, besides removing soluble proteins, is the only proteolytic pathway capable of degrading large protein assemblies, including those observed in neurodegenerative conditions. In this chapter, we will focus on the selective clearance of protein aggregates by autophagy, a process referred to as aggrephagy (aggregates + autophagy). We will first give an overview of the mechanism underlying aggrephagy and the impact of this process on neurodegenerative diseases, followed by a discussion on the various therapeutic strategies that are available or under clinical trials to modulate the aggrephagy pathway. Lastly, we will present methodologies widely used to study the autophagic clearance of protein aggregates *in vivo* and *in vitro*.

9.2 CYTOPROTECTIVE MECHANISMS TO PROTECT AGAINST ACCUMULATION OF PROTEIN AGGREGATES

9.2.1 COMPARTMENTALIZATION OF AGGREGATED PROTEINS INTO INCLUSION BODIES

Sequestering of protein aggregates into a single, large foci is an evolutionarily conserved pathway observed in bacterial and eukaryotic cells [27–35] (Figure 9.1). These spatially distinct inclusions are found to be enriched in amyloid-like proteins, and are formed under acute and chronic stress conditions that favor destabilization of protein folding. In bacterial cells, formation of such inclusions is often observed during recombinant protein production, where accelerated synthesis of polypeptides by the ribosomes exceeds the coping ability of molecular chaperones to keep the

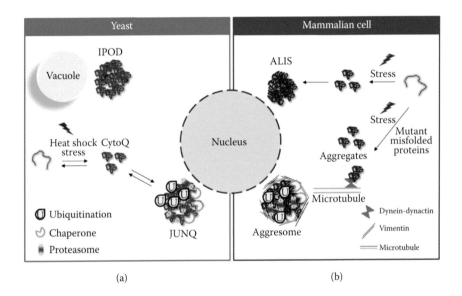

(a)   (b)

FIGURE 9.1 Different types of inclusion compartments observed in yeast and mammalian cells. (a) Heat shock leads to protein misfolding and aggregation to form transient CytoQ bodies. CytoQ can be ubiquitinated and transported along the microtubules to the perinuclear region to sequester into JUNQ compartment together with proteasomes and molecular chaperones. IPOD is a structure found near the vacuole in yeast. Unlike JUNQ, IPOD is free of proteasomes and molecular chaperones. (b) Terminally misfolded or mutant proteins can assemble to form aggregates that may interact and impair proteasomal function. Impairment in the proteasome is a signal to induce compartmentalization of cytosolic aggregates into perinuclear aggresome inclusion. Ubiquitinated protein aggregates are transported along microtubules, mediated by the dynein–dynactin motor complex to the MTOC region. Similar to yeast, proteasomes and molecular chaperones are sequestered into the aggresome. The aggresome is caged by vimentin, making the structure relatively inert. Aggresome-like induced structures (ALIS) also form in mammalian cells. They help to compartmentalize ubiquitinated proteins in response to bacterial infection, and can be induced by puromycin treatment associated with accumulation of defective newly synthesized proteins.

proteins in the native, folded forms [27]. In yeast cells, heat stress causes major protein unfolding accompanied with appearance of protein deposits intracellularly [28,30]. In mammalian cells, compromised proteasomal function drives the formation of inclusion foci as a way to unload the substrate burden on the proteasome [33]. Hence, these cellular compartments represent quality control strategies to protect the cellular proteome from aggregation-induced toxicity.

In the yeast, three types of inclusion bodies have been identified and characterized: (1) cytosolic Q bodies (CytoQ), (2) juxtanuclear quality control deposit (JUNQ), and (3) insoluble protein deposit (IPOD) [30,36,37] (Figure 9.1a). CytoQ are multiple cytosolic aggregates that appear upon heat stress and can be sequestered into perinuclear JUNQ compartment [38,39]. JUNQ is enriched with proteasomes and molecular chaperones. Studies have shown that JUNQ is a dynamic deposition site where misfolded proteins can be extracted by JUNQ-localized chaperones. These damaged protein species will subsequently be targeted to JUNQ-associated proteasomes for degradation once the condition is favorable to dissolve the structure. The IPOD in contrast is free of proteasomes and molecular chaperones, and is localized next to the vacuole. Its close vicinity to the vacuole suggests that the compartment serves as a staging area for bulk degradation of insoluble protein aggregates by autophagy [30,37]. Sorting of the aggregated proteins to JUNQ appears to be dependent on ubiquitination, while proteins routing to IPOD do not require such modification. It has been shown that JUNQ precursors, such as the CytoQ, are ubiquitinated [40]. A study reported that overexpression of deubiquitinating enzyme (DUB) Ubp4 or ablation of E2 ubiquitin-conjugating enzyme Ubc4 and Ubc5 inhibited the formation of JUNQ, and fusion of the IPOD substrate Rnq1 with ubiquitin rechanneled it to the JUNQ compartment [30]. These lines of evidence highlight the importance of ubiquitination as a sorting signal for protein aggregates compartmentalization to the JUNQ in the yeast system.

Similar to yeast, the concept of spatial compartmentalization of cytosolic aggregate-prone proteins is also observed in the mammalian systems. In mammalian cells, the most well-characterized type of protein quality control compartment is known as the aggresome (Figure 9.1b), which shares striking similarities to many neurodegenerative-linked protein inclusions such as LB in α-synucleopathies [41]. The formation of aggresome represents a stress response by the cells to cope with proteasomal failure or overloading, where aggregated proteins throughout the cells are retrograde transported along the microtubule network to the microtubule-organizing center (MTOC) region to be concentrated into the aggresome structure [33,34]. Spatial localization of the aggresome to the perinuclear region has a functional significance. During proteasomal stress, microtubule-mediated transport of lysosomes around the perinuclear region is reduced, thus creating an entrapment zone enriched with the proteolytic organelles. This factor helps to promote the degradation of perinuclearly localized aggresome by the lysosomes [42]. Aggresome shares similar features with the JUNQ compartment in the yeast where it selectively sequesters ubiquitinated aggregation-prone proteins, as well as proteasomes and molecular chaperones [34]. In contrast to JUNQ, aggresome is also enriched with autophagy cargo receptors known as p62 and NBR1, which promote targeting of cytosolic aggregated proteins into the aggresome [43] for subsequent in-bulk removal by the autophagy-lysosomal pathway [44]. Different types of aggresome-like induced structures (ALIS) have also been observed in certain mammalian cell types such as macrophages and dendritic cells [45,46]. Formation of ALIS is induced by puromycin treatment or oxidative stress [47]. The differences between the aggresome and ALIS are that the latter tend to be transient and the formation does not depend on transport along the microtubules [45,46].

9.2.2 AGGREPHAGY: SELECTIVE CLEARANCE OF PROTEIN AGGREGATES

The importance of autophagy for the turnover of protein inclusion is demonstrated by the association of neurodegenerative pathologies with various autophagy-deficient systems. Studies have shown that ablation of essential autophagic genes such as *atg5* and *atg7* resulted in motor deficit, degeneration of neurons accompanied with accumulation of cytosolic ubiquitinated protein inclusions in these mice [48,49]. This evidence clearly links the loss of autophagic function to neurodegeneration [50]. In contrast, pharmacological activation of autophagy has shown to successfully alleviate aggregation-induced neurotoxicity [51]. These evidence highlights that autophagy is an important quality control pathway to degrade protein aggregates to prevent their toxic accumulation, thereby protecting the cell from proteotoxicity.

Once thought as a bulk-removal pathway to provide energy during starvation [52], it is now demonstrated that the autophagy can also perform selective removal of specific cargoes in response to different stressors. Among the selective forms of autophagy, aggrephagy is most extensively studied. Emerging evidence suggests that the cell copes with different types of protein inclusions by targeting them for clearance by different autophagic pathways. Upon proteasomal stress, cells form the aggresome to facilitate high-volume removal of aggregating proteins by stress-induced aggrephagy, whereas small cytosolic protein aggregates formed basally are constitutively turned over by basal quality control aggrephagy [39,53]. This specificity in autophagic targeting is observed in both yeast and mammalian systems [39,53]. In aggrephagy, selectivity is conferred by the ability of the cell to differentiate between normal cytoplasmic protein constituents and unwanted aggregated proteins to only target the latter for autophagic disposal. The selectivity requires the presence of molecular determinants and players to mediate recognition of protein substrates by the autophagy pathway. Here, we will discuss the aggrephagy machinery observed in mammalian cells.

9.3 AGGREPHAGY CASCADE

9.3.1 AGGRESOME ASSEMBLY

Aggrephagy involves the decision to compartmentalize misfolded proteins into the aggresome for subsequent handling by the autophagy pathway. The formation of aggresome is an active process in response to proteasomal impairment and involves retrograde transportation of ubiquitinated aggregated proteins along the microtubules to the MTOC region for sequestration [33] (Figure 9.2a). Selectivity in this process is determined by the presence of specific ubiquitination signal on the aggregate-prone

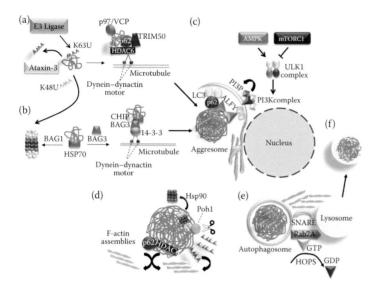

FIGURE 9.2 Aggrephagy cascade in mammalian cells. A myriad of molecular players and determinants regulate the aggrephagy pathway. (a–b) Formation of perinuclear aggresome can be via HDAC6-mediated ubiquitination-dependent (a) or BAG3 chaperone-mediated ubiquitination-independent (b) pathways. In HDAC6-mediated formation of aggresome, cytosolic aggregates are ubiquitinated by tug-of-war between E3 ligases (e.g., Parkin) and DUB (e.g., ataxin-3). While Parkin mediates K63U on cytosolic protein aggregates, ataxin-3 specifically cleaves off the modification and leads to an enrichment of K48U which directs the aggregates to the proteasome for degradation. Meanwhile, K63U-modified protein substrates are loaded onto the dynein–dynactin motor complex, and transported along the microtubules to the perinuclear region with the help of p97, TRIM50, and p62. BAG3-mediated aggresome formation does not require ubiquitination of protein substrates, but only targets Hsp70-associated protein species to the perinuclear aggresome structure. However, a high BAG1 to BAG3 ratio will favor the targeting to the proteasome. Movement of the Hsp70-associated substrates along the microtubules to form the aggresome is aided by the dynein–dynactin motor complex, CHIP, and 14-3-3 proteins. (c) The aggresome is a staging area to facilitate formation and linking of autophagosome membrane to the protein aggregates. p62 mediates the tethering of ubiquitinated protein aggregates to LC3 on autophagosome membrane. Corecruitment of scaffolding protein Alfy to the protein aggregates promotes efficient degradation of the latter by facilitating recruitment of PI3-kinase complex important for elongation of autophagosome membrane around the protein aggregates. (d) Aggresome disassembly has been observed prior to degradation. Cleavage of K63U chains from the aggresome mediated by Hsp90 and Poh1 is important to activate HDAC6-mediated F-actin remodeling to facilitate efficient fusion between aggresome-containing autophagosome with the lysosome. Here, p62 in the aggresome promotes perinuclear clustering of F-actin assemblies for efficient remodeling by HDAC6. (e–f) Fusion between autophagosome and lysosome is mediated by endosomal protein Rab7A and HOPS complex. HOPS complex converts Rab7A to its active GTP-bound form to facilitate membrane tethering between autophagosome and lysosome. SNARE proteins increase membrane permeability and induce membrane opening to facilitate fusion between the two organelles.

proteins for specific recruitment of machineries involved in aggresome biogenesis and aggrephagy pathway. In contrast to K48-linked polyubiquitination that targets misfolded proteins to proteasome for degradation, aggregation-prone proteins that are modified by K63-linked polyubiquitination (K63U) are preferentially recognized by histone deacetylase 6 (HDAC6). HDAC6 interacts with the dynein–dynactin motor complex on microtubules to transport these K63U modified proteins into the aggresome [54,55]. E3 ligases and DUB are important molecular players for editing ubiquitin codes on protein substrates to facilitate aggrephagy. Parkin and TRAF6 are the two E3 ligases that have been implicated in catalyzing K63U on protein aggregates to influence formation and clearance of aggresomes [56,57]. Conversely, ataxin-3, a DUB associated with protein aggregates, has been shown to remove K63U to inhibit aggresome formation [58]. Recently, a study has shown that ataxin-3 can generate unanchored C-terminal ubiquitin chains from polyubiquitin linkages on protein aggregates to serve as a recognition signal for HDAC6, thus suggesting an alternative pathway for selection of ubiquitinated protein aggregates for aggresome formation [59].

Regulation of HDAC6 activity profoundly affects aggresome formation. Different factors that influence HDAC6 activity include: (1) PTMs such as phosphorylation by casein kinase II (CKII) [60] and protein kinase C (PKC) zeta [61], (2) interaction with cargo receptor protein p62 [62], (3) interaction with molecular chaperone p97/VCP [63], and (4) modifications by E3 ligase TRIM50 (tripartite motif containing 50), [64] and ubiquitin-like modifier FAT10 [65]. Hence, different PTMs and interacting partners of HDAC6 participate in modifying its deacetylase activity and thereby influence aggresome formation as well as clearance.

Recently, a ubiquitination-independent pathway for aggresome formation has been identified (Figure 9.2b). This process requires the presence of molecular chaperones to shuffle misfolded proteins into the aggresome. Briefly, Hsp70 chaperone binds to misfolded proteins and the complex is recognized by a co-chaperone BAG3 [66]. The ATPase domain of BAG3 interacts with the dynein motor, facilitated by 14-3-3 regulatory protein, and promotes the transportation of proteins along the microtubules to the aggresome [67]. Surprisingly, although ubiquitination is dispensable in this process, interaction between Hsp70 and E3 ligase CHIP is important for BAG3-mediated aggresome formation [68].

9.3.2 Targeting Protein Inclusions to Autophagic Compartments

Special autophagic adaptors known as cargo receptor proteins facilitate tethering of protein inclusions to the autophagosomal membrane [44]. In mammalian cells, p62 and NBR1 are the most well-characterized autophagic receptors [69]. These receptor proteins contain (1) a short, linear LC3-interacting region (LIR) motif with the consensus sequence D/E-D-W-x-x-L (x = any amino acid) that interacts with LC3, the autophagosomal membrane protein [70] and (2) a C-terminal ubiquitin-associated (UBA) domain that binds to ubiquitinated protein aggregates. Hence, cargo receptor proteins help to bridge the protein aggregates with the autophagosomal membrane to facilitate loading of the protein aggregates into autophagosomes. However, NBR1 and p62 engage different types of ubiquitinated protein aggregates; p62 has preferential binding to K63U substrates [54,71], whereas NBR1 does not have selectivity against specific ubiquitin linkage [72]. In addition, structural differences in the UBA domains between p62 and NBR1 account for much higher affinity of NBR1 for ubiquitin chains than p62 [72]. Despite the differences, p62 and NBR1 have been shown to act cooperatively to promote aggregation and selective degradation of ubiquitinated protein inclusions [69]. In addition to being a cargo receptor, p62 has also been shown to initiate formation of autophagosomal membrane around protein aggregates and hence, facilitates sequestration of protein substrates into autophagosomes (Figure 9.2c). In this model, p62 is shown to interact with unc-51-like autophagy-activating kinase (ULK1) and mammalian target of rapamycin (mTOR) at the autophagosome initiation site called the omegasome. This favors the recruitment and assembly of upstream autophagic proteins to the initiation sites to promote autophagosome formation [73,74].

The activities of p62 and NBR1 are regulated by phosphorylation. CKII-mediated phosphorylation of the UBA domain in p62 has been shown to increase its binding affinity for ubiquitinated misfolded proteins to enhance formation and degradation of the protein inclusions [75]. On the contrary, phosphorylation of NBR1 by GSK3 perturbs formation and autophagic clearance of ubiquitinated protein aggregates [76]. Scaffolding proteins have also been shown to interact with cargo receptor proteins to facilitate formation of autophagosomal membrane around the protein inclusions. In mammalian cells, autophagy-linked FYVE (named after the four cysteine-rich proteins: Fab1, YOTB, Vac1 and EEA1) protein (Alfy) [77] interacts with p62, Atg5, and phosphatidylinositol 3-phosphate (PI3P) found in autophagic membranes to promote autophagosome biogenesis around the p62-labeled protein cargoes [78]. Recently, HTT protein implicated in HD has been identified as a novel scaffolding protein that not only promotes interactions between p62, LC3, and K63U signals in protein inclusions for cargo recognition, but also serves a second function of activating ULK to initiate autophagy [79].

9.3.3 Fusion between Autophagosome and Lysosome

The protein cargoes in the autophagosome are degraded by fusion of the organelle with the lysosome to form autophagolysosome (Figure 9.2e). The fusion process involves remodeling of F-actin-cortactin network by HDAC6 [80]. p62 in protein aggregates plays a role in this step by recruiting and enriching the perinuclear region with F-actin network assemblies for remodeling [62]. Microtubule-mediated transport of lysosomes within the perinuclear region is also reduced to enhance the lysosomal population in the region for efficient fusion with autophagosomes [42]. Wild-type HTT also plays a role in the trafficking of lysosomes to the perinuclear region to further enhance the lysosomal population in the region [81]. Tethering of the autophagosomal membrane with the lysosomal membrane prior to fusion is mediated by late endosome protein Rab7A and homotypic fusion and protein sorting (HOPS) complex. Briefly, Rab7A is converted to its active Guanosine-5'-triphosphate (GTP)-bound form by HOPS complex to facilitate membranes tethering. Meanwhile, soluble N-ethylmaleimide-sensitive factor (NSF) attachment protein receptor (SNARE) proteins will increase membrane permeability and induce membrane opening to facilitate fusion [82,83].

9.3.4   Disassembly of Aggresome for Degradation

In yeast cells, a disaggregase triad consisting of Hsp70-Hsp40-Hsp110 is able to solubilize and extract polypeptides from inclusion bodies, and target them for refolding or degradation by the proteasome system under chronic heat stress. This shows that large insoluble protein inclusions are dissolved prior to degradation. In mammalian system, emerging evidence also shows that the aggresomal structure is disassembled prior to degradation by the autophagy pathway (Figure 9.2d). Proteasome-associated DUB enzyme Poh1 cleaves off K63U chains from the protein aggregates, releasing free unanchored K63U chains to activate HDAC6 to promote cortactin-dependent F-actin remodeling to facilitate fusion between autophagosome and lysosome [84]. This process requires the dissociation of 26S proteasomal subunit from the aggresome, and is mediated by molecular chaperone Hsp90 [85].

9.4   FAILURE OF AGGREPHAGY IN NEURODEGENERATIVE DISEASES

Failure of aggrephagy in neurodegeneration can be due to perturbation at various steps along the autophagy pathway or due to inefficient priming of protein inclusions for clearance. Here, we examine how the autophagic clearance of protein aggregates has gone awry in neurodegeneration, with particular focus on how the key aspects of aggrephagy such as appropriate labeling and recognition of cargoes are altered in neurodegenerative disorders.

9.4.1   Defects in Cargo Recognition

Cargo receptor proteins are important for selective recognition of ubiquitinated protein aggregates by the autophagy pathway for degradation. Mutations in p62 gene have been observed in Paget's disease of the bone (PDB), as well as familial and sporadic forms of amyloid lateral sclerosis (ALS) [86,87]. In PDB, mutations in p62 occur mainly in the UBA domain, while in ALS, mutations affect several p62 domains, including the UBA, Phox, and Bem1 (PB1) domain, and LIR regions [86,88]. Mutations in the UBA domain account for the inability of p62 to bind ubiquitinated protein cargoes to target them to the autophagosomes in the disease state. In HD, electron microscopy revealed that the autophagosomes of HD cells are devoid of cellular cargoes, showing that there is a defect in the sequestration of cellular cargoes by the autophagic vacuoles [89]. Recently, wild-type HTT has been shown to act as a physiological scaffold linking ULK1 and p62 together, to form autophagic initiation complex to induce autophagosome formation around sequestered ubiquitinated proteins in response to aggregation-induced proteotoxicity [79]. Thus, the loss of wild-type HTT function in HD might partially account for the appearance of empty autophagosomes observed with mutant HTT in HD.

9.4.2   Inefficient Priming of Aggresome for Clearance

Failure in protein cargo recognition can be due to dysfunctional autophagic cargo receptor or scaffolding proteins mentioned above, or caused by inappropriate labeling of the protein cargoes to engage the aggrephagy machineries [90]. Although the presence of p62 and NBR1 cargo receptors in protein inclusions is crucial for its susceptibility to selective autophagy, emerging studies have shown that presence of p62 is not an absolute criterion for aggrephagy. First, not all p62-containing inclusion bodies are susceptible to autophagic clearance, and these include aggresomes formed by PD-linked p38 protein, myopathy-linked mutant desmin protein, and exogenous administered α-synuclein fibrils [90,91]. These inclusions are not able to recruit the autophagy-lysosome machinery and are not susceptible to autophagic degradation.

Converging lines of evidence now show that the presence of "extra" determinants plays a role in further fine-tuning the competency of protein inclusions toward autophagic amenability. We have recently identified synphilin-1, a protein that promotes protein aggregation and highly represented in various LBs of α-synucleinopathies, to be a novel molecular determinant of successful aggrephagy [53]. Synphilin-1 protein can render autophagic-resistant PD-linked p38 inclusions amenable to autophagy, and ameliorates neurotoxicity induced by α-synuclein fibrils by sequestering them into aggresome and promoting autophagic degradation [92–94]. Specifically, the presence of the ankyrin-1 (ANK1) domain of synphilin-1 in these inclusions is sufficient to promote recruitment of autophagosomes and lysosomes. It appears that the ANK1 provides a platform for favorable K63U modification, which helps to stabilize protein mobility on the surface of the aggresome to promote efficient recruitment of the class III PI3-kinase complex for initiating autophagosome biogenesis to engulf the protein inclusions [53]. It remains unknown whether the aggrephagy-promoting role of synphilin-1 is restricted to LB associated with α-synucleinopathies or has a more general influence on all types of inclusions. Taken together, the persistence of protein lesions in diseases can also be due to failure of the aggresome to sequester key aggrephagy-promoting factor that regulates competency of the inclusion to be recognized by the autophagy machinery.

9.4.3 DEFECTS IN INITIATION OF AUTOPHAGOSOME BIOGENESIS

A few neurodegenerative disorders have been implicated with defective autophagosomal membrane formation, which results in failure to engulf misfolded and damaged protein substrates into the autophagosomes. In PD, wild-type α-synuclein has been shown to inhibit the activity of Rab1 GTPase, an important protein to recruit Atg1 and downstream Atg9, to the omegasome. Atg9 complex (Atg9-Atg2-Atg18) shuffles between cytoplasmic vesicle pool, the Golgi apparatus, and the omegasome to supply lipid components for the formation of autophagosomal membrane. Inhibition of Rab1 by α-synuclein causes Atg9 mislocalization and inhibits omegasome formation [95]. Another example is Lafora disease, an autosomal recessive genetic disorder caused by mutation in phosphatase laforin that leads to accumulation of hyperphosphorylated glycogens in the form of Lafora bodies [96]. Reduced autophagosome number was observed in this disorder where aberrant mTOR activity has been proposed to cause deregulation in autophagy induction [96]. In AD, aberrant caspase 3-mediated cleavage of beclin-1 significantly reduced beclin-1 level, which forms part of the class III PI3-kinase complex that initiates formation of omegasome [97].

9.4.4 COMPROMISED LYSOSOME PROTEOLYSIS

Degradation of protein cargoes by the lysosome is the final step in the aggrephagy pathway. Lysosomal dysfunction is closely associated with neurodegeneration in a group of diseases known as lysosomal storage disorders (LSD). A distinct pathology of LSD is the abnormal accumulation of undigested or partially digested macromolecules in the lysosomal compartments [98,99]. An acidic pH environment is essential for the proper activation of lysosomal proteases for proteolytic digestion. Compromised lysosomal pH has been identified as a leading cause of LSD pathogenesis. Mutations in *MCOLN1* gene, which encodes a nonselective cation channel mucolipin1, causes lysosomal pH elevation and is responsible for the pathogenesis of mucolipidosis type IV congenital disorder, a type of LSD [100]. Mutations in gene encoding various lysosomal proteases are also a cause for defective lysosomal proteolysis observed in LSD. For instance, mutation in cathepsin D gene leads to the development of neuronal ceroid lipofuscinoses, also known as Batten's disease [101].

LSD-like lysosomal pathology is also seen in AD and PD, thus highlighting that the disruption of lysosomal proteolysis could be a common causative factor in the pathogenesis of many neurodegenerative disorders. In early-onset familial AD, presenilin-1 protein is mutated, which leads to impairment in targeting of vacuolar-ATPase (v-ATP) to the lysosome. As v-ATPase plays a crucial role in maintaining the acidification of lysosomal lumen, the reduced presence of v-ATP causes lysosomal dysfunction in mutant presenilin-1- expressing cells [102]. Furthermore, mutation in apolipoprotein (APOE4) gene encoding for a cholesterol protein transporter is observed in late-onset AD. This mutation destabilizes lysosomal membrane and disrupts lysosomal integrity [103]. In addition, an abnormally high level of calpain is reported in AD pathology. This enzyme cleaves off Hsp70 and contributes to the loss of lysosomal integrity and promotes necrotic neuronal cell death [104]. In late-onset PD, leucine-rich repeat kinase-2 (LRRK2) is found to reduce acidity of lysosomes by affecting v-ATPase function in the lysosomes [105].

9.5 TARGETING AGGREPHAGY AS THERAPEUTIC INTERVENTION FOR PROTEINOPATHIES

Currently, there are a variety of drugs and compounds that act on different stages of the autophagy pathway to upregulate autophagy. These drugs have shown great potential in reducing aggregated protein loads and alleviating neurotoxicity in various cell and animal neurodegenerative models [106–108]. However, it is now known that besides impairment in cargo recognition and autophagy induction, defects in the terminal organelle lysosomes as those seen in LSD can also contribute to neurodegeneration. Enhancing the levels of autophagy in disorders with primary lysosomal defects may be deleterious rather than protective, by causing further accumulation of neuropathological autophagic vacuoles and lysosomes containing undigested cytoplasmic materials. Thus, adopting autophagy upregulation as a therapeutic strategy for neurodegeneration should be carefully evaluated, with assessment of the origin of defect in the autophagy-lysosomal pathway represented in the disease of interest.

9.5.1 Enhancing Global Autophagy Activity: Targeting mTOR and AMPK Signaling Pathways

Enhancing the autophagy levels by modulating the mTOR and AMPK signaling pathways is the most common intervention to alleviate intracellular burden and toxicity from protein aggregates. Although these modulations target and influence the general, nonselective autophagy pathway rather than the selective clearance of protein aggregates, modulating the signaling pathways has shown to be beneficial to fight against neurodegeneration.

Upregulation of autophagy can be achieved by inhibiting the mTOR pathway. Rapamycin, a lipophilic macrolide antibiotic that binds to mTORC1 and inactivates its activity [109], has been shown to reduce the levels of intracytoplasmic protein aggregates and relieve aggregation-induced toxicity in AD, HD, PD, frontotemporal dementia, SCA type III, and prion disease models [110–115]. However, given the varied physiological functions of mTOR signaling which also encompass roles in ribosome biogenesis and cellular metabolism, a key caveat is the possibility of off-target effects [116]. Hence, in lieu of the side effects, identification of mTORC1 inhibitors with fewer side effects is favorable. Alternatively, targeting mTOR-independent signaling pathway to modulate autophagy, like the phosphatidylinositol or AMPK pathways, might be a more ideal cellular route.

The phosphoinositol cycle replenishes the levels of inositol 1,4,5-trisphosphate (IP_3) from inositol, and enhanced cellular levels of IP_3 inhibits autophagy [117]. Mood-stabilizing drugs, such as lithium, sodium valproate, and carbamazepine [118], and hypertension drugs, such as rilmenidine [119], have been shown to induce autophagy by disrupting the phosphoinositol cycle and reducing the levels of IP_3. These drugs have been shown to be effective against toxic effects of mutant HTT, and promote reduction as well as clearance of the polyQ-expanded proteins. Combined treatment of lithium and valproate has also been shown to be more effective than a single compound in delaying disease onset and prolonging survival of ALS mouse model [120]. Trehalose, a disaccharide, has been reported to upregulate autophagy independently of mTOR signaling pathway. Although the induction mechanism is unknown, emerging studies have shown that trehalose can ameliorate the behavioral symptoms and pathological damage associated with the accumulation of mutant pathogenic proteins such as A53T α-synuclein mutant, G93A superoxide dismutase 1 (SOD1) mutant, and HTT mutants [121–125].

Besides this, metformin, an antidiabetic medication, has been shown to be neuroprotective against HD and tauopathies by activating AMPK signaling and autophagy [126–128]. In AD cell model, besides promoting autophagy, metformin also increases protein phosphatase 2A level that reduces tau phosphorylation and aggregation [127,128]. Another AMPK activator, AICAR (5-amino-4-carboxamide-1-β-D-ribonucleoside) also helps to reduce the levels of phosphorylated tau in neuronal cell line under stress-induced condition, and may be a potential compound against tauopathies [129].

Targeting mTOR-dependent and independent pathways synergistically has also been shown to have enhanced benefits for relieving intracellular protein aggregate burden. Administration of rapamycin along with lithium was shown to provide greater protection against neurodegeneration in HD cell and Drosophila models than the use of a single compound [130].

9.5.2 ENHANCING AGGREPHAGY

Targeting the aggrephagy process rather than global autophagy activity as a strategy against neurodegeneration may be more favorable as it increases the specific removal of protein aggregates without overwhelming the basal, housekeeping autophagy. Cargo receptor proteins and HDAC6 are the most appealing therapeutic targets for this purpose, as they are key components that influence the selectivity of protein aggregates. Currently, pharmacological compounds that upregulate the activities of these group of proteins are limited, and those attempting to influence their actions focus on altering their expression levels or PTM. Deprenyl, a candidate drug for PD, induces nuclear accumulation and activation of transcription factor Nrf2, which upregulates the expression of p62 [131]. This can potentially increase the pool of cargo receptor proteins available to match the elevated levels of protein aggregates in proteinopathies for more efficient autophagic targeting.

9.5.3 RESTORATION OF LYSOSOME PROTEOLYTIC ACTIVITY

Therapeutic interventions to enhance efficiency of lysosomal degradation and stability have shown some promising outcomes against LSD. Cysteine proteases are enzymes that target and inactivate lysosomal protease cathepsin. Administration of cysteine protease inhibitors like cystatin B has been reported to alleviate Aβ-peptide-induced toxicity in transgenic AD mice model expressing mutant amyloid precursor protein [132]. Similarly, GSK-3β inhibitors such as lithium restores lysosomal acidification and ameliorate Aβ pathology [133]. Lysosome membrane instability can be rescued by administrating calpain inhibitors, which are shown to have therapeutic effects against AD [134].

9.6 METHODOLOGIES TO MONITOR AGGREPHAGY

Various methods established to monitor general autophagy pathway are also applicable to study aggrephagy. However, it is important to establish selectivity as a prerequisite of the process during experimental studies. Here, we will start by discussing the methods that have been used to create protein inclusions *in vivo* and *in vitro*, and then evaluate the approaches that have been established to study the downstream clearance process. Lastly, we will describe the control assays that are required to determine the selective nature of the autophagic process.

9.6.1 INCLUSION MODEL PARADIGM

Under normal physiological conditions, cells do not usually express disease-associated or aggregation-prone mutant proteins. Even if misfolded or damaged proteins are formed, molecular chaperones and the proteasome activity will keep the levels in check before the proteins can accumulate and aggregate. Hence, studying clearance of protein aggregates in normal cellular physiology is hard as their appearances are infrequent and rare. To circumvent this problem, an overexpression system *in vivo* or *in vitro* is commonly used to artificially induce the synthesis of disease-associated mutant proteins. This includes the use of transgenic model organisms, transient transfection, or stable expression of the mutant proteins. In addition, the mutant protein of interest is often expressed with a fluorescence tag to facilitate detection of the protein *in vivo* and *in vitro*.

Although overexpression of these mutant proteins can lead to spontaneous protein aggregation, the number of protein inclusions observed often remains low due to the active UPS quality control mechanism. Hence, proteasome inhibitors such as MG132, bortezomib, and lactacystin are commonly applied to accelerate the formation of inclusions. These inhibitors form covalent bonds with the N-terminal threonine on the catalytic subunits of the proteasomal active site and inactivate its chymotrypsin-like activity [135]. Perinuclear aggresomes are the most common type of inclusions formed under proteasome inhibition. Hence, treatment of cells with proteasome inhibitors can be used to induce aggresome formation to study their amenabilities to autophagy [33]. The aggresomal nature of the perinuclear structures can be verified by staining for aggresome markers, including γ-tubulin, ubiquitin, and vimentin [33]. Other aggresome-like inclusions such as ALIS or p62 bodies can be artificially induced in cell cultures by treatment with antibiotic puromycin. Generally, puromycin causes premature termination of protein translation at the ribosome and subsequent accumulation of the defective ribosomal translation products that form ALIS or p62 bodies [136].

9.6.2 Interaction between Inclusion Bodies and Autophagy

Examining the association of protein inclusions described above with the autophagy-lysosomal compartments will give a preliminary indication on the susceptibility of the protein inclusions toward aggrephagy. Electron microscopy is routinely employed to visualize autophagic structures in cells and tissue samples [137,138]. Morphologically, autophagosomes appear as double membrane-bound vesicles that can further be classified into early- or late-stage structures. Early autophagosome often contains intact organelles or more electron-dense cytoplasmic materials, while late autophagosomes, which include amphisome or autophagolysosome, consist of partially digested products [139,140]. Extra attention is required to determine whether a cellular vacuole is an autophagic compartment, as presence of multivesicular bodies (MVB) characterized by multiple membrane-bound organelles can be mistaken as the former. A guideline to distinguish between an MVB and the autophagosome is the presence of cytoplasmic contents enclosed within the membranes for the latter [139]. The nature of the autophagosomal structures that accumulative often provides useful information on the step in the autophagy pathway that is defective in various neurodegenerative diseases. For instance, the "empty" autophagosomes observed in HD cells suggest a defect in cargo loading into autophagosomes [89], while the accumulation of late-stage autophagosomes in AD is indicative of a problem with lysosome proteolysis [140].

Immunostaining and fluorescence microscopy provide alternative ways to look at the interaction between protein inclusions and components of the autophagy-lysosomal machinery. Autophagosome and lysosome can be visualized through immunostaining for LC3 and LAMP1, respectively. If a protein inclusion is predisposed for clearance, activation of autophagy will lead to enhanced presence of autophagic markers in the inclusion. Hence, by comparing the percentages of cells exhibiting colocalization under basal and autophagic induction conditions, it can indicate whether the protein inclusions recruit the autophagic structures during autophagy activation to facilitate removal.

The use of microscopy techniques to determine whether protein aggregates are targeted to the autophagic compartments may not offer a global representation of the autophagy activity in the cells or tissues. This is because only a few random areas of the tissue or cell layer were analyzed each time. Hence, as a complementary method to microscopy, biochemical method such as sodium dodecyl sulfate polyacrylamide gel electrophoresis (SDS-PAGE) provides a non-biased way for the analysis of protein aggregate accumulation in autophagic structures [141]. Briefly, autophagic compartments such as the autophagosomes, autophagolysosomes, and the lysosomes can be differentially fractionated from the cytosol using a metrizamide gradient and immunoblot for the levels of aggregation-prone proteins associated with the different organelle fractions [89]. This will also provide information on whether the accumulation of the protein aggregates occurs in the autophagic structures or in the cytosolic fraction.

9.6.3 MONITORING CLEARANCE OF PROTEIN INCLUSIONS

Both immunofluorescence and SDS-PAGE have been used to study autophagic turnover of protein inclusions. Microscopic monitoring of changes in the levels of protein inclusion under autophagic induction and/or inhibition will demonstrate the amenability of the inclusion bodies toward aggrephagy. One can either quantify the changes in the number or size of the protein inclusions within each cell or analyze in a population of cells, the changes in the percentage of cells harboring protein inclusions. For the latter study, typically large sample size of more than 100 cells is analyzed. Inhibition of proteasomal activity is a way to artificially induce formation of protein inclusions to assess their turnover [90]. With the presence of an adequate pre-existing pool of protein inclusions, one can monitor changes in the percentage of protein inclusions in response to autophagic activation or inhibition to measure the turnover (if any) by aggrephagy. If there is an aggrephagy flux, autophagic upregulation (e.g., rapamycin treatment or starvation) will reduce the percentage of protein inclusions as compared to untreated control condition. This reduction represents the protein inclusions turnover by autophagy.

Accumulation of protein inclusions upon autophagy inhibition is another readout for autophagy turnover. The accumulation would represent the pool of inclusion bodies that would have been cleared if the autophagic process is intact. This method is particularly useful for studying basal aggrephagy [53].

Sometimes, it is not clear whether the observed decline in levels of protein inclusion is due to a defect in formation, or the clearance of the pre-existing pool of inclusion bodies. To circumvent this issue, a tetracycline-inducible system can be adopted to temporally regulate the stable expression of the aggregation-prone protein [142]. Briefly, stress-induced synthesis of proteins can be switched on by treating with tetracycline (*tet-on* system), or removal of tetracycline (*tet-off* system). After this, clearance of protein inclusions can be monitored at various time points after antibiotic addition or wash-out. This method allows *de novo* synthesis to be halted, and the subsequent clearance process to be monitored in the absence of synthesis variations.

Turnover of protein inclusions by autophagy can also be monitored biochemically using SDS-PAGE and western blotting to examine changes in the levels of insoluble proteins upon autophagy activation or inhibition. This technique provides a non-bias analysis to complement protein turnover analysis by microscopy techniques. Differential detergent extraction method can be used to separate the soluble and the insoluble protein fractions for analysis of changes in each fraction. Soluble proteins are extracted under mild detergent conditions (e.g., NP-40, Triton X-100), while the insoluble protein aggregates can only be extracted under harsh detergent conditions (e.g., 1% SDS). In addition, a filter-trap experiment can be further utilized to complement the SDS-PAGE. Briefly, SDS-insoluble fraction is extracted from the protein lysates and applied to cellulose acetate filter where they will be retained. By immunoblotting the filter paper with antibody against the protein of interest, the levels of SDS-insoluble protein inclusions can be determined [143].

9.6.4 PARAMETERS TO DETERMINE SELECTIVITY IN AUTOPHAGIC CLEARANCE OF PROTEIN INCLUSIONS

To show that macroautophagy is the pathway responsible for any clearance profiles observed, it is important to determine the readout under autophagy inhibition. Genetic ablation of key components of the autophagic machinery such as Atg5, Atg7, or ULK1 using knockout or siRNA/shRNA-mediated knockdown in cell or animal models, or chemical inhibition using 3-methyladenine (3-MA), Bafilomycin A1 (Baf A1), or combination of ammonium chloride and leupeptin (NL), can be used to create an autophagy-deficient environment. 3-MA is a class III PI3-kinase complex inhibitor that prevents autophagosome formation, and Baf A1 and NL disrupt lysosomal acidification by inhibiting v-ATPase proton pump and inactivating the lysosomal proteolytic enzymes. Perturbed turnover of protein aggregates under macroautophagy inhibition will suggest that the clearance is due to the latter's activity.

To pinpoint that clearance of protein inclusion is due to selective targeting of aggregated proteins to autophagy and not due to bulk degradation, it is important to establish that the enhanced interaction between the protein inclusion and autophagosomes or lysosomes is also accompanied by higher association with cargo receptor proteins for selective targeting. Colocalization between inclusion bodies and receptors like p62 and NBR1 is an indication for selective autophagy. It is also essential to perform experiments in cargo receptor–deficient cellular environment. Accumulation of protein aggregates under this condition will indicate the role of cargo receptors in the removal of the protein inclusions. In addition to monitoring protein clearance, checking the turnovers of p62 and NBR1 under normal and autophagy inhibition conditions by SDS-PAGE can also be a readout for selective autophagy, as cargo receptors themselves are degraded along with their cargoes.

ACKNOWLEDGMENTS

Work in our laboratory is supported by grants from MOE Tier 2 M4020161.080 (ARC 25/13), MOE Tier 1 M4011565.080 (RG139/15), and SUG M4080753.080.

REFERENCES

1. Dobson, C.M. Protein folding and misfolding. *Nature*, 2003. **426**(6968): 884–90.
2. Soto, C. Unfolding the role of protein misfolding in neurodegenerative diseases. *Nat Rev Neurosci*, 2003. **4**(1): 49–60.
3. Chiti, F. and C.M. Dobson. Protein misfolding, functional amyloid, and human disease. *Annu Rev Biochem*, 2006. **75**: 333–66.
4. Ross, C.A. and M.A. Poirier. Protein aggregation and neurodegenerative disease. *Nat Med*, 2004. **10 Suppl**: S10–7.
5. Rubinsztein, D.C. The roles of intracellular protein-degradation pathways in neurodegeneration. *Nature*, 2006. **443**(7113): 780–6.
6. Meredith, S.C. Protein denaturation and aggregation: Cellular responses to denatured and aggregated proteins. *Ann N Y Acad Sci*, 2005. **1066**: 181–221.
7. Strnad, P., et al. Mallory-Denk-bodies: Lessons from keratin-containing hepatic inclusion bodies. *Biochim Biophys Acta*, 2008. **1782**(12): 764–74.
8. Zatloukal, K., et al. From Mallory to Mallory-Denk bodies: What, how and why? *Exp Cell Res*, 2007. **313**(10): 2033–49.
9. Hull, R.L., et al. Islet amyloid: A critical entity in the pathogenesis of type 2 diabetes. *J Clin Endocrinol Metab*, 2004. **89**(8): 3629–43.
10. Knowles, T.P., M. Vendruscolo, and C.M. Dobson. The amyloid state and its association with protein misfolding diseases. *Nat Rev Mol Cell Biol*, 2014. **15**(6): 384–96.
11. Chhangani, D. and A. Mishra. Protein quality control system in neurodegeneration: A healing company hard to beat but failure is fatal. *Mol Neurobiol*, 2013. **48**(1): 141–56.
12. Nixon, R.A. and D.S. Yang. Autophagy failure in Alzheimer's disease—Locating the primary defect. *Neurobiol Dis*, 2011. **43**(1): 38–45.
13. Spillantini, M.G. and M. Goedert. The alpha-synucleinopathies: Parkinson's disease, dementia with Lewy bodies, and multiple system atrophy. *Ann N Y Acad Sci*, 2000. **920**: 16–27.
14. Shao, J. and M.I. Diamond. Polyglutamine diseases: Emerging concepts in pathogenesis and therapy. *Hum Mol Genet*, 2007. **16 Spec No. 2**: R115–23.
15. Lee, S.J., et al. Cell-to-cell transmission of non-prion protein aggregates. *Nat Rev Neurol*, 2010. **6**(12): 702–6.
16. Polymenidou, M. and D.W. Cleveland. The seeds of neurodegeneration: Prion-like spreading in ALS. *Cell*, 2011. **147**(3): 498–508.
17. Collinge, J. and A.R. Clarke. A general model of prion strains and their pathogenicity. *Science*, 2007. **318**(5852): 930–6.
18. Jucker, M. and L.C. Walker. Self-propagation of pathogenic protein aggregates in neurodegenerative diseases. *Nature*, 2013. **501**(7465): 45–51.
19. Walker, L.C., et al. Mechanisms of protein seeding in neurodegenerative diseases. *JAMA Neurol*, 2013. **70**(3): 304–10.
20. Caughey, B. and P.T. Lansbury. Protofibrils, pores, fibrils, and neurodegeneration: Separating the responsible protein aggregates from the innocent bystanders. *Annu Rev Neurosci*, 2003. **26**: 267–98.
21. Walsh, D.M. and D.J. Selkoe. Oligomers on the brain: The emerging role of soluble protein aggregates in neurodegeneration. *Protein Pept Lett*, 2004. **11**(3): 213–28.

22. Baglioni, S., et al. Prefibrillar amyloid aggregates could be generic toxins in higher organisms. *J Neurosci*, 2006. **26**(31): 8160–7.

23. Haass, C. and D.J. Selkoe. Soluble protein oligomers in neurodegeneration: Lessons from the Alzheimer's amyloid beta-peptide. *Nat Rev Mol Cell Biol*, 2007. **8**(2): 101–12.

24. Winner, B., et al. In vivo demonstration that alpha-synuclein oligomers are toxic. *Proc Natl Acad Sci U S A*, 2011. **108**(10): 4194–9.

25. Campioni, S., et al. A causative link between the structure of aberrant protein oligomers and their toxicity. *Nat Chem Biol*, 2010. **6**(2): 140–7.

26. Cremades, N., et al. Direct observation of the interconversion of normal and toxic forms of alpha-synuclein. *Cell*, 2012. **149**(5): 1048–59.

27. Ventura, S. and A. Villaverde. Protein quality in bacterial inclusion bodies. *Trends Biotechnol*, 2006. **24**(4): 179–85.

28. Miller, S.B., A. Mogk, and B. Bukau. Spatially organized aggregation of misfolded proteins as cellular stress defense strategy. *J Mol Biol*, 2015. **427**(7): 1564–74.

29. Sontag, E.M., W.I. Vonk, and J. Frydman. Sorting out the trash: The spatial nature of eukaryotic protein quality control. *Curr Opin Cell Biol*, 2014. **26**: 139–46.

30. Kaganovich, D., R. Kopito, and J. Frydman. Misfolded proteins partition between two distinct quality control compartments. *Nature*, 2008. **454**(7208): 1088–95.

31. Tyedmers, J., A. Mogk, and B. Bukau. Cellular strategies for controlling protein aggregation. *Nat Rev Mol Cell Biol*, 2010. **11**(11): 777–88.

32. Markossian, K.A. and B.I. Kurganov. Protein folding, misfolding, and aggregation. Formation of inclusion bodies and aggresomes. *Biochemistry (Mosc)*, 2004. **69**(9): 971–84.

33. Johnston, J.A., C.L. Ward, and R.R. Kopito. Aggresomes: A cellular response to misfolded proteins. *J Cell Biol*, 1998. **143**(7): 1883–98.

34. Kopito, R.R. Aggresomes, inclusion bodies and protein aggregation. *Trends Cell Biol*, 2000. **10**(12): 524–30.

35. Corboy, M.J., P.J. Thomas, and W.C. Wigley. Aggresome formation. *Methods Mol Biol*, 2005. **301**: 305–27.

36. Miller, S.B., et al. Compartment-specific aggregases direct distinct nuclear and cytoplasmic aggregate deposition. *EMBO J*, 2015. **34**(6): 778–97.

37. Bagola, K. and T. Sommer. Protein quality control: On IPODs and other JUNQ. *Curr Biol*, 2008. **18**(21): R1019–21.

38. Specht, S., et al. Hsp42 is required for sequestration of protein aggregates into deposition sites in Saccharomyces cerevisiae. *J Cell Biol*, 2011. **195**(4): 617–29.

39. Escusa-Toret, S., W.I. Vonk, and J. Frydman. Spatial sequestration of misfolded proteins by a dynamic chaperone pathway enhances cellular fitness during stress. *Nat Cell Biol*, 2013. **15**(10): 1231–43.

40. Shiber, A., et al. Ubiquitin conjugation triggers misfolded protein sequestration into quality control foci when Hsp70 chaperone levels are limiting. *Mol Biol Cell*, 2013. **24**(13): 2076–87.

41. McNaught, K.S., et al. Aggresome-related biogenesis of Lewy bodies. *Eur J Neurosci*, 2002. **16**(11): 2136–48.

42. Zaarur, N., et al. Proteasome failure promotes positioning of lysosomes around the aggresome via local block of microtubule-dependent transport. *Mol Cell Biol*, 2014. **34**(7): 1336–48.

43. Komatsu, M., et al. Homeostatic levels of p62 control cytoplasmic inclusion body formation in autophagy-deficient mice. *Cell*, 2007. **131**(6): 1149–63.

44. Pankiv, S., et al. p62/SQSTM1 binds directly to Atg8/LC3 to facilitate degradation of ubiquitinated protein aggregates by autophagy. *J Biol Chem*, 2007. **282**(33): 24131–45.

45. Lelouard, H., et al. Transient aggregation of ubiquitinated proteins during dendritic cell maturation. *Nature*, 2002. **417**(6885): 177–82.

46. Lelouard, H., et al. Dendritic cell aggresome-like induced structures are dedicated areas for ubiquitination and storage of newly synthesized defective proteins. *J Cell Biol*, 2004. **164**(5): 667–75.

47. Szeto, J., et al. ALIS are stress-induced protein storage compartments for substrates of the proteasome and autophagy. *Autophagy*, 2006. **2**(3): 189–99.

48. Hara, T., et al. Suppression of basal autophagy in neural cells causes neurodegenerative disease in mice. *Nature*, 2006. **441**(7095): 885–9.

49. Komatsu, M., et al. Essential role for autophagy protein Atg7 in the maintenance of axonal homeostasis and the prevention of axonal degeneration. *Proc Natl Acad Sci U S A*, 2007. **104**(36): 14489–94.

50. Wong, E. and A.M. Cuervo. Autophagy gone awry in neurodegenerative diseases. *Nat Neurosci*, 2010. **13**(7): 805–11.

51. Rubinsztein, D.C., et al. Potential therapeutic applications of autophagy. *Nat Rev Drug Discov*, 2007. **6**(4): 304–12.

52. Kuma, A., et al. The role of autophagy during the early neonatal starvation period. *Nature*, 2004. **432**(7020): 1032–6.

53. Wong, E., et al. Molecular determinants of selective clearance of protein inclusions by autophagy. *Nat Commun*, 2012. **3**: 1240.

54. Olzmann, J.A., et al. Parkin-mediated K63-linked polyubiquitination targets misfolded DJ-1 to aggresomes via binding to HDAC6. *J Cell Biol*, 2007. **178**(6): 1025–38.

55. Kirkin, V., et al. A role for NBR1 in autophagosomal degradation of ubiquitinated substrates. *Mol Cell*, 2009. **33**(4): 505–16.

56. Lim, K.L., V.L. Dawson, and T.M. Dawson. Parkin-mediated lysine 63-linked polyubiquitination: A link to protein inclusions formation in Parkinson's and other conformational diseases? *Neurobiol Aging*, 2006. **27**(4): 524–9.

57. Moscat, J., M.T. Diaz-Meco, and M.W. Wooten. Signal integration and diversification through the p62 scaffold protein. *Trends Biochem Sci*, 2007. **32**(2): 95–100.

58. Burnett, B.G. and R.N. Pittman. The polyglutamine neurodegenerative protein ataxin 3 regulates aggresome formation. *Proc Natl Acad Sci U S A*, 2005. **102**(12): 4330–5.

59. Ouyang, H., et al. Protein aggregates are recruited to aggresome by histone deacetylase 6 via unanchored ubiquitin C termini. *J Biol Chem*, 2012. **287**(4): 2317–27.

60. Watabe, M. and T. Nakaki. Protein kinase CK2 regulates the formation and clearance of aggresomes in response to stress. *J Cell Sci*, 2011. **124**(Pt 9): 1519–32.

61. Du, Y., et al. aPKC phosphorylation of HDAC6 results in increased deacetylation activity. *PLoS One*, 2015. **10**(4): e0123191.

62. Yan, J., et al. SQSTM1/p62 interacts with HDAC6 and regulates deacetylase activity. *PLoS One*, 2013. **8**(9): e76016.

63. Boyault, C., et al. HDAC6-p97/VCP controlled polyubiquitin chain turnover. *EMBO J*, 2006. **25**(14): 3357–66.

64. Fusco, C., et al. The E3-ubiquitin ligase TRIM50 interacts with HDAC6 and p62, and promotes the sequestration and clearance of ubiquitinated proteins into the aggresome. *PLoS One*, 2012. **7**(7): e40440.

65. Kalveram, B., G. Schmidtke, and M. Groettrup. The ubiquitin-like modifier FAT10 interacts with HDAC6 and localizes to aggresomes under proteasome inhibition. *J Cell Sci*, 2008. **121**(Pt 24): 4079–88.

66. Gamerdinger, M., et al. BAG3 mediates chaperone-based aggresome-targeting and selective autophagy of misfolded proteins. *EMBO Rep*, 2011. **12**(2): 149–56.

67. Xu, Z., et al. 14-3-3 protein targets misfolded chaperone-associated proteins to aggresomes. *J Cell Sci*, 2013. **126**(Pt 18): 4173–86.

68. Zhang, X. and S.B. Qian. Chaperone-mediated hierarchical control in targeting misfolded proteins to aggresomes. *Mol Biol Cell*, 2011. **22**(18): 3277–88.

69. Lamark, T., et al. NBR1 and p62 as cargo receptors for selective autophagy of ubiquitinated targets. *Cell Cycle*, 2009. **8**(13): 1986–90.
70. Johansen, T. and T. Lamark. Selective autophagy mediated by autophagic adapter proteins. *Autophagy*, 2011. **7**(3): 279–96.
71. Wooten, M.W., et al. Essential role of sequestosome 1/p62 in regulating accumulation of Lys63-ubiquitinated proteins. *J Biol Chem*, 2008. **283**(11): 6783–9.
72. Walinda, E., et al. Solution structure of the ubiquitin-associated (UBA) domain of human autophagy receptor NBR1 and its interaction with ubiquitin and polyubiquitin. *J Biol Chem*, 2014. **289**(20): 13890–902.
73. Itakura, E. and N. Mizushima. p62 Targeting to the autophagosome formation site requires self-oligomerization but not LC3 binding. *J Cell Biol*, 2011. **192**(1): 17–27.
74. Duran, A., et al. p62 is a key regulator of nutrient sensing in the mTORC1 pathway. *Mol Cell*, 2011. **44**(1): 134–46.
75. Matsumoto, G., et al. Serine 403 phosphorylation of p62/SQSTM1 regulates selective autophagic clearance of ubiquitinated proteins. *Mol Cell*, 2011. **44**(2): 279–89.
76. Nicot, A.S., et al. Phosphorylation of NBR1 by GSK3 modulates protein aggregation. *Autophagy*, 2014. **10**(6): 1036–53.
77. Isakson, P., P. Holland, and A. Simonsen. The role of ALFY in selective autophagy. *Cell Death Differ*, 2013. **20**(1): 12–20.
78. Filimonenko, M., et al. The selective macroautophagic degradation of aggregated proteins requires the PI3P-binding protein Alfy. *Mol Cell*, 2010. **38**(2): 265–79.
79. Rui, Y.N., et al. Huntingtin functions as a scaffold for selective macroautophagy. *Nat Cell Biol*, 2015. **17**(3): 262–75.
80. Lee, J.Y., et al. HDAC6 controls autophagosome maturation essential for ubiquitin-selective quality-control autophagy. *EMBO J*, 2010. **29**(5): 969–80.
81. Caviston, J.P., et al. Huntingtin coordinates the dynein-mediated dynamic positioning of endosomes and lysosomes. *Mol Biol Cell*, 2011. **22**(4): 478–92.
82. Zucchi, P.C. and M. Zick. Membrane fusion catalyzed by a Rab, SNAREs, and SNARE chaperones is accompanied by enhanced permeability to small molecules and by lysis. *Mol Biol Cell*, 2011. **22**(23): 4635–46.
83. Itakura, E., C. Kishi-Itakura, and N. Mizushima. The hairpin-type tail-anchored SNARE syntaxin 17 targets to autophagosomes for fusion with endosomes/lysosomes. *Cell*, 2012. **151**(6): 1256–69.
84. Hao, R., et al. Proteasomes activate aggresome disassembly and clearance by producing unanchored ubiquitin chains. *Mol Cell*, 2013. **51**(6): 819–28.
85. Nanduri, P., et al. Chaperone-mediated 26S proteasome remodeling facilitates free K63 ubiquitin chain production and aggresome clearance. *J Biol Chem*, 2015. **290**(15): 9455–64.
86. Rea, S.L., et al. SQSTM1 mutations—Bridging Paget disease of bone and ALS/FTLD. *Exp Cell Res*, 2014. **325**(1): 27–37.
87. Chen, Y., et al. SQSTM1 mutations in Han Chinese populations with sporadic amyotrophic lateral sclerosis. *Neurobiol Aging*, 2014. **35**(3): 726 e7–9.
88. Fecto, F., et al. SQSTM1 mutations in familial and sporadic amyotrophic lateral sclerosis. *Arch Neurol*, 2011. **68**(11): 1440–6.
89. Martinez-Vicente, M., et al. Cargo recognition failure is responsible for inefficient autophagy in Huntington's disease. *Nat Neurosci*, 2010. **13**(5): 567–76.
90. Wong, E.S., et al. Autophagy-mediated clearance of aggresomes is not a universal phenomenon. *Hum Mol Genet*, 2008. **17**(16): 2570–82.
91. Tanik, S.A., et al. Lewy body-like alpha-synuclein aggregates resist degradation and impair macroautophagy. *J Biol Chem*, 2013. **288**(21): 15194–210.
92. Hernandez-Vargas, R., et al. Synphilin suppresses alpha-synuclein neurotoxicity in a Parkinson's disease Drosophila model. *Genesis*, 2011. **49**(5): 392–402.

93. Smith, W.W., et al. Synphilin-1 attenuates neuronal degeneration in the A53T alpha-synuclein transgenic mouse model. *Hum Mol Genet*, 2010. **19**(11): 2087–98.

94. Engelender, S., et al. Synphilin-1 associates with alpha-synuclein and promotes the formation of cytosolic inclusions. *Nat Genet*, 1999. **22**(1): 110–4.

95. Winslow, A.R., et al. alpha-Synuclein impairs macroautophagy: Implications for Parkinson's disease. *J Cell Biol*, 2010. **190**(6): 1023–37.

96. Aguado, C., et al. Laforin, the most common protein mutated in Lafora disease, regulates autophagy. *Hum Mol Genet*, 2010. **19**(14): 2867–76.

97. Rohn, T.T., et al. Depletion of Beclin-1 due to proteolytic cleavage by caspases in the Alzheimer's disease brain. *Neurobiol Dis*, 2011. **43**(1): 68–78.

98. Futerman, A.H. and G. van Meer. The cell biology of lysosomal storage disorders. *Nat Rev Mol Cell Biol*, 2004. **5**(7): 554–65.

99. Platt, F.M., B. Boland, and A.C. van der Spoel. The cell biology of disease: Lysosomal storage disorders: The cellular impact of lysosomal dysfunction. *J Cell Biol*, 2012. **199**(5): 723–34.

100. Soyombo, A.A., et al. TRP-ML1 regulates lysosomal pH and acidic lysosomal lipid hydrolytic activity. *J Biol Chem*, 2006. **281**(11): 7294–301.

101. Siintola, E., et al. Cathepsin D deficiency underlies congenital human neuronal ceroid-lipofuscinosis. *Brain*, 2006. **129(Pt 6)**: 1438–45.

102. Lee, J.H., et al. Lysosomal proteolysis and autophagy require presenilin 1 and are disrupted by Alzheimer-related PS1 mutations. *Cell*, 2010. **141**(7): 1146–58.

103. Ji, Z.S., et al. Reactivity of apolipoprotein E4 and amyloid beta peptide: Lysosomal stability and neurodegeneration. *J Biol Chem*, 2006. **281**(5): 2683–92.

104. Yamashima, T. Hsp70.1 and related lysosomal factors for necrotic neuronal death. *J Neurochem*, 2012. **120**(4): 477–94.

105. Gomez-Suaga, P., et al. Leucine-rich repeat kinase 2 regulates autophagy through a calcium-dependent pathway involving NAADP. *Hum Mol Genet*, 2012. **21**(3): 511–25.

106. Rubinsztein, D.C., C.F. Bento, and V. Deretic. Therapeutic targeting of autophagy in neurodegenerative and infectious diseases. *J Exp Med*, 2015. **212**(7): 979–90.

107. Harris, H. and D.C. Rubinsztein. Control of autophagy as a therapy for neurodegenerative disease. *Nat Rev Neurol*, 2012. **8**(2): 108–17.

108. Rubinsztein, D.C., P. Codogno, and B. Levine. Autophagy modulation as a potential therapeutic target for diverse diseases. *Nat Rev Drug Discov*, 2012. **11**(9): 709–30.

109. Noda, T. and Y. Ohsumi. Tor, a phosphatidylinositol kinase homologue, controls autophagy in yeast. *J Biol Chem*, 1998. **273**(7): 3963–6.

110. Caccamo, A., et al. Molecular interplay between mammalian target of rapamycin (mTOR), amyloid-beta, and Tau: Effects on cognitive impairments. *J Biol Chem*, 2010. **285**(17): 13107–20.

111. Spilman, P., et al. Inhibition of mTOR by rapamycin abolishes cognitive deficits and reduces amyloid-beta levels in a mouse model of Alzheimer's disease. *PLoS One*, 2010. **5**(4): e9979.

112. Berger, Z., et al. Rapamycin alleviates toxicity of different aggregate-prone proteins. *Hum Mol Genet*, 2006. **15**(3): 433–42.

113. Menzies, F.M., et al. Autophagy induction reduces mutant ataxin-3 levels and toxicity in a mouse model of spinocerebellar ataxia type 3. *Brain*, 2010. **133(Pt 1)**: 93–104.

114. Cortes, C.J., et al. Rapamycin delays disease onset and prevents PrP plaque deposition in a mouse model of Gerstmann-Straussler-Scheinker disease. *J Neurosci*, 2012. **32**(36): 12396–405.

115. Wang, I.F., et al. Autophagy activators rescue and alleviate pathogenesis of a mouse model with proteinopathies of the TAR DNA-binding protein 43. *Proc Natl Acad Sci U S A*, 2012. **109**(37): 15024–9.

116. Sarbassov, D.D., S.M. Ali, and D.M. Sabatini. Growing roles for the mTOR pathway. *Curr Opin Cell Biol*, 2005. **17**(6): 596–603.
117. Sarkar, S. and D.C. Rubinsztein. Inositol and IP3 levels regulate autophagy: Biology and therapeutic speculations. *Autophagy*, 2006. **2**(2): 132–4.
118. Sarkar, S., et al. Lithium induces autophagy by inhibiting inositol monophosphatase. *J Cell Biol*, 2005. **170**(7): 1101–11.
119. Rose, C., et al. Rilmenidine attenuates toxicity of polyglutamine expansions in a mouse model of Huntington's disease. *Hum Mol Genet*, 2010. **19**(11): 2144–53.
120. Feng, H.L., et al. Combined lithium and valproate treatment delays disease onset, reduces neurological deficits and prolongs survival in an amyotrophic lateral sclerosis mouse model. *Neuroscience*, 2008. **155**(3): 567–72.
121. Sarkar, S., et al. Trehalose, a novel mTOR-independent autophagy enhancer, accelerates the clearance of mutant huntingtin and alpha-synuclein. *J Biol Chem*, 2007. **282**(8): 5641–52.
122. Castillo, K., et al. Trehalose delays the progression of amyotrophic lateral sclerosis by enhancing autophagy in motoneurons. *Autophagy*, 2013. **9**(9): 1308–20.
123. He, Q., et al. Treatment with trehalose prevents behavioral and neurochemical deficits produced in an AAV alpha-synuclein rat model of Parkinson's disease. *Mol Neurobiol*, 2015. **53**(4): 2258–68.
124. Lan, D., et al. Proteasome inhibitor-induced autophagy in PC12 cells overexpressing A53T mutant alpha-synuclein. *Mol Med Rep*, 2015. **11**(3): 1655–60.
125. Zhang, X., et al. MTOR-independent, autophagic enhancer trehalose prolongs motor neuron survival and ameliorates the autophagic flux defect in a mouse model of amyotrophic lateral sclerosis. *Autophagy*, 2014. **10**(4): 588–602.
126. Ma, T.C., et al. Metformin therapy in a transgenic mouse model of Huntington's disease. *Neurosci Lett*, 2007. **411**(2): 98–103.
127. Kickstein, E., et al. Biguanide metformin acts on tau phosphorylation via mTOR/protein phosphatase 2A (PP2A) signaling. *Proc Natl Acad Sci U S A*, 2010. **107**(50): 21830–5.
128. Salminen, A., et al. AMP-activated protein kinase: A potential player in Alzheimer's disease. *J Neurochem*, 2011. **118**(4): 460–74.
129. Kim, J., et al. AMPK activation inhibits apoptosis and tau hyperphosphorylation mediated by palmitate in SH-SY5Y cells. *Brain Res*, 2011. **1418**: 42–51.
130. Sarkar, S., et al. A rational mechanism for combination treatment of Huntington's disease using lithium and rapamycin. *Hum Mol Genet*, 2008. **17**(2): 170–8.
131. Nakaso, K., et al. Novel cytoprotective mechanism of anti-parkinsonian drug deprenyl: PI3K and Nrf2-derived induction of antioxidative proteins. *Biochem Biophys Res Commun*, 2006. **339**(3): 915–22.
132. Yang, D.S., et al. Reversal of autophagy dysfunction in the TgCRND8 mouse model of Alzheimer's disease ameliorates amyloid pathologies and memory deficits. *Brain*, 2011. **134**(Pt 1): 258–77.
133. Avrahami, L., et al. Inhibition of glycogen synthase kinase-3 ameliorates beta-amyloid pathology and restores lysosomal acidification and mammalian target of rapamycin activity in the Alzheimer disease mouse model: *In vivo* and *in vitro* studies. *J Biol Chem*, 2013. **288**(2): 1295–306.
134. Trinchese, F., et al. Inhibition of calpains improves memory and synaptic transmission in a mouse model of Alzheimer disease. *J Clin Invest*, 2008. 118(8): 2796–807.
135. Pellom, S.T., Jr. and A. Shanker. Development of proteasome inhibitors as therapeutic drugs. *J Clin Cell Immunol*, 2012. **S5**: 5.
136. Vazquez, D. Inhibitors of protein synthesis. *FEBS Lett*, 1974. 40(0): suppl:S63–84.
137. Eskelinen, E.L., et al. Seeing is believing: The impact of electron microscopy on autophagy research. *Autophagy*, 2011. **7**(9): 935–56.

138. Yla-Anttila, P., et al. Monitoring autophagy by electron microscopy in Mammalian cells. *Methods Enzymol*, 2009. **452**: 143–64.
139. Eskelinen, E.L. Fine structure of the autophagosome. *Methods Mol Biol*, 2008. **445**: 11–28.
140. Eskelinen, E.L. Maturation of autophagic vacuoles in Mammalian cells. *Autophagy*, 2005. **1**(1): 1–10.
141. Marzella, L., J. Ahlberg, and H. Glaumann. Isolation of autophagic vacuoles from rat liver: Morphological and biochemical characterization. *J Cell Biol*, 1982. **93**(1): 144–54.
142. Yamamoto, A., M.L. Cremona, and J.E. Rothman. Autophagy-mediated clearance of huntingtin aggregates triggered by the insulin-signaling pathway. *J Cell Biol*, 2006. **172**(5): 719–31.
143. Xu, G., V. Gonzales, and D.R. Borchelt. Rapid detection of protein aggregates in the brains of Alzheimer patients and transgenic mouse models of amyloidosis. *Alzheimer Dis Assoc Disord*, 2002. **16**(3): 191–5.

10 Chaperone-Mediated Autophagy and Neurodegeneration

Moumita Rakshit and Esther Wong
Nanyang Technological University
Singapore

CONTENTS

10.1 INTRODUCTION: BASICS OF CHAPERONE-MEDIATED AUTOPHAGY

In the early years, autophagy (degradation in the lysosomes) was thought to be a nonselective bulk degradation pathway. However, later experiments demonstrated that some cytosolic proteins upon microinjection into cultured fibroblast cells underwent differential rates of degradation in the lysosomes [1]. Furthermore, these proteins were degraded more rapidly under conditions of serum starvation compared to others [1]. This process of selective degradation of a subset of long-lived soluble cytosolic proteins in the lysosomes was henceforth identified as chaperone-mediated autophagy (CMA).

Molecular characterization of CMA using *in vitro* systems with purified lysosomes, cultured cells, and different organs from rodents led to the identification of selection prerequisite in the substrate proteins and the subset of cytosolic chaperones and lysosomal proteins that govern the selectivity of this pathway. CMA is a multi-step process as shown in Figure 10.1a. The first step involves the recognition of a CMA-targeting motif in the amino acid sequence of substrate protein by cytosolic hsc70 (heat shock-cognate protein of 70 kDa) chaperone which forms a complex with other co-chaperones like heat shock protein 40 (hsp40), hsp90, hsc70-interacting protein (hip), hsp70-hsp90 organizing protein (Hop), and BCL2-associated athanogene 1 (BAG1) [2,3]. The CMA-targeting motif was initially identified in RNase A [4,5], the first documented CMA substrate, whereby a pentapeptide motif "KFERQ" was found to be necessary and sufficient for its rapid catabolism by lysosomes [6,7]. A consensus CMA-targeting motif biochemically related to "KFERQ" found in RNase A was subsequently uncovered in about 30% of cytosolic proteins via sequence analysis [8–10]. Various substrate characterization studies and recent proteomic analysis revealed that these proteins are involved in diverse biological processes such as transcriptional regulation, neuronal survival and function, stress responses, metabolism, immune response, and cytoskeleton and protein degradation/synthesis [8–11]. The CMA motif is degenerated and allows a series of amino acid combinations guided by an invariant glutamine (Q) residue flanked on either side by four amino acid residues consisting of a basic (lysine [K] or arginine [R]), an acidic (glutamic acid [E] or aspartic acid [D]), a bulky hydrophobic (phenylalanine [F], valine [V], leucine [L], or isoleucine [I]), and a repeated basic or bulky hydrophobic amino acid residue [12]. Exposure of CMA motif that may often time be buried in the core of the substrate protein is required for recognition by hsc70 [13]. Therefore, conditions that either enhance or obscure the accessibility of CMA motif will directly impact CMA proteolysis. Additionally, post-translational modifications of an amino acid stretch missing just one of the residues of the CMA motif, such as phosphorylation when the acidic residue is missing or acetylation when a basic residue is lacking, could also give rise to a functional noncanonical CMA tag [12].

Upon recognition by cytosolic hsc70-chaperone complex, the substrate is delivered to the lysosomal surface where it binds to the cytosolic tail of lysosome-associated membrane protein type 2A (LAMP-2A), the receptor responsible for CMA [14]. LAMP-2A acts as a docking and entry point for the substrate into the lysosome. It is one of the three splice variants of the *lamp-2* gene. Binding of the

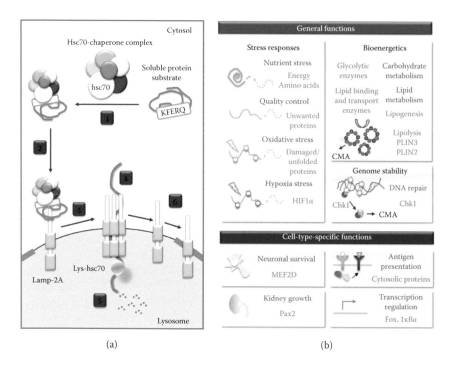

(a)   (b)

FIGURE 10.1 Different steps of CMA and its physiological importance. (a) CMA step by step: (1) Binding of hsc70 and its associated co-chaperones to the CMA-targeting motif in substrate protein, (2) Delivery of substrate-chaperone complex to LAMP-2A, the CMA receptor at lysosomal membrane, (3) Substrate binding to the cytosolic tail of LAMP-2A promotes its multimerization to form an active translocation complex, (4) Unfolding of the substrate protein prior to translocation across the lysosomal membrane assisted by a lysosomal resident form of hsc70 (Lys-hsc70), (5) Degradation by lysosomal proteases, (6) Disassembly of LAMP-2A translocation complex into LAMP-2A monomers to return to basal level. (b) General and cell-type specific functions of CMA.

substrate to the LAMP-2A receptor is the rate-limiting step in CMA and initiates the multimerization of LAMP-2A into an active translocation complex in the lysosomal membrane to allow substrate to transverse from the cytosol into the lysosomal lumen after substrate unfolding (Figure 10.1a) [15]. After CMA degradation, the multimeric LAMP-2A translocation complex disassemblies into its monomers and transits into cholesterol-rich lipid microdomains of the lysosomal membrane for degradation to return to the basal CMA state. Thus, CMA activity is determined by the levels of LAMP-2A at the lysosomal membrane. When CMA is less active, most of the LAMP-2A are concentrated at the lipid microdomains of the lysosomal membrane where LAMP-2A continuously undergoes turnover by a pair of lysosomal proteases, cathepsin A, and a yet unidentified metalloprotease [16,17]. During CMA activation, the level of LAMP-2A at the lysosomal membrane increases. This is achieved by migration of LAMP-2A out of the lipid microdomains upon substrate binding to prevent its degradation. Additionally, the level of LAMP-2A at the lysosomal membrane is also enhanced through transcriptional upregulation of the *lamp-2* gene and mobilization of a pool of full length LAMP-2A residing in the lysosomal lumen to the surface of the lysosomes [16].

LAMP-2A conformation at the lysosomal membrane is highly dynamic. It undergoes continuous cycles of assembly and disassembly to form translocation complex active for CMA in response to substrate availability (Figure 10.2a). This dynamicity is conferred by the chaperones hsc70 and hsp90 [18,19]. The effect of hsc70 on LAMP-2A is substrate dependent. When hsc70 is in substrate-bound form, it promotes the multimerization of the LAMP-2A complex while in the absence of substrate, it predominantly promotes LAMP-2A complex disassembly. Hsp90 drives LAMP-2A assembly by stabilizing LAMP-2A intermediate conformers during the transition from monomeric to multimeric form (Figure 10.2a). The hsp90 responsible for this function is localized at the luminal side of the lysosomal membrane [19,20]. Apart from hsp90 and hsc70, LAMP-2A is stabilized in the active complex form by glial fibrillary acidic protein (GFAP) and elongation factor 1α (EF1α) at the lysosomal surface. Under basal low CMA condition, high level of GTP in the cell activates EF1α and causes it to dissociate from GFAP leading to GFAP self-dimerization and release from LAMP-2A complex. This induces the disassembly and migration of LAMP-2A into the lipid microdomains for degradation [18–20]. Only a subset of lysosomes is competent to perform CMA. The CMA-active lysosomes are those that contain in their lumen a lysosomal variant of hsc70 (Lys-hsc70) which is stable at the low lysosomal pH of 5.1–5.5 [21]. Although the exact mechanism remains unclear, Lys-hsc70 may aid substrate translocation by binding to and actively pulling the substrate inside the lysosomal lumen or help to secure the substrate portion already transversed into the lysosomal lumen to prevent its retraction back to the cytosol (Figure 10.2a). The level of Lys-hsc70 also changes with the activation of CMA where starvation and oxidative stress, the two potent stimuli of CMA, have shown to increase the amount of Lys-hsc70 per lysosome [13]. After the substrate enters the lysosome for degradation by the lysosomal proteases, the LAMP-2A translocation complex rapidly dissociates into monomers where substrates can bind again. Therefore, the rate of CMA can be modulated by the rate of assembly and disassembly of LAMP-2A translocation complex [19].

10.2 REGULATION OF CMA BY mTORC2/PHLPP1/AKT SIGNALING PATHWAY IN THE LYSOSOME

CMA signaling mechanism has remained elusive until recent identification of a lysosomal mTORC2/PHLPP1/Akt signaling pathway that regulates active LAMP-2A complex formation at the lysosomal membrane [22,23]. Arias and coworkers have shown that a pair of kinase and phosphatase, mammalian targets of rapamycin complex 2 (mTORC2) and Pleckstrin homology domain and a leucine-rich repeat protein phosphatase 1 (PHLPP1), regulate the activity of lysosomal Akt to control the dynamics of LAMP-2A assembly and disassembly into CMA translocation complex in CMA-active lysosomes (Figure 10.2b) [22,23]. While mTORC1 is a negative regulator of macroautophagy, mTORC2 acts as an endogenous inhibitor of

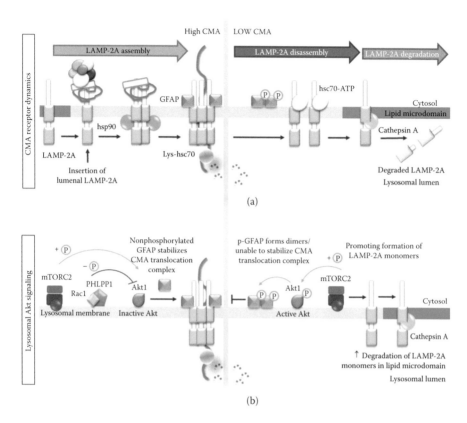

FIGURE 10.2 Regulation of CMA receptor dynamics by phosphorylation state of lysosomal Akt. (a) Dynamics of CMA receptor at the lysosomal membrane. LAMP-2A undergoes continuous cycle of assembly and disassembly upon demand for substrate binding and uptake. Activation of CMA is associated with mobilization of LAMP-2A monomers out of discrete lipid microdomain in the lysosomal membrane to prevent its degradation, resulting in a net increase in LAMP-2A levels at the lysosomal membrane. Increased LAMP-2A levels are also achieved via lysosomal membrane insertion of a pool of full size LAMP-2A residing in lysosomal lumen. Under basal condition, a percentage of LAMP-2A monomers are continuously turned over in the lipid microdomain by cathepsin A and an unidentified metalloprotease. (b) CMA receptor dynamics is regulated by the levels of phosphorylated Akt at the lysosomal membrane. Stress-induced activation of CMA elicits recruitment of phosphatase PHLPP1 and its stabilizing co-factor, GTPase Rac1, to the lysosomes to inactivate Akt via dephosphorylation. Reduced Akt activity leads to decreased phosphorylation of its substrate GFAP thereby enhancing the stability of CMA translocation complex at the lysosomal membrane to promote CMA. Under basal condition, CMA activity is constitutively inhibited by a resident pool of lysosomal-associated mTORC2 that activates lysosomal Akt to phosphorylate GFAP to form dimers. This represses stable assembly of CMA translocation complex, leading to accumulation of LAMP-2A monomers that get degraded in the lipid microdomain.

CMA by phosphorylating lysosomal Akt to negatively modulate CMA translocation complex formation [22,24]. Activated Akt phosphorylates GFAP to promote dimerization of phospho-GFAP thereby abrogating GFAP interaction with LAMP-2A and destabilizes the active translocation complex (Figure 10.2b). The dual kinase inhibitory effects are neutralized during stress conditions that induce CMA through the recruitment of PHLPP1 and GTPase Rac1 to the lysosomal membrane. Rac1 stabilizes PHLPP1 at the lysosomal membrane to allow dephosphorylation of lysosomal Akt to render it inactive. Thus, nonphosphorylated GFAP monomers are able to associate with and stabilize LAMP-2A multimeric complex to upregulate CMA (Figure 10.2b) [22]. The elucidation of this lysosomal mTORC2/PHLPP1/Akt signaling axis that modulates basal CMA activity and its activation by stress offers an exciting avenue to restore CMA in CMA-deficient conditions such as in aging and neurodegeneration.

10.3 PHYSIOLOGICAL IMPORTANCE OF CMA

10.3.1 CMA—A QUALITY CONTROL AND STRESS RESPONSE PATHWAY

Basal CMA activity has been detected in most cell types for maintenance of cellular homeostasis [4,5]. At the basal level, CMA plays an important role in protein quality control. CMA degrades a wide range of substrates ranging from glycolytic enzymes [25], transcription factors [10], inhibitors of transcription factors [9], calcium binding proteins (specifically annexins) [26], lipid-binding proteins (alpha-2-microglobulin) [27], proteins involved in vesicular trafficking [28,29] to even catalytic subunits of the proteasome (Figure 10.1b) [30]. CMA helps to maintain protein homeostasis by degrading proteins that are not required and generate raw materials like amino acids for cellular functioning. Apart from basal activity, CMA is highly induced under different stress conditions. The very first evidence supporting the role of CMA as a stress response pathway came from early studies where prolonged starvation was shown to maximally activate CMA [4]. When the cells are deprived of nutrient supply, CMA is activated to degrade unnecessary proteins to recycle amino acids as raw materials for energy supply [4]. Till date, nutrient starvation is the best demonstrated and most widely used stimulus for activating CMA in animal models and cultured cells [4,5]. In contrast to macroautophagy, which is maximally activated after 4–6 hours of nutrient depletion and only sustains up to 8 hours of nutrient starvation, CMA requires 16–18 hours of starvation to be maximally induced and can stay activated for up to 3 days upon induction [31]. This temporal difference in the activation of CMA during nutrient stress serves a regulatory role in modulating other cellular pathway responses to cope with nutrient scarcity. During initial starvation, bulk removal mechanism of macroautophagy is favored to facilitate rapid mass degradation of cellular components to quickly meet metabolic demand. Under prolonged period of starvation, such bulk self-eating strategy is no longer sustainable. For continuous survival, the cell switches to CMA degradation so as to selectively target only those proteins that are not required for its immediate survival for energy. For instance, a well-known substrate of CMA is glyceraldehyde 3-phosphate dehydrogenase

(GAPDH), which is the principal component of the glycolysis pathway. During starvation, glycolysis is not required and so GAPDH can be degraded to stop glycolysis and also generate raw materials for survival [32,33]. Likewise IkBα is also degraded during starvation by CMA, thereby causing starvation-induced changes in the transcriptome and may affect survivability against nutrient stress [9]. This increase in CMA activity is due to the increase in the levels of lysosomal hsc70 [34] and LAMP-2A [15,35]. The increase in LAMP-2A levels is primarily due to enhanced stability of the LAMP-2A multimers at the lysosomal membrane as well as relocation of the LAMP-2A resident in the lysosomal lumen into the lysosomal membrane [15,35]. There is no change in the rates of synthesis of LAMP-2A detected under starvation [15,35].

Apart from starvation, the next most well characterized stimulus for CMA activation is oxidative stress (Figure 10.1b). It was first observed that in the presence of antioxidant, the rates of degradation of CMA substrates reduced considerably, whereas upon inducing oxidative stress, the same substrates were degraded much more efficiently [9]. It was further revealed that oxidized proteins were found in the lysosomal matrix. Mild oxidative stress induces protein unfolding that helps to uncover hidden CMA-targeting motifs in the unfolded proteins, thereby facilitating increased lysosomal uptake of oxidized proteins to boost CMA activity [9]. This increase in CMA activity during mild oxidative stress is brought about by an increase in the levels of lysosomal hsc70, Hip, and LAMP-2A in the lysosomes [13]. Unlike starvation where LAMP-2A levels are stabilized at the lysosomal membrane, under oxidative stress there is no significant stabilization of LAMP-2A, instead the *lamp-2a* mRNA level increases leading to enhanced *de novo* synthesis of LAMP-2A [13].

Further, it has been observed that CMA activity and LAMP-2A level increases in cells under hypoxia stress [36]. This increase is due to a higher level of *lamp-2a* mRNA synthesis while hsp90 and hsc70 levels remain unchanged. Induction of CMA in hypoxic cells protects the cells against apoptotic cell death. *In vivo* studies also showed increase in neural expression of LAMP-2A after ischemic shock in rat brains, resulting in CMA upregulation and mitigation of ischemic stress [36]. Thus, CMA plays a cellular protective role under hypoxic conditions (Figure 10.1b). Interestingly, hypoxia inducible factor-1α (HIF-1α) is a CMA substrate. HIF-1α forms part of the HIF-1 heterodimer that transcriptionally regulates a repertoire of proteins involved in the adaptive response toward reduced oxygen availability and hypoxic stress. However, the functional relationship between CMA upregulation and the degradation of HIF-1α during hypoxia remains unclear, especially since recent studies observed that excessive degradation of HIF-1α by CMA compromises cellular response to hypoxic stress [37,38].

10.3.2 CMA AND BIOENERGETICS

Apart from being a stress response pathway, CMA also plays an imperative role in maintaining metabolic balance in more than one way (Figure 10.1b). CMA has been implicated in lipid and carbohydrate metabolism pathways [11,39]. The development of the hepatic LAMP-2A knockout (KO) mouse model has shed light on the role of CMA in maintaining bioenergetic balance [11]. Disruption of CMA in the liver of

LAMP-2A KO mouse showed distinct metabolic phenotype with reduced glycogen storage and negative energy balance. The animal showed higher overall dependency on carbohydrate oxidation for energy production with an elevated rate of glycolysis. Many glycolytic and tricarboxylic acid (TCA) cycle enzymes are CMA substrates. Thus, these enzymes accumulate in high levels in LAMP-2A KO liver due to loss of CMA, thereby leading to enhanced rate of glycolysis [11]. CMA is also crucial for lipid metabolism. LAMP-2A KO liver developed severe hepatosteatosis with overt lipid accumulation and reduced oxidation of fatty acids [11]. The study shows that many enzymes involved in lipid binding and transport, and triglyceride synthesis pathway are also CMA substrates. Therefore, CMA failure results in increase in lipid uptake and lipogenesis. Furthermore, CMA also degrades lipid droplet–associated proteins, perilipin 2 (PLIN2) and 3 (PLIN3), and their degradation is necessary for cytosolic lipases and autophagy-related proteins to assess the lipid droplets to carry out lipolysis [40]. Thus, CMA-incompetent liver also suffers from a reduced rate of lipolysis.

CMA metabolic regulation plays a significant role in promoting tumor growth and metastasis. Many cancer cells possess abnormal high basal CMA that has been shown to fuel the Warburg effect, the aerobic glycolysis in cancer cells [41]. Unlike loss of CMA in normal cells, blockade of CMA in cancer cells surprisingly caused marked reduction of glycolytic enzymes like GAPDH, 3-phospho glycerate kinase (3PGK), and aldolase levels. CMA regulates glycolysis in cancer cells via different mechanism from normal cells. In cancer cells, CMA degrades p53 which transcriptionally represses the expression of glycolytic enzymes. Thus, a functional CMA is required to release p53 transcriptional brake on glycolysis to augment Warburg effect to support cancer growth and progression [41]. Another glycolytic enzyme pyruvate kinase M2 (PKM2) is important for cancer cell growth. The end product of PKM2 is pyruvate which can be further reduced to lactate or oxidized to acetyl-coA and fed to the TCA cycle for energy production. PKM2 undergoes acetylation on its lysine K305 residue in cancer cells turning it into a CMA substrate and resulting in its degradation. This leads to accumulation of glycolytic intermediates and enhances the glycolytic flux in cancer cells to support tumor growth [42].

10.3.3 CMA in DNA Damage Response

Besides alleviating proteotoxicity, CMA also protects against genotoxicity (Figure 10.1b). This role of CMA is executed by regulated degradation of phosphorylated checkpoint kinase 1 (Chk1). Chk1 is a *bona fide* CMA substrate [43]. During cell division when there is DNA damage, Chk1 delays cell cycle progression to allow DNA repair. Once repair is done, release of cell cycle arrest is necessary. Chk1 is then phosphorylated and tagged for CMA degradation [44]. CMA-deficient cells showed nuclear accumulation of Chk1, resulting in deregulated cell cycle progression with persistence of DNA damage upon genotoxic stress. CMA blockade also perturbed the stability of early step DNA repair complex, thereby making CMA-deficient cells more vulnerable to etoposide-mediated genotoxicity and genome instability due to inefficient DNA repair [43].

10.3.4 OTHER ROLES OF CMA

CMA also performs specialized functions related to the client substrates it regulates in certain cell types (Figure 10.1b). In neurons, CMA maintains the functional pool of myocyte enhancer factor 2D (MEF2D), a neuronal transcription factor important for neuronal survival, by the timely removal of dysfunctional MEF2D to prevent the damaged transcription proteins from interfering with MEF2D function. It has been reported that high level of inactive MEF2D accumulates in the cytoplasm of the neuronal cells from CMA compromised models [8]. CMA activity has also been shown to contribute to organogenesis and development of the kidney [45]. CMA determines kidney growth by regulating the level of Pax2, another transcription factor that modulates kidney tubular cell proliferation [45,46]. In antigen-presenting cells, CMA participates in antigen presentation by the major histocompatibility complex (MHC) class II molecules [47]. CMA also controls the regulation of transcription factor NFkB by controlling the degradation of its inhibitor IkB which is a CMA substrate [9].

10.4 CROSS TALK BETWEEN CMA AND OTHER DEGRADATIVE PATHWAYS

In order to ensure adequate protein quality control, recycling of raw materials, and maintenance of cellular homeostasis, coordination between the various proteolytic pathways is necessary, particularly in eliciting compensatory responses when one of the degradative pathways fails to work properly. There are increasing evidences showing that the ubiquitin–proteasome system (UPS) and autophagy are functionally coupled processes [48]. Studies show that pharmacological inhibition of autophagy using chloroquine which prevents lysosomal degradation enhances IPS activity [49]. Similarly, abrogation of macroautophagy by knockdown of autophagy-related genes (ATG) such as phosphatidylinositol 3-kinase catalytic subunit type 3 (PIK3C3), ATG5, or ATG7 also enhanced UPS activity although induction of autophagy by rapamycin did not reduce UPS activity [49]. In the fly model, macroautophagy compensates for UPS failure in a histone deacetylase 6 (HDAC6)-dependent manner to control Endoplasmic Reticulum (ER) stress and susceptibility to cell death [50]. UPS also collaborates with CMA to degrade the level of regulator of calcineurin 1 (RCAN1) which is overexpressed in pathological conditions such as Alzheimer's disease (AD) and Down syndrome [51]. Likewise, cross talk exists between UPS and CMA where acute inhibition of proteasome by chemical inhibitors significantly activates CMA [52]. In addition, some of the proteasomal subunits such as 19S regulatory components are *bona fide* substrates of CMA [30]. Enhanced accumulation of 19S regulatory components has been observed in LAMP-2A KO liver and proposed to lead to increased proteasomal activity to compensate for the loss of proteolytic flux through CMA [53].

Extensive communication also exists between macroautophagy and CMA. Apart from sharing the crucial organelle for degradation-lysosome, they also share a number of common substrates. For example, LRRK2 and α-synuclein implicated in the pathology of Parkinson's disease (PD) are degraded by both the lysosomal pathways [54–56]. Furthermore, blockade of macroautophagy either chemically or genetically (via knockout of ATG5) upregulates CMA activity [57]. Under macro-autophagy-deficient condition, an increase in the total number of lysosomes active for CMA was observed [57]. In the absence of ATG5, the fusion of lysosomes with autophagosomes is blocked and as such there is less dissipation of the lysosomal acidic pH. As a result, the lysosomal pH remains more acidic and this property serves to further stabilize the Lys-hsc70 leading to more CMA-competent lysosomes and greater CMA activity [57]. The reverse is also true where ablation of CMA by siRNA interference or KO of LAMP-2A in cells and animals resulted in compensatory activation of macroautophagy [31,53]. In this case, upregulation of macro-autophagy is due to the stabilization of transcription factor EB (TFEB), a master transcriptional regulator of a plethora of ATG and lysosomal genes, in the absence of CMA activity. It turns out that TFEB is also a CMA substrate [53].

10.5 CMA AND AGING

Numerous studies have unequivocally demonstrated that CMA activity declines with age and contributes to the proteotoxicity that characterizes the cells and tissues of aging organisms [58–61]. Further, recent developments support a role of progressive loss of CMA function in precipitating age-associated alterations in organismal metabolism, stress resistance, and immunity [11,39,53,62]. Thus, it is helpful to understand the mechanisms underlying CMA age-dependent demise to derive ways to halt the deterioration of this pathway in order to delay the aging process.

During aging, distinct changes in the terminal lysosomal degradative compartment and the autophagic processes, macroautophagy and CMA, which deliver substrates to the lysosomes have been reported [63–66]. The most evident alteration in the lysosomes is the progressive accumulation of undigested material in the form of yellowish-brown autofluorescent pigment known as lipofuscin [66–68]. Lysosomal accumulation of lipofuscin reduces the ability of lysosomes to fuse and/or degrade the autophagosomal contents delivered via macroautophagy, leading to accumulation of autophagic vacuoles in the cytosol. Besides perturbed removal of autophagosomes, problems in induction of macroautophagy and the biogenesis of autophagosomes have also been observed with age that collectively result in macroautophagic failure [69,70]. In the case of CMA, a decrease in CMA activity with age was first described in the senescent primary fibroblasts [58] and in the lysosomes isolated from different tissues of old rodents [59] that reported a lower translocation and degradation rate of *bona fide* CMA substrates. Analysis of the pool of CMA-active lysosomes did not show overt morphological changes with age nor display visible accumulation of lipofuscin in their lumen [59]. CMA substrate recognition by hsc70-chaperone complex and targeting to the lysosomal membrane, as well as their degradation once inside the lysosomal lumen are also well preserved until late in life [59]. The major factor responsible for CMA decline in older organisms is the dramatic decrease in its capacity to bind and translocate CMA substrates into the lysosomes for degradation [59].

The translocation defect observed in old CMA lysosomes originates from a drastic reduction in the levels of LAMP-2A at the lysosomal membrane to form active CMA translocation complex (Figure 10.3a) [59]. The decline in LAMP-2A levels

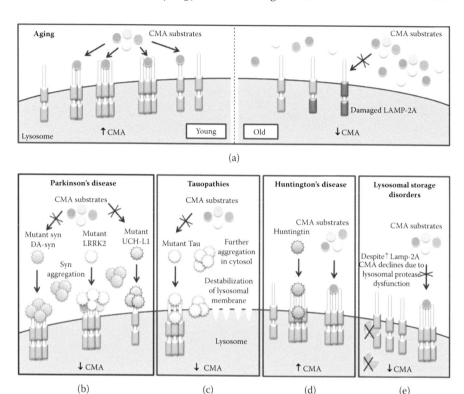

FIGURE 10.3 CMA dysfunction in aging and neurodegeneration. (a) Age-dependent decline in LAMP-2A levels at the lysosomal membrane due to altered degradation contributes to CMA decline in aging. (b) CMA failure in Parkinson's disease (PD). Many PD-related proteins bear CMA-targeting motifs: α-synuclein, LRRK2, and UCH-LI shown here. Mutant and undesirably post-translationally modified forms of these proteins bind abnormally to LAMP-2A, albeit via different mechanisms, leading to blockage of their own degradation as well as degradation of other CMA substrates. CMA failure causes accumulation and aggregation of these toxic proteins that could contribute to Lewy body formation in PD. (c) Perturbation of CMA by mutant tau in tauopathies. Tau protein is a CMA substrate but mutant tau fails to translocate fully into the lysosomal lumen for degradation, leading to tau oligomerization at the lysosome that destabilizes lysosomal membrane to cause lysosomal leakage. Release of lysosomal tau oligomers into the cytosol further promotes tau aggregation. (d) Upregulation of CMA activity has been observed in Huntington's disease (HD) to compensate for failure in other degradative pathways. (e) Compromised CMA in lysosomal storage disorders (LSD) is due to reduced lysosomal degradation as a result of protease dysfunction. Syn: α-synuclein; DA-syn: dopamine modified α-synuclein.

happens prior to the manifestation of age-dependent alterations in CMA activity. This decrease in LAMP-2A levels does not result from age-dependent changes in the transcriptional regulation of LAMP-2A, its splicing patterns, nor its trafficking from Golgi to lysosome [20]. Instead, the compromise in CMA activity is due to age-related changes in the lysosomal membrane that significantly alter the dynamics and stability of LAMP-2A in the lysosomes [20]. Aging results in a diminished ability to mobilize lysosomal lumenal LAMP-2A to the lysosomal membrane through reinsertion when CMA is activated, leading to enhanced accumulation of LAMP-2A in the lysosomal lumen where LAMP-2A undergoes massive unregulated degradation. Additionally, comparative lipodomics studies of lysosomes from young and old mice revealed an age-dependent expansion of the lipid microdomains in the lysosomal membrane [71]. This factor may further destabilize the levels of LAMP-2A at the lysosomal membrane by promoting aberrant LAMP-2A degradation by cathepsin A and metalloprotease associated with the microdomains. The resultant overall deleterious turnovers of LAMP-2A with age significantly deplete the presence of LAMP-2A at the lysosomal membrane to form active translocation complex for uptake of substrates, thereby reducing CMA activity in aging (Figure 10.3a).

10.6 FUNCTIONAL FAILURE OF CMA IN NEURODEGENERATIVE DISEASES

In physiological aging, studies have shown a causative relationship between the gradual loss of CMA functions and the manifestation of aging phenotypes [53,72]. On the contrary, pathological aging is associated with severe CMA failure which has been observed in the pathogenesis of age-related diseases such as neurodegeneration. Similar to the reduced LAMP-2A levels seen in physiological aging, the mechanisms underlying CMA dysfunction in neurodegenerative pathologies also involve different facets of alterations in LAMP-2A levels or functions in the CMA translocation complex (Figure 10.3b–e). Furthermore, many pathogenic proteins associated with neurodegenerative diseases are degraded by CMA and shown to interact aberrantly with CMA components during their degradation leading to CMA blockade [12,61]. Such failure of CMA in the presence of continuous toxic protein production leads to a vicious cycle of increasing CMA impairment and cellular accumulation of pathogenic proteins that are intimately linked to the neuronal demise seen in various neurodegenerative conditions. Here, we briefly look at a few neurodegenerative conditions where CMA alterations have been characterized and how CMA failure contributes to the pathogenesis of these disorders.

10.6.1 PARKINSON'S DISEASE

Parkinson's disease (PD) is the most common age-related neurodegenerative movement disorder characterized by selective loss of dopaminergic neurons in the *substantia nigra* of the midbrain. Perturbed proteostasis plays a central role in PD pathogenesis as exemplified by the presence of intra-neuronal protein inclusions known as Lewy bodies (LBs) in the affected brain regions of PD patients [73]. Notably, a highly unstable presynaptic terminal-enriched protein known as α-synuclein formed the major component of LBs, and altered α-synuclein turnover was seen in all forms of PD. α-synuclein and numerous other PD-linked gene products turned out to be substrates of CMA, thus affirming the importance of CMA in modulating intracellular levels of toxic proteins that need to be tightly controlled. CMA dysfunction has been well characterized in both familial and idiopathic PD. CMA has been shown experimentally to regulate the levels of both wild-type (WT) and mutated α-synuclein and leucine-rich repeat kinase 2 (LRRK2), the two most frequently mutated proteins in familial PD [28,74,75]. During CMA proteolysis, mutant α-synuclein and LRRK2 proteins bind aberrantly to the LAMP-2A translocation complex with abnormally high affinity that prevent their translocation into the lysosomal lumen thereby obstructing the translocation step (Figure 10.3b). This blockage affects not only the translocation and degradation of α-synuclein or LRRK2 mutants but also other CMA substrates [55]. Likewise, I93M ubiquitin C-terminal hydrolase L1 (UCH-L1), another PD-linked mutant, also shows enhanced and aberrant binding to LAMP-2A, resulting in perturbed organization and function of the translocation complex (Figure 10.3b) [76]. Notably, inhibition of CMA by either mutant UCH-L1 or LRRK2 also accentuates intracellular levels of α-synuclein favoring its aggregation. Thus, failed CMA proteolysis of a single toxic protein is enough to potentiate the harmful effects of other CMA substrates to cause severe CMA inhibition and increase the misfolded protein load inside the cell.

Apart from mutations, environmental and cellular stressors, evident in many human pathologies and during the normal aging process, can also negatively impact CMA function by altering CMA substrate properties through unfavorable post-translational modifications (PTM). Dopamine (DA)-induced PTM of WT α-synuclein have been shown to cause similar CMA blockage during its degradation by CMA in idiopathic PD, the major form of PD (Figure 10.3b) [55]. DA-induced CMA inhibition is more prominent in dopaminergic neurons than in non-DA-producing cortical neurons. Importantly, α-synuclein appears to be the primary mediator of DA-induced blockade of CMA as dopaminergic neurons derived from α-synuclein null mice were relatively spared from the inhibitory effects of DA on CMA activity [59]. DA-modified α-synuclein is prone to form protofibrils and oligomers that can assemble into pore-like structures capable of destabilizing membrane. Thus, an accumulation of this modified form of α-synuclein on the lysosomal surface as a result of CMA translocation failure could promote lysosome destabilization and toxic α-synuclein aggregation, resulting in considerable neuronal cell toxicity (Figure 10.3b). Furthermore, dopamineric neurons are particularly vulnerable to oxidative stress due to the oxidative nature of dopamine metabolism. Thus, the loss of CMA-mediated stress response mechanisms will further sensitize DA neurons to oxidative stress and may explain why DA-producing neurons are especially susceptible to degeneration in PD.

Besides toxic interactions between mutant proteins and LAMP-2A translocation complex, overt changes in the levels of CMA components were also seen in the postmortem brain samples from PD patients [54,77]. An increase in LAMP-2A levels was initially observed during the early development of PD in both mouse models and PD patient brain samples. This could be an early attempt to upregulate LAMP-2A to counteract the ongoing CMA insults. In contrast, a drastic reduction in LAMP-2A and hsc70 levels has been observed in dopaminergic neurons of advanced stage PD patients [54,77]. Furthermore, there exists a distinct correlation between deficiency in LAMP-2A and selective vulnerability of the brain regions to α-synuclein aggregation. More recent studies suggest that microRNA deregulation and/or sequence variation in *lamp-2* gene promoter region in PD patients may account for the downregulation of CMA components in some forms of PD [78–80].

10.6.2 ALZHEIMER'S DISEASE

CMA modulates two important proteins, tau and RCAN-1, involved in Alzheimer's disease (AD). Tau is a CMA substrate (Figure 10.3c). It has two CMA-targeting motifs and interacts with hsc70 [81]. Mutant tau also binds to hsc70 and is transported to the lysosomal membrane where it binds to LAMP-2A but does not undergo complete internalization [29]. The partially translocated mutant tau undergoes partial cleavage by cathepsin L inside the lysosomal lumen, leading to generation of pathogenic fragments of tau on the lysosomal membrane surface. These fragments oligomerize at the lysosomal membrane and affect the membrane integrity. The destabilized lysosomal membrane is prompt to leakage and lysis ultimately resulting in the release of the lysosomal toxic tau oligomers into the cytosol to seed further tau aggregation, thereby potentiating tau toxicity (Figure 10.3c). The partially internalized tau also causes CMA translocation blockade to prevent degradation of other CMA substrates further aggravating the disease condition [29]. RCAN-1 is another protein which is overexpressed in the brains of AD patients [82] which results in neuronal cell death [51]. RCAN-1 has two CMA-targeting motifs and is a CMA substrate [82]. Thus, blockade of CMA due to mutant Tau oligomerization can further aggravate the AD condition by promoting the accumulation of RCAN-1.

10.6.3 HUNTINGTON'S DISEASE

Huntington's disease (HD) is yet another age-related neurodegenerative disease which is caused by abnormal expansion of the highly unstable polyglutamine (polyQ) repeat in the N-terminus of the huntingtin (Htt) protein [83]. This causes misfolding of the Htt protein making it more prone to oligomerization and aggregation [84]. The aggregates accumulate in the cytosol triggering toxicity and neurodegeneration. These aggregates cause the failure of both UPS and macroautophagy pathways [85,86]. An overt upregulation of CMA has been detected in HD cells and tissues to compensate for the loss of UPS and MA proteolytic flux which may account for the balanced proteostasis observed in young HD animals before the onset of HD phenotype (Figure 10.3d) [52]. CMA activity is upregulated through the increased expression and stability of LAMP-2A. Additionally, Htt protein also contains a non-canonical CMA-targeting motif, which undergoes post-translational phosphorylation by Ikk to convert it into a functional CMA-targeting motif [87]. Thus, the increased CMA may also help to keep the level of Htt in check. However, the ability of CMA to compensate for the failure of other proteolytic pathways in HD decreases with age due to the age-dependent deterioration that occurs in CMA pathway, thereby contributing in part to the onset of HD pathology with age [52].

10.6.4 LYSOSOMAL STORAGE DISORDERS

Lysosomal storage disorders (LSD) are a family of genetic diseases resulting from alterations in lysosomal enzyme functions that cause dysfunction in lysosomes and massive accumulation of undigested substrates in the lysosomal compartment [33].

CMA failure has been implicated in at least two forms of LSDs, namely galactosialidosis and mucolipidosis type IV (Figure 10.3e). Galactosialidosis is caused by the lack of the lysosomal protease cathepsin A. Cathepsin A, besides acting as a chaperone for other lysosomal proteases, also controls the levels of LAMP-2A by cleaving it at the lipid microdomains [18]. In the absence of this protease, LAMP-2A levels are abnormally high in the lysosomes and result in abnormally high CMA translocation. However, due to the overall reduction in the lysosomal degradative capacity, CMA degradation is overall reduced [18,57,88]. In mucolipidosis type IV, perturbation in lysosomal pH has been proposed to destabilize the lumenal hsc70 chaperone, thereby impacting CMA functions [89].

10.7 THE LAMP-2A-DEFICIENT MODELS

10.7.1 Hepatic LAMP-2A Knockout Model

The development of the liver-specific conditional LAMP-2A KO mice was a significant step in understanding the physiological roles of CMA [11]. These mice showed increased accumulation of cholesterol esters, triglycerides, free fatty acids, and diacylglycerol in the liver. They were also more sensitive to chronic high fat diet, resulting in larger liver size and accumulation of lipid droplets. Thus, the hepatic CMA-deficient mice showed altered lipid metabolism making the liver more susceptible to lipid stress. Additionally, glycogen storage capacity was also reduced in the hepatic LAMP-2A KO animals where overnight fasting caused a significant drop in blood glucose levels. Analysis of the livers showed a shift from carbohydrate synthesis and storage toward carbohydrate hydrolysis and utilization. It was already known that GAPDH, a major glycolytic enzyme, is a *bona fide* CMA substrate [32]. Now in the recent KO studies, 8 out of the 10 glycolytic enzymes were found to contain the CMA-targeting motif. Comparative proteomic analysis of the liver confirmed 30% of the proteins were CMA substrates which include proteins involved in carbohydrate and lipid metabolism. It is interesting to note that the conditional KO of LAMP-2A in the liver not only affected the liver but also caused changes in the peripheral organs. Thus, abrogation of CMA activity in liver not only perturbs liver protein quality control and functions but also compromises the overall metabolic homeostasis and adaptation to changing energy demands of the whole animal as what has been observed in aging. In a way, the hepatic LAMP-2A KO mice model behaves more like the aging models at the metabolic level [11,53]. This promotes the idea that upregulation of CMA can be of therapeutic benefit in delaying aging which will be discussed in the next section.

10.7.2 Brain LAMP-2A Knockdown Model

Currently, whole brain KO model for CMA is not available. However, a recent study demonstrates an obligate role of functional CMA in the proper functioning and maintenance of the dopaminergic neuronal system in the brain [90]. Specifically, downregulation of LAMP-2A level using adeno-associated virus-mediated short hairpin RNA interference in the dopaminergic neurons in rat brain inhibited CMA functions in this class of neurons. This resulted in accumulation of poly-ubiquitinated α-synuclein aggregates in the DA neurons with progressive loss of dopaminergic neurons, reduction in dopamine production, and significant motor deficits as seen in PD pathology [90]. Thus, CMA is essential for the function of DA neurons, and CMA impairment is a major factor underscoring PD pathogenesis.

10.8   METHODOLOGY TO STUDY CMA

CMA is an autophagic pathway found only in mammals and is not found in lower organisms such as yeast, nematode worm, or fish [91]. Consequently, experimental models and techniques used to study CMA activity have been developed in mammals. Basal and stress-induced forms of CMA activity are detectable in a plethora of mammalian cell types, including neurons, astrocytes, hepatocytes, retinal cells, fibroblasts, and different types of immune cells and cancer cells [92]. At the tissue level, CMA activity has been demonstrated through the analysis of lysosomes isolated from organs such as liver, spleen, kidney, and brain [92]. In this section, we describe a series of methods commonly used to study two broad aspects of CMA in cultured cells and tissues: (1) deciphering whether a protein of interest is a CMA substrate and (2) measuring CMA activity in various physiological, pathological, or therapeutic interventions.

10.8.1   PARAMETERS THAT DEFINE *BONA FIDE* CMA SUBSTRATES

An important aspect of studying CMA is to verify whether a protein of interest is a CMA substrate. About 30% of the cytosolic proteins are CMA substrates [7]. There are certain criteria that a protein must possess in order to qualify as a CMA substrate.

1. *The CMA-targeting motif*—the most important criterion for a candidate protein to qualify as a CMA substrate is to possess in its primary sequence a CMA-targeting motif [7]. This pentapeptide motif has to be exposed to the surface of the protein for efficient recognition by the chaperones. Often, the proteins that do not have the CMA-targeting motif can undergo post-translational modifications like acetylation or phosphorylation and be rendered as a CMA target [7,93].
2. *Long-lived cytosolic protein*—CMA substrates usually have long half-lives and their half-lives are dependent on CMA activity. Metabolic labeling of the substrates followed by pulse and chase can determine the half-life of the proteins.
3. *Interaction with cytosolic hsc70 and LAMP-2A*—The mere presence of the CMA-targeting motif is not enough to conclude that a substrate is degraded by CMA. The CMA-targeting motif needs to be recognizable by the cytosolic chaperone hsc70 to direct it to LAMP-2A on the lysosomal membrane for lysosomal translocation [14,35]. Hence, it is essential to determine interaction of the substrate with hsc70 and LAMP-2A which can be studied by co-immunoprecipitation or immunofluorescence colocalization studies.

4. *Entry into the lysosomal lumen*—It is important to demonstrate binding and uptake of the potential candidate into the lysosomal lumen for degradation.

5. *Competition with other known CMA substrates*—Most of the CMA substrates known till date compete with other substrates to enter the lysosomal lumen for degradation. The only known exception is LRRK2 which is an unusual substrate as well as regulator of the pathway. LRRK2 translocation has shown to increase in the presence of other *bona fide* substrates like GAPDH and RNase [93].

10.8.2 Measurement of Key CMA Components

The levels of different components of CMA change with CMA activity. The most important marker is LAMP-2A the level of which level correlates with CMA activity. An increase in the level of Lys-hsc70 may also be indicative of increased CMA activity due to the expansion of the number of CMA-active lysosomes [15,21]. It is thus plausible to make a rough estimate of CMA activity by examining the levels of LAMP-2A and hsc70 in the lysosomes. This can be achieved by isolating total lysosomal fractions for immunoblotting analysis. Further, CMA-active lysosomes can be separated out from the rest of the lysosomal pool and immunoblotted against specific CMA markers (more details in the next section). Immunofluorescence is another technique usually employed to track levels of different CMA components together with co-staining for a lysosomal marker such as LAMP-1.

Apart from the levels of different CMA components, stability of the LAMP-2A multimeric translocation complex is pertinent for forming the active translocation channel to transport substrates into the lysosomal lumen. The stability of the LAMP-2A multimer depends on various factors such as membrane lipid composition [88], phosphorylation state of GFAP [19,22], and concentrations of hsc70 and hsp90 [19]. We can measure the stability of the 700kDa translocation complex using biochemical methods such as native gel electrophoresis, sucrose density gradient centrifugation, or *in vitro* LAMP-2A stability assay. The first two methods rely on the difference in the molecular weights of the monomeric and the multimetic LAMP-2A while the LAMP-2A stability assay measures the stability of the LAMP-2A against lysosomal degradation by lysosomal proteases which degrade the monomeric LAMP-2A.

10.8.3 Purification of CMA-Active Lysosomes

CMA-active lysosomes can be purified from different cultured cells or tissue samples [94]. These lysosomes can then be used to study the various aspects of CMA. The following protocol has been developed by Cuervo et al. (1997). Briefly, the cells or tissues are resuspended and washed two to three times in ice-cold 0.25M sucrose solution. The key point here is to keep everything cold and in 0.25M sucrose solution which makes the cell membranes more fragile due to hyperosmotic shock. The cells are then disrupted in nitrogen cavitation chamber or a manual/

mechanical homogenizer. Once the cells are disrupted, they are subjected to a series of serial fractionations to get the crude mitochondria-lysosome fraction. This fraction can be used directly for analysis or can be further purified to yield CMA-active lysosomes by loading onto a variable metrizamide gradient to separate out the CMA-active and -inactive lysosomes by centrifugation. This pool of pure CMA-active lysosomes can be discriminated from the CMA-active lysosomes by the presence of lumenal hsc70.

10.8.4 MEASUREMENT OF CMA ACTIVITY

1. *Metabolic labeling followed by pulse and chase.*
   Most CMA substrates are long-lived proteins and thus measuring the half-lives of these long-lived proteins can give us an estimate of lysosomal functions [1]. This process relies on two stages, pulse—where radioactively labeled amino acids are incorporated into the proteins synthesized over a fixed period of time. The next stage is chase—the degradation rates of these proteins are tracked by measuring the amount of radioactivity released into the media. Amino acids and small peptides are separated by acid precipitation of the intact proteins in the medium [95].

2. *Binding and uptake assay*
   This is an *in vitro* technique of monitoring CMA activity using isolated lysosomes. It analyzes the lysosomal binding and uptake into the lumen steps in CMA [96]. It is an improvised version of the protease protection assay [32]. In this technique, purified lysosomes are incubated with CMA substrates, for example, GAPDH or α-synuclein, in the presence or absence of protease inhibitors (PI). At the end of the incubation, lysosomes are pelleted along with the associated substrates by centrifugation for immunoblotting of the amount of lysosome-associated substrate protein in the presence and absence of PI. The amount of protein associated with lysosomes upon PI inhibition represents the total pool of substrates bound and the uptake into the lysosomes for degradation. The level in the absence of PI represents only the bound substrate level. Substrate lysosomal uptake rate is calculated by subtracting the levels of bound proteins from the total proteins associated with lysosomes under protease inhibition [32]. It is important to ensure the purity of the lysosomal fraction and the integrity of the lysosomal membrane in this assay [94].

3. *CMA reporter*
   The limitation in studying CMA in all the above methods is the difficulty in monitoring CMA activity in intact cells. The above methods are either *in vitro* assays which require isolation of lysosomes or involves radioactivity. A simple and efficient technique to measure rate of CMA activity *in situ* is via the use of CMA reporter. As previously discussed, the KFERQ motif is necessary and sufficient for a protein to be degraded by CMA [7]. This concept was put to use while developing the CMA reporter. The first 20 amino acids of RNase A containing the KFERQ motif were fused to a monomeric photoswitchable cyan fluorescent protein. The CMA reporter

fusion protein photoconverts from cyan to green upon photoactivation to permit tracking of only the pre-synthesized CMA reporter proteins in a similar manner as the pulse and chase approach. All the fusion protein synthesized after photoconversion has a different excitation and emission spectra, thus allowing easy visualization and tracking of the pre-synthesized CMA reporter protein in association with lysosomes to distinguish from protein synthesis effect [52].

10.9   CONCLUDING REMARKS

CMA performs a plethora of physiological functions and its failure has been implicated in aging as well as in various neurodegenerative diseases. Various studies have shown that the levels of pathogenic proteins like α-synuclein are intimately regulated by CMA activity [55,74,75]. Half-life of endogenous α-synuclein was prominently reduced in LAMP-2A overexpressing cells with concomitant improvement in cell survival rate [56]. Most significantly, enhancing CMA *in vivo* through overexpression of LAMP-2A has shown to ameliorate α-synuclein-induced DA neuron degeneration [97]. In physiological aging, the restoration of CMA activity in mouse liver of aged mice through overexpression of a single copy of LAMP-2A gene significantly delayed the aging phenotypes in the transgenic animals [72], providing proof of principle rationale for targeting CMA pathway as an anti-aging strategy. Thus, restoration of CMA is an important therapeutic approach in delaying aging and mitigation of many neurodegenerative disorders.

Therefore, ways to specifically control CMA activity is highly desirable. Although genetic augmentation of LAMP-2A has proven beneficial in counteracting aging and disease state in animal models, such manipulation may be highly challenging in humans, especially in aged individuals. Recent studies have shown that retinoic acid receptor (RAR) inhibits CMA activity endogenously. Strategy targeted at design of synthetic derivatives of all-trans-retinoic acid to specifically neutralize RAR inhibitory effect on CMA has proven successful [98]. The RAR derivatives are small in size and hydrophobic which make them easier to be transported across the lipid bilayers. The antagonists selectively activate the CMA pathway and have been shown to protect cells from oxidative stress, thus providing a potential therapeutic application [98].

Another approach is the generation of synthetic peptides which are fusion proteins containing the CMA-targeting motif. This approach has been shown useful in overcoming the HD pathology in animal models [99]. In this technique, two polyQ-binding proteins (QBP1) have been fused to two CMA-targeting motifs [99,100]. The QBP1 binds to the expanded polyQ tract of mutant Htt while the CMA-targeting motif directs the QBP1 bound mutant Htt protein toward CMA degradation [99].

ACKNOWLEDGMENTS

Work in our laboratory is supported by grants from MOE Tier 2 M4020161.080 (ARC 25/13), MOE Tier 1 M4011565.080 (RG139/15), and SUG M4080753.080.

REFERENCES

1. Auteri, J.S., et al., Regulation of intracellular protein degradation in IMR-90 human diploid fibroblasts. *J Cell Physiol*, 1983. **115**(2): 167–74.
2. Callahan, M.K., et al., Differential acquisition of antigenic peptides by Hsp70 and Hsc70 under oxidative conditions. *J Biol Chem*, 2002. **277**(37): 33604–9.
3. Agarraberes, F.A. and J.F. Dice, A molecular chaperone complex at the lysosomal membrane is required for protein translocation. *J Cell Sci*, 2001. **114**(Pt 13): 2491–9.
4. Cuervo, A.M., et al., Activation of a selective pathway of lysosomal proteolysis in rat liver by prolonged starvation. *Am J Physiol*, 1995. **269**(5 Pt 1): C1200–8.
5. Dice, J.F., et al., Regulation of catabolism of microinjected ribonuclease A. Identification of residues 7–11 as the essential pentapeptide. *J Biol Chem*, 1986. **261**(15): 6853–9.
6. Backer, J.M., L. Bourret, and J.F. Dice, Regulation of catabolism of microinjected ribonuclease A requires the amino-terminal 20 amino acids. *Proc Natl Acad Sci U S A*, 1983. **80**(8): 2166–70.
7. Dice, J.F., Peptide sequences that target cytosolic proteins for lysosomal proteolysis. *Trends Biochem Sci*, 1990. **15**(8): 305–9.
8. Smith, P.D., et al., Calpain-regulated p35/cdk5 plays a central role in dopaminergic neuron death through modulation of the transcription factor myocyte enhancer factor 2. *J Neurosci*, 2006. **26**(2): 440–7.
9. Cuervo, A.M., et al., IkappaB is a substrate for a selective pathway of lysosomal proteolysis. *Mol Biol Cell*, 1998. **9**(8): 1995–2010.
10. Aniento, F., et al., Selective uptake and degradation of c-Fos and v-Fos by liver lysosomes. *FEBS Lett*, 1996. **390**: 47–52.
11. Schneider, J.L., Y. Suh, and A.M. Cuervo, Deficient chaperone-mediated autophagy in liver leads to metabolic dysregulation. *Cell Metab*, 2014. **20**(3): 417–32.
12. Martinez-Vicente, M. and E. Wong, Chaperone-mediated autophagy and Parkinson's disease, in *Protein chaperones and protection from neurodegenerative diseases*, S.N. Witt, Editor 2011, John Wiley & Sons, Inc.: New Jersey. p. 101–138.
13. Kiffin, R., et al., Activation of chaperone-mediated autophagy during oxidative stress. *Mol Biol Cell*, 2004. **15**: p. 4829–40.
14. Chiang, H.L., et al., A role for a 70-kilodalton heat shock protein in lysosomal degradation of intracellular proteins. *Science*, 1989. **246**(4928): 382–5.
15. Cuervo, A.M. and J.F. Dice, Unique properties of lamp2a compared to other lamp2 isoforms. *J Cell Sci*, 2000. **113**(Pt 24): 4441–50.
16. Cuervo, A.M. and J.F. Dice, Regulation of lamp2a levels in the lysosomal membrane. *Traffic*, 2000. **1**(7): 570–83.
17. Cuervo, A.M., et al., Cathepsin A regulates chaperone-mediated autophagy through cleavage of the lysosomal receptor. *EMBO J*, 2003. **22**(1): 47–59.
18. Bandyopadhyay, U., et al., The chaperone-mediated autophagy receptor organizes in dynamic protein complexes at the lysosomal membrane. *Mol Cell Biol*, 2008. **28**(18): 5747–63.
19. Bandyopadhyay, U., et al., Identification of regulators of chaperone-mediated autophagy. *Mol Cell*, 2010. **39**(4): 535–47.
20. Kiffin, R., et al., Altered dynamics of the lysosomal receptor for chaperone-mediated autophagy with age. *J Cell Sci*, 2007. **120**(Pt 5): 782–91.
21. Agarraberes, F.A., S.R. Terlecky, and J.F. Dice, An intralysosomal hsp70 is required for a selective pathway of lysosomal protein degradation. *J Cell Biol*, 1997. **137**(4): 825–34.
22. Arias, E., et al., Lysosomal mTORC2/PHLPP1/Akt regulate chaperone-mediated autophagy. *Mol Cell*, 2015. **59**(2): 270–84.
23. Arias, E., Lysosomal mTORC2/PHLPP1/Akt axis: A new point of control of chaperone-mediated autophagy. *Oncotarget*, 2015. **6**(34): 35147–8.

24. Pattingre, S., et al., Regulation of macroautophagy by mTOR and Beclin 1 complexes. *Biochimie*, 2008. **90**(2): 313–23.
25. Cuervo, A.M., et al., Selective binding and uptake of ribonuclease A and glyceraldehyde-3-phosphate dehydrogenase by isolated rat liver lysosomes. *J Biol Chem*, 1994. **269**(42): 26374–80.
26. Cuervo, A.M., et al., Selective degradation of annexins by chaperone-mediated autophagy. *J Biol Chem*, 2000. **275**(43): 33329–35.
27. Cuervo, A.M., et al., Direct lysosomal uptake of alpha 2-microglobulin contributes to chemically induced nephropathy. *Kidney Int*, 1999. **55**(2): 529–45.
28. Vogiatzi, T., et al., Wild type alpha-synuclein is degraded by chaperone-mediated autophagy and macroautophagy in neuronal cells. *J Biol Chem*, 2008. **283**(35): 23542–56.
29. Wang, Y., et al., Tau fragmentation, aggregation and clearance: The dual role of lysosomal processing. *Hum Mol Genet*, 2009. **18**(21): 4153–70.
30. Cuervo, A.M., et al., Degradation of proteasomes by lysosomes in rat liver. *Eur J Biochem*, 1995. **227**(3): 792–800.
31. Massey, A.C., et al., Consequences of the selective blockage of chaperone-mediated autophagy. *Proc Natl Acad Sci U S A*, 2006. **103**(15): 5805–10.
32. Aniento, F., et al., Uptake and degradation of glyceraldehyde-3-phosphate dehydrogenase by rat liver lysosomes. *J Biol Chem*, 1993. **268**(14): 10463–70.
33. Bejarano, E. and A.M. Cuervo, Chaperone-mediated autophagy. *Proc Am Thorac Soc*, 2010. **7**(1): 29–39.
34. Cuervo, A.M., J.F. Dice, and E. Knecht, A population of rat liver lysosomes responsible for the selective uptake and degradation of cytosolic proteins. *J Biol Chem*, 1997. **272**(9): 5606–15.
35. Cuervo, A.M. and J.F. Dice, A receptor for the selective uptake and degradation of proteins by lysosomes. *Science*, 1996. **273**(5274): 501–3.
36. Dohi, E., et al., Hypoxic stress activates chaperone-mediated autophagy and modulates neuronal cell survival. *Neurochem Int*, 2012. **60**(4): 431–42.
37. Hubbi, M.E., et al., Chaperone-mediated autophagy targets hypoxia-inducible factor-1alpha (HIF-1alpha) for lysosomal degradation. *J Biol Chem*, 2013. **288**(15): 10703–14.
38. Ferreira, J.V., et al., STUB1/CHIP is required for HIF1A degradation by chaperone-mediated autophagy. *Autophagy*, 2013. **9**(9): 1349–66.
39. Tasset, I. and A.M. Cuervo, Role of chaperone-mediated autophagy in metabolism. *FEBS J*, 2016. **283**(13): 2403–13.
40. Kaushik, S. and A.M. Cuervo, Degradation of lipid droplet-associated proteins by chaperone-mediated autophagy facilitates lipolysis. *Nat Cell Biol*, 2015. **17**(6): 759–70.
41. Kon, M., et al., Chaperone-mediated autophagy is required for tumor growth. *Sci Transl Med*, 2011. **3**(109): 109ra117.
42. Lv, L., et al., Acetylation targets the M2 isoform of pyruvate kinase for degradation through chaperone-mediated autophagy and promotes tumor growth. *Mol Cell*, 2011. **42**(6): 719–30.
43. Park, C., Y. Suh, and A.M. Cuervo, Regulated degradation of Chk1 by chaperone-mediated autophagy in response to DNA damage. *Nat Commun*, 2015. **6**: 6823.
44. Zhang, Y.W., et al., Genotoxic stress targets human Chk1 for degradation by the ubiquitin-proteasome pathway. *Mol Cell*, 2005. **19**(5): 607–18.
45. Franch, H.A., et al., A mechanism regulating proteolysis of specific proteins during renal tubular cell growth. *J Biol Chem*, 2001. **276**(22): 19126–31.
46. Franch, H.A., Chaperone-mediated autophagy in the kidney: The road more traveled. *Semin Nephrol*, 2014. **34**(1): 72–83.
47. Zhou, D., et al., Lamp-2a facilitates MHC class II presentation of cytoplasmic antigens. *Immunity*, 2005. **22**(5): 571–81.

48. Wong, E. and A.M. Cuervo, Integration of clearance mechanisms: The proteasome and autophagy. *Cold Spring Harb Perspect Biol*, 2010. **2**(12): a006734.
49. Wang, X.J., et al., A novel crosstalk between two major protein degradation systems: Regulation of proteasomal activity by autophagy. *Autophagy*, 2013. **9**(10): 1500–8.
50. Pandey, U.B., et al., HDAC6 rescues neurodegeneration and provides an essential link between autophagy and the UPS. *Nature*, 2007. **447**(7146): 859–63.
51. Liu, H., et al., Degradation of regulator of calcineurin 1 (RCAN1) is mediated by both chaperone-mediated autophagy and ubiquitin proteasome pathways. *FASEB J*, 2009. **23**(10): 3383–92.
52. Koga, H., et al., A photoconvertible fluorescent reporter to track chaperone-mediated autophagy. *Nat Commun*, 2011. **2**: 386.
53. Schneider, J.L., et al., Loss of hepatic chaperone-mediated autophagy accelerates proteostasis failure in aging. *Aging Cell*, 2015. **14**(2): 249–64.
54. Orenstein, S.J., et al., Interplay of LRRK2 with chaperone-mediated autophagy. *Nat Neurosci*, 2013. **16**(4): 394–406.
55. Martinez-Vicente, M., et al., Dopamine-modified alpha-synuclein blocks chaperone-mediated autophagy. *J Clin Invest*, 2008. **118**(2): 777–88.
56. Xilouri, M., et al., Abberant alpha-synuclein confers toxicity to neurons in part through inhibition of chaperone-mediated autophagy. *PLoS One*, 2009. **4**(5): e5515.
57. Kaushik, S., et al., Constitutive activation of chaperone-mediated autophagy in cells with impaired macroautophagy. *Mol Biol Cell*, 2008. **19**(5): 2179–92.
58. Dice, J.F., Altered degradation of proteins microinjected into senescent human fibroblasts. *J Biol Chem*, 1982. **257**: 14624–7.
59. Cuervo, A.M. and J.F. Dice, Age-related decline in chaperone-mediated autophagy. *J Biol Chem*, 2000. **275**: 31505–13.
60. Cuervo, A.M., Autophagy and aging: Keeping that old broom working. *Trends Genet*, 2008. **24**(12): 604–12.
61. Cuervo, A.M. and E. Wong, Chaperone-mediated autophagy: Roles in disease and aging. *Cell Res*, 2014. **24**: 92–104.
62. Cuervo, A.M. and F. Macian, Autophagy and the immune function in aging. *Curr Opin Immunol*, 2014. **0**: 97–104.
63. Martinez-Vicente, M., G. Sovak, and A.M. Cuervo, Protein degradation and aging. *Exp Gerontol*, 2005. **40**(8–9): 622–33.
64. Rajawat, Y.S., Z. Hilioti, and I. Bossis, Aging: Central role for autophagy and the lysosomal degradative system. *Ageing Res Rev*, 2009. **8**(3): 199–213.
65. Terman, A., B. Gustafsson, and U.T. Brunk, Autophagy, organelles and ageing. *J Pathol*, 2007. **211**(2): 134–43.
66. Brunk, U.T. and A. Terman, Lipofuscin: Mechanisms of age-related accumulation and influence on cell function. *Free Radic Biol Med*, 2002. **33**(5): 611–9.
67. Cuervo, A.M., et al., Autophagy and aging: The importance of maintaining "clean" cells. *Autophagy*, 2005. **1**(3): 131–40.
68. Cuervo, A.M. and J.F. Dice, When lysosomes get old. *Exp Gerontol*, 2000. **35**(2): 119–31.
69. Donati, A., et al., Age-related changes in the autophagic proteolysis of rat isolated liver cells: Effects of antiaging dietary restrictions. *J Gerontol A Biol Sci Med Sci*, 2001. **56**(9): B375–83.
70. Wong, E. and A.M. Cuervo, Neuronal autophagy gone awry: Many fixings for the autophagic wrong-doing. *Nat Neurosci*, 2010. **13**: 805–11.
71. Rodriguez-Navarro, J.A., et al., Inhibitory effect of dietary lipids on chaperone-mediated autophagy. *Proc Natl Acad Sci U S A*, 2012. **109**(12): E705–14.
72. Zhang, C. and A.M. Cuervo, Restoration of chaperone-mediated autophagy in aging liver improves cellular maintenance and hepatic function. *Nat Med*, 2008. **14**(9): 959–65.

73. Lees, A.J., The Parkinson chimera. *Neurology*, 2009. **72**(7 Suppl): S2–11.
74. Cuervo, A.M., et al., Impaired degradation of mutant alpha-synuclein by chaperone-mediated autophagy. *Science*, 2004. **305**(5688): 1292–5.
75. Mak, S.K., et al., Lysosomal degradation of alpha-synuclein in vivo. *J Biol Chem*, 2010. **285**(18): 13621–9.
76. Kabuta, T., et al., Aberrant interaction between Parkinson disease-associated mutant UCH-L1 and the lysosomal receptor for chaperone-mediated autophagy. *J Biol Chem*, 2008. **283**(35): 23731–8.
77. Alvarez-Erviti, L., et al., Chaperone-mediated autophagy markers in Parkinson disease brains. *Arch Neurol*, 2010. **67**(12): 1464–72.
78. Alvarez-Erviti, L., et al., Influence of microRNA deregulation on chaperone-mediated autophagy and alpha-synuclein pathology in Parkinson's disease. *Cell Death Dis*, 2013. **4**: e545.
79. Pang, S., et al., Genetic analysis of the LAMP-2 gene promoter in patients with sporadic Parkinson's disease. *Neurosci Lett*, 2012. **526**(1): 63–7.
80. Rothaug, M., et al., LAMP-2 deficiency leads to hippocampal dysfunction but normal clearance of neuronal substrates of chaperone-mediated autophagy in a mouse model for Danon disease. *Acta Neuropathol Commun*, 2015. **3**: 6.
81. Petrucelli, L., et al., CHIP and Hsp70 regulate tau ubiquitination, degradation and aggregation. *Hum Mol Genet*, 2004. **13**(7): 703–14.
82. Harris, C.D., G. Ermak, and K.J. Davies, RCAN1-1L is overexpressed in neurons of Alzheimer's disease patients. *FEBS J*, 2007. **274**(7): 1715–24.
83. Zoghbi, H.Y. and H.T. Orr, Glutamine repeats and neurodegeneration. *Annu Rev Neurosci*, 2000. **23**: 217–47.
84. Wong, E.S., et al., Autophagy-mediated clearance of aggresomes is not a universal phenomenon. *Hum Mol Genet*, 2008. **17**(16): 2570–82.
85. Martinez-Vicente, M., et al., Cargo recognition failure is responsible for inefficient autophagy in Huntington's disease. *Nat Neurosci*, 2010. **13**(5): 567–76.
86. Hipp, M.S., et al., Indirect inhibition of 26S proteasome activity in a cellular model of Huntington's disease. *J Cell Biol*, 2012. **196**(5): 573–87.
87. Thompson, L.M., et al., IKK phosphorylates Huntingtin and targets it for degradation by the proteasome and lysosome. *J Cell Biol*, 2009. **187**(7): 1083–99.
88. Kaushik, S., A.C. Massey, and A.M. Cuervo, Lysosome membrane lipid microdomains: Novel regulators of chaperone-mediated autophagy. *EMBO J*, 2006. **25**(17): 3921–33.
89. Cuervo, A.M., Chaperone-mediated autophagy: Selectivity pays off. *Trends Endocrinol Metab*, 2010. **21**: 142–50.
90. Xilouri, M., et al., Impairment of chaperone-mediated autophagy induces dopaminergic neurodegeneration in rats. *Autophagy*, 2016: **12**(11): 2230–47.
91. Gough, N.R., C.L. Hatem, and D.M. Fambrough, The family of LAMP-2 proteins arises by alternative splicing from a single gene: Characterization of the avian LAMP-2 gene and identification of mammalian homologs of LAMP-2b and LAMP-2c. *DNA Cell Biol*, 1995. **14**(10): 863–7.
92. Patel, B. and A.M. Cuervo, Methods to study chaperone-mediated autophagy. *Methods*, 2015. **75**: 133–40.
93. Orenstein, S.J. and A.M. Cuervo, Chaperone-mediated autophagy: Molecular mechanisms and physiological relevance. *Semin Cell Dev Biol*, 2010. **21**(7): 719–26.
94. Storrie, B. and E.A. Madden, Isolation of subcellular organelles. *Methods Enzymol*, 1990. **182**: 203–25.
95. Kaushik, S. and A.M. Cuervo, Methods to monitor chaperone-mediated autophagy. *Methods Enzymol*, 2009. **452**: 297–324.
96. Salvador, N., et al., Import of a cytosolic protein into lysosomes by chaperone-mediated autophagy depends on its folding state. *J Biol Chem*, 2000. **275**(35): 27447–56.

97. Xilouri, M., et al., Boosting chaperone-mediated autophagy in vivo mitigates alpha-synuclein-induced neurodegeneration. *Brain*, 2013. **136**(Pt 7): 2130–46.

98. Anguiano, J., et al., Chemical modulation of chaperone-mediated autophagy by retinoic acid derivatives. *Nat Chem Biol*, 2013. **9**(6): 374–82.

99. Bauer, P.O., et al., Harnessing chaperone-mediated autophagy for the selective degradation of mutant huntingtin protein. *Nat Biotechnol*, 2010. **28**(3): 256–63.

100. Nagai, Y., et al., Induction of molecular chaperones as a therapeutic strategy for the polyglutamine diseases. *Curr Pharm Biotechnol*, 2010. **11**(2): 188–97.

11 Autophagy in Synaptic Structure and Function

Guomei Tang, Sheng-Han Kuo, and David Sulzer
Columbia University Medical Center
New York, NY

CONTENTS

11.1 METHODOLOGY

The major function of the neuronal synapse is neurotransmission. This is classically measured by electrophysiology or electrochemistry in neuronal cultures, brain slices, and *in vivo* live animal brain. Investigators also study the morphological and physiological features of axonal presynaptic and dendritic postsynaptic sites in these model systems using fluorescent labeling, immunohistochemistry, and electron microscopy. The study of neuronal autophagy is challenging because traditional biochemical approaches used in heterologous cell systems are poorly adapted for neurons, particularly in synapses which constitute a relatively small fraction of the total cellular material. Recently, live imaging of the biogenesis and turnover of fluorescent-labeled autophagosomes has been adapted for the study of autophagy flux in neurons and neurites, in combination with electrophysiological, neuroanatomical methods, and genetic rodent models, that are deficient for essential autophagy genes. These technologies promise further advances in autophagy research in synapses *in vitro* and *in vivo*.

11.2 INTRODUCTION

The human brain contains billions of neurons and more than 100 trillion (10^{14}) synaptic connections that form its neural circuits. Neuroscientists have long been interested in how this complex synaptic network is built during development and how it is remodeled during learning and disease. Recent advances suggest that appropriate levels of autophagy are necessary for neuronal survival and function. Dysregulation of autophagy has been implicated in synaptic dysfunction and neuronal cell death in acute and chronic neurological disorders[1,2,3] including stroke, Parkinson's disease (PD), Alzheimer's disease (AD), Huntington's disease (HD), motor neuron disorders, prion disease, lysosomal storage disorders, and in neurodevelopmental diseases.[4–6] Elucidation of how autophagy regulates synaptic development and function will offer insight into underlying mechanisms of disease pathogenesis and provide new targets for the treatment of these disorders. Here, we describe recent findings that illustrate roles for autophagy in the formation, function, plasticity, and elimination of synapses in the nervous system.

11.3 NEURONS AND SYNAPSES

The form and function of neurons are extremely variable, ranging from photoreceptors that are activated by light or its absence and release neurotransmitter at the end of a short axon from a single release site, to midbrain dopamine neurons in which axonal length in a mouse can approach a meter and release neurotransmitter from hundreds of thousands boutons, sometimes independently of cell body activation, in patterns that diffuse over perhaps millions of synapses. A "typical" neuron is a highly polarized postmitotic cell composed of a soma (cell body), often a complex pattern of dendrites and a single, long, and often highly branched axon. Communication between neurons is achieved by neurotransmission at specialized junctions known as synapses.[7] The nerve impulse or action potential is generated near the cell body and travels to the presynaptic axonal terminals. Electrochemical signals are classically triggered by release from synapses of small molecule neurotransmitters, including glutamate, γ-aminobutyric acid (GABA), acetylcholine, dopamine, and serotonin (Figure 11.1). These neurotransmitters are typically produced by the presynaptic neuron and stored in synaptic vesicles at presynaptic boutons. At classical synapses, synaptic vesicles make contact with a thickening of the presynaptic plasma membrane, known as the active zone, where vesicle fusion and exocytosis of neurotransmitters occur. In the central nervous system (CNS), axonal boutons typically possess many synaptic vesicles, each containing thousands of neurotransmitter molecules. Classically, neurotransmitters are released upon neuronal activity from presynaptic boutons into the synaptic cleft between neurons, and bind to neurotransmitter receptors in the dendrites of the receptive postsynaptic neuron. These events result in changes in the permeability of the postsynaptic cell membrane to specific ions, opening up receptors that express ion channels which gate calcium,

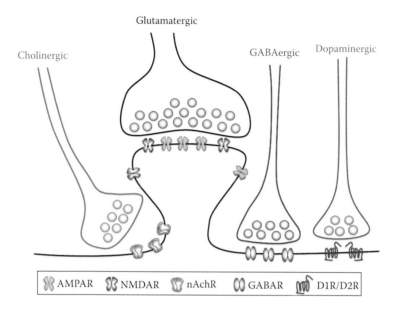

FIGURE 11.1 Schematic representation of a striatopallidal medial spiny neuron synapses with glutamatergic, GABAergic, dopaminergic, and cholinergic synaptic inputs. AMPAR, α-amino-3-hydroxy-5-methyl-4-isoxazolepropionic acid (AMPA) receptor; NMDAR, N-methyl-D-aspartate (NMDA) receptor; nAchR, nicotinic acetylcholine (nAch) receptor; D1R/D2R, D1, or D2-type dopamine receptor.

sodium, potassium, or chloride, or bind to modulate the activity of G protein–coupled receptors that affect second messenger systems. Some neurotransmitters tend to excite (e.g., glutamate) or inhibit (e.g., GABA) postsynaptic neurons, while some (e.g., dopamine) modulate the function of excitatory and inhibitory synapses.

11.4 POSTNATAL SYNAPSE DEVELOPMENT

Postnatal synaptic development in mammalian brain, including humans as discovered by the late Peter Huttenlocher,[7,8] is characterized by an initial overproduction of synapses followed by a large-scale net synaptic pruning that extends from childhood through adolescence (Figure 11.2). During early stages of postnatal development, synapse formation exceeds elimination, resulting in an excess of excitatory synapses, a process that may be essential for the assembly of neural circuits. Excessive or inappropriate synapses are subsequently eliminated in late childhood and adolescent brain.[7,8] The synaptic pruning process provides the precise selection and maturation of functional synaptic connections and neural circuits and is hypothesized to be required for forms of learning and memory.[9,10] Synaptic pruning during early postnatal development appears to occur in many brain regions, including the cerebral cortex, cerebellum, olfactory bulb, and hippocampus.[11] Sensory experience, learning, and neuronal activity can stabilize or eliminate excitatory synapses and sculpt mature neuronal circuits.[12–14] The cellular mechanisms that underlie activity

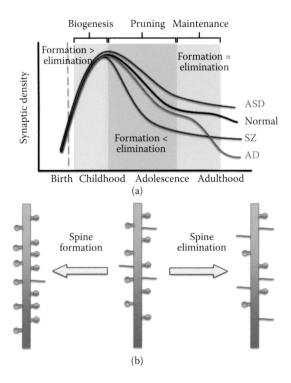

(a)

(b)

FIGURE 11.2 Model of postnatal synapse development in mammalian brain. (a) Dendritic spine synapse density in normal subjects (black), in autism (ASD, red), in schizophrenia (SZ, Blue), and in Alzheimer's disease (AD, brown) patients. (b) Schematic representation of dendritic spine synapse formation and elimination. Figure modified from Penzes et al. 2011.[15]

or experience-dependent synapse elimination in the CNS remain a focus of much investigation, with roles of autophagy and phagocytosis by astrocytes or inflammatory cells implicated in various paradigms.

The majority of morphological studies on synaptic development have measured dendritic spines, the postsynaptic structural component of excitatory synapses. Consistent with the changes in synapses during development, the number of dendritic spines increases before and after birth, and decreases during childhood and adolescence to adult levels.[15] Dendritic spine structural plasticity is closely coordinated with synaptic function and plasticity, with spine enlargement often associated with long-term potentiation (LTP) of the synapse, and spine shrinkage has been associated with long-term depression (LTD)[16] (Figure 11.2).

Abnormalities in dendritic spine shape, size, or number have been identified in many psychiatric and neurologic diseases, including mental retardation, schizophrenia, PD, autistic spectrum disorders (ASD), AD, obsessive–compulsive disorders, and addiction.[17] Changes in dendritic spines in neurodevelopmental disorders presumably result from the disruptions of normal spine pruning events during neural development and circuit maturation.[18–21] Insufficient spine pruning in the cortex, cerebellum, and limbic system is correlated with increases in white matter and has been linked to epilepsy and ASD, while excessive cortical neurites and spine pruning during puberty are correlated with early-onset schizophrenia. In neurodegenerative disorders, for example, AD, synaptic loss is to date the best pathologic correlate of cognitive decline, and synaptic dysfunction is evident long before synapses and neurons are lost.[17] The synapse thus constitutes an important target for treatments to slow disease progression and preserve cognitive and functional abilities in these disorders. It is not clear whether an abnormal spine number or morphology initiates symptom cascades, is a secondary effect, or is a compensatory response. Whatever the etiology, synaptic alterations are prominent pathology features of these disorders and rescue of pathophysiological synaptic changes may provide opportunities for therapeutic intervention.

11.5 NEURONAL AUTOPHAGY AND SYNAPSE DEVELOPMENT

Recent evidence implicates roles for proteolysis in synaptic plasticity and synapse elimination. In hippocampal neuronal cultures, neuronal activity decreases synapse number in part through activation of the myocyte enhancer factor 2 (MEF2) family of transcription factors,[22–24] which triggers synapse elimination by promoting ubiquitin–proteasome system (UPS)-dependent degradation of the postsynaptic scaffolding protein, PSD95.[25] In addition to the UPS, which is responsible for the degradation of many short-lived cytosolic proteins, neurons utilize lysosomal-dependent degradation mechanism for turnover of long-lived synaptic proteins and damaged organelles not degraded by UPS. At least three forms of protein degradation via the lysosome have been identified: macroautophagy, microautophagy, and chaperone-mediated autophagy (CMA).[26,27] These forms differ in physiological function and in the cargoes they deliver to the lysosome.

Macroautophagy (often referred to as autophagy) is the primary mechanism responsible for the turnover of organelles, further provides a major means to degrade cytoplasmic proteins and lipids, and is an important regulator of cellular energy homeostasis. Basal autophagy is important for protecting neurons, particularly from the toxic effects of misfolded proteins, as murine neuronal knockouts of two key autophagy genes result in neurodegeneration with protein aggregate accumulation.[28,29] The autophagic process involves the sequestration of cytoplasmic material, including organelles, to form a nascent autophagosome. The nascent autophagosomes rapidly fuse with pre-existing lysosomes that contain lysosomal enzymes to form degradative autophagosomes. Although autophagy was previously considered an "in-bulk" process, overwhelming evidence now supports selectivity in the sequestration of autophagic cargo. Recognition of certain post-translational modifications, mostly polyubiquitination, by molecules that bind both cargo and components of the autophagic machinery mediates this selectivity. p62 is the best characterized cargo-recognition molecule, and binds preferentially to a particular ubiquitin linkage (Lys63) on the surface of protein aggregates. This assists in autophagosome formation and sequestration of protein aggregates through interaction with LC3.[30] Cargo recognition by p62 is not only limited to protein aggregates but also includes polyubiquitinated proteins and organelles.

Previous studies *in vivo* and *in vitro*, using the autophagosome marker green fluorescent protein (GFP) / microtubule-associated protein 1 light chain 3 (LC3), showed that autophagosomes are scarce in healthy neurons under nutrient-rich conditions.[31] However, inhibiting lysosomal degradation under nutrient-rich conditions causes rapid accumulation of autophagosomes in primary cortical neurons, indicating the presence of constitutive autophagy in neurons.[32] Under basal conditions, autophagosomes appear to be constitutively generated in the axon terminals or boutons, likely often fuse with endosomes to form "amphisomes," and are trafficked on microtubules toward the cell body where they *fuse with lysosomes*. Biogenesis of autophagosomes clearly can occur not only in distal axons but also in axonal boutons. At axonal terminals, the double FYVE domain-containing protein 1 (DFCP1)- positive subdomain of the membranes from the endoplasmic reticulum may provide the source of autophagosome membrane, whereas plasma- or mitochondrial-derived membranes appear to not be incorporated into nascent autophagosomes.[32] While studies have shown that constitutive autophagy is critical for neuronal survival, and autophagy deficiency may cause pathological and behavioral changes associated with many neurodegenerative diseases,[33,34] the detailed relationship between autophagy and synaptic physiology remains incompletely understood. It is becoming increasingly appreciated that the removal of misfolded proteins, protein aggregates, and damaged organelles is of crucial importance for proper synaptic function.

In support of autophagy as a regulatory mechanism of synaptic structure and function, various autophagic proteins are found at synapses,[35] including LC3 and a functionally similar protein, GABA receptor–associated protein (GABARAP). GABARAP regulates aspects of GABA receptor degradation in an invertebrate model during neuronal development.[36] Autophagy proteins further interact with both UPS[37,38] and the endocytic/lysosomal system,[39,40] suggesting an interaction between autophagy and canonical proteasome and endo-lysosomal degradation systems at synapses that are responsible for the degradation and sorting of synaptic vesicles and neurotransmitter receptors.[41]

The few extant reports on autophagy and synaptic structure and function indicate powerful regulation, but divergent effects. Autophagy can potentiate synapse growth at the Drosophila neuromuscular junction,[42] as driving overexpression of the autophagy gene *Atg1* in neurons increased synapse size, while reducing autophagy reduces synapse size. These studies suggest that autophagy regulates synapse growth both at the transcriptional level by activation of c-Jun N terminal kinase (JNK)/Activator protein -1 (AP-1) and locally by regulating levels of synaptic adhesion effectors such as extracellular-signal-regulated kinase (ERK)/fasciclin II. In contrast, recent studies in mouse brain[43] indicate that chronic autophagy deficiency increases the size of dopaminergic synaptic terminal profiles and striatal dopaminergic innervation, consistent with additional studies that implicate autophagy in retraction of neuronal processes[44] and neurite growth during development.[45] Thus, role of autophagy in synaptic morphology and size may be complex.

A role for autophagy in mammalian synaptic development is revealed by a recent report that demonstrates autophagy is essential for synapse pruning during postnatal development in cerebral cortex.[6] Autophagy remodels synapse maturation downstream of mTOR, and autophagy deficiency in response to overactive mTOR contributes to synaptic alterations, for example, increased synaptic density. Dysregulated mTOR signaling has been identified in multiple neurodevelopmental disorders including autism, fragile X syndrome, tuberous sclerosis (TSC), neurofibro-matosis, and phosphatase and tension homolog (PTEN)-mediated macroencephaly,[46] each of which features altered dendritic spine densities. Consistently, Tang et al. reported a higher spine density in basal dendrites of layer V pyramidal neurons in ASD patients than controls. The increased spine density was associated with a defect in net postnatal spine pruning that was correlated with hyperactivated mTOR and impaired autophagy. Using *Tsc1/2*-mutant mice in which mTOR activity is consti-tutively high, the authors found that impaired autophagy is responsible for autism-like synaptic pathology. Pharmacological inhibition of mTOR activity normalized ASD-like spine pruning deficits and ASD-like behaviors in *Tsc1/2* mice, largely by activating neuronal autophagy.

Synapse development in cultured primary hippocampal neurons follows a similar trajectory as that *in vivo*, as it is characterized by initial synapse overgrowth and subse-quent synapse elimination: The number of synapses increased 6–10 days *in vitro* (DIV), peaked at DIV 14–21, and then decreased at DIV 21–28. Using this *in vitro* model sys-tem, Tang et al.[6] (2014) found that autophagy regulates spine elimination but not spine formation during the developmental period when dendritic spines are pruned. Silencing the autophagy gene *Atg7* on DIV 14 increased the spine density at DIV 19–20. The rate of synapse formation was equivalent to control neurons during this developmental period, while autophagy deficiency suppressed synapse elimination without affecting synapse formation. The regulatory effect of constitutive autophagy on synapse elimi-nation was confirmed *in vivo* in an *Atg7*-deficient mouse line, in which *Atg7* is deleted on postnatal days 23 and 24 in forebrain excitatory neurons, a time when net dendritic spine synapse elimination begins and synapse formation rate drops dramatically.

11.6 NEURONAL AUTOPHAGY AND SYNAPTIC DEGENERATION

Pathological increases in autophagosomes and lysosomes have been reported in PD, AD, HD, prion disease, and acute brain injuries.[47] Increased autophagosomes observed in these disease states could reflect impaired autophagosome clearance, as suggested for AD and lysosomal storage disorders.[1] The number of autophago-somes correlates with the degree of synaptic and neurite degeneration, a prominent feature of neurodegenerative diseases,[48] such as the Lewy neurites in PD/Lewy body diseases,[49] dystrophic neurites in AD,[50] huntingtin protein in cortical neurites,[51] and spongiform features in prion diseases.[52]

In AD, dystrophic axons are particularly abundant in the hippocampal fiber systems originating from the subiculum, CA1 hippocampus, and the entorhinal cortex.[53–56] These abnormalities in axonal and presynaptic structures represent an early manifes-tation of axonal damage preceding synaptic and neuronal loss, and could significantly contribute to AD pathology in the preclinical stages. Synapse loss in the neocor-tex and the hippocampus provides the best pathological correlate of early cognitive decline in AD. Autophagosomes accumulate within dystrophic axons and synapses in the brains of humans with AD and AD mouse models,[53] suggesting that defects in the autophagy may contribute to AD synaptic pathology. Interestingly, autopha-gic compartments also participate in Amyloid precursor protein (APP) processing and Aβ peptide production,[57] indicative of a causal relationship between autophagy, plaque formation, and neuritic dystrophy. Remarkably, restoring autophagy ameliorates disease progression and cognition deficits in an AD mouse model,[58] indicating a potential therapeutic value of autophagy induction in early stages of the disease to provide neuronal function recovery.

Multiple genetic risk factors associated with PD have been associated with different forms of autophagy, each of which can be implicated in the synaptic dysfunction in PD. For example, α-synuclein is degraded by both autophagy and CMA,[59,60] and CMA can be disrupted by A53T-mutant α-synuclein, mutant G2019S or R1441C-mutant leucine-rich repeat kinase 2 (LRRK2), and mutant I93M ubiquitin C-terminal hydrolase L1 (UCH-L1).[61] In addition, Parkin and Pink can be relocalized to mitochondria following mitochondrial damage, which promotes mitochondrial autophagy (mitophagy).[62]

Mutant LRRK2 disrupts the morphology and likely the function of the neuritic/synaptic compartments that precede neuronal cell death.[63,64] These effects can be mediated through mutant LRRK2 effects on autophagic neurite catabolism or via secondary autophagic responses to neurite injury. Suppressing autophagy via siRNA knockdown of LC3 or *Atg7* reverses the neurite injury and retraction induced by the overexpression of mutant LRRK2, suggesting that autophagy may underlie LRRK2 mutation–mediated neurite injury.[64,65]

Parkin is an E3 ubiquitin ligase expressed in neurons that impacts synaptic function via the UPS.[66] In the postsynaptic compartment, parkin regulates ubiquitination of postsynaptic protein endophilin-A, and thereby its endocytosis.[67] In addition to catalyzing the formation of K48 polyubiquitin chains that target substrates for proteasomal degradation, parkin targets substrates for autophagy via K63-linked polyubiquitin tagging.[68] It is possible that loss of parkin function affects synapses by impairing autophagic proteolysis of synaptic proteins and mitochondria.

The discovery of α-synuclein as a component of Lewy bodies in postmortem PD human brain has promulgated research into the role of α-synuclein aggregation in PD pathogenesis. Endogenous α-synuclein resides in presynaptic terminals and is involved in synaptic vesicle docking, fusion/release, and the dynamics of the reserve pool of synaptic vesicles and synaptic release.[69,70] α-Synuclein in its native form is partly degraded by CMA,[60] while mutant α-synuclein and dopamine-modified α-synuclein block the lysosomal/CMA degradation pathway. Abnormal α-synuclein inhibits not only CMA but also autophagy through RAB1A and omegasome formation.[71]

11.7 AUTOPHAGY IN GLUTAMATE RECEPTOR– MEDIATED NEUROEXCITOTOXICITY

Excitotoxicity, a phenomenon induced by overactivation of excitatory amino acid (glutamate) receptors, is the major mediator of neuronal death in cerebral ischemia and traumatic brain injury (TBI). An induction of autophagy has been demonstrated in several acute excitotoxicity model systems, including organotypic hippocampal slices treated with *N*-methyl-D-aspartate (NMDA),[72] mice injected with kainate directly into the brains,[73] and in cerebral ischemia mouse models.[74]

The role of autophagy in excitotoxic neuronal death was originally explored in the *Lurcher (Lc)*-mutant mice, a chronic excitotoxicity model in which a mutant glutamate receptor causes continuous ion flow and kills cerebellar Purkinje cells (PCs). PCs start to degenerate by postnatal day 8 (P8), and most die within the first month. Constitutive activation of the GluRδ2Lc triggered a rapid and robust accumulation of autophagosomes in dystrophic axonal swellings and terminals.[75] It was suggested that autophagy in *Lc* PCs could be directly activated by an interaction between the postsynaptic GluRδ2Lc, the postsynaptic density/Drosophila disc tumor suppressor/zonula occludes 1 (PDZ)-containing protein nPIST, and the highly conserved autophagy protein Beclin 1. The constitutive activation of the autophagy pathway may contribute to PC death in *Lc* mice.

It remains uncertain how activation of local postsynaptic signaling pathway in dendrites initiates autophagosome formation in axon compartments. Nishiyama et al.[76] (2010) found that the continuous ion flux through GluRδ2Lc channels decreases intracellular ATP levels, leading to the activation of AMP-activated protein kinase (AMPK) and consequently autophagy. In this study, however, induced autophagy appeared not to cause PC death, but rather, PCs died from ATP depletion with necrotic volume increase. Autophagy may possess a homeostatic protective role to maintain intracellular ATP in cells suffering from ion overload.

Autophagy activation is also implicated in NMDA receptor (NMDAR) hyperactivation-medicated neuronal death in TBI,[77] through the protein interactions among Beclin-1, the NR2B receptor, and the synaptic scaffolding proteins postsynaptic density protein (PSD) 95, Shank, and Homer within membrane raft microdomains in synapses. Acute brain injury induced a rapid recruitment of NR2B into membrane rafts but translocation of Beclin-1 out from the raft microdomains. The release of Beclin-1 or PSD95 Shank-Homer-Beclin-1 from the complex may be a critical event required for inducing autophagy in response to excessive stimulation of NR2B following acute brain injury. Wang et al.[78] used $50\,\mu M$ NMDA to induce acute injury in hippocampal neurons and elicit autophagy-dependent cell death. NMDA suppressed the phosphorylation of the serine/threonine kinase AKT and mTOR, which precedes the increased expression of Beclin-1 and the ratio of LC3-II/LC3-I. These data suggest that NMDA may induce autophagy by inhibiting the phosphatidylinositol 3-kinase (PI3K)-AKT-mTOR pathway. IGF-1 pretreatment can effectively suppress autophagy induced by NMDA via the inhibition of PI3K-AKT-mTOR pathway, thus protecting against the early phase of excitotoxicity.

11.8 NEURONAL AUTOPHAGY AND SYNAPTIC PLASTICITY

Initial evidence for the role of autophagy in synaptic plasticity comes from the study on mTOR regulated autophagy and dopamine synaptic transmission by Hernandez et al.[43] To investigate the possible contribution of autophagy on neurotransmission, the authors generated mice that lack *Atg7* specifically in dopamine neurons by crossing *Atg7flox/flox* mice with mice expressing Cre recombinase under the control of the dopamine transporter (DAT) gene promoter, producing *Atg7-Cre-DAT* conditional knockout mice. In the acute striatal slice preparation, dopamine axons are severed from their cell bodies but continue to synthesize, release, and reaccumulate neurotransmitter for up to 10 hours. In this model system, electrochemical recordings of evoked dopamine release and reuptake in the striatum can provide a unique means to measure dopaminergic neurotransmission with millisecond resolution that is independent of postsynaptic response.

At many synapses, when stimulated with two action potentials in rapid succession, the second postsynaptic response is larger than the first, a phenomenon termed "paired-pulse facilitation." Compared to controls, striatal slices from *Atg7DAT-CreCKO* mice demonstrated increased evoked dopamine release and faster presynaptic recovery following paired-pulse stimulation. Both the initial and subsequent pulses showed increased amplitudes relative to control mice, indicating a higher release probability. Rapamycin depressed evoked dopamine release in control mice but had no effect on *Atg7DAT-CreCKO* mice, suggesting that the enhanced release and recovery in the mutant mice were autophagy-dependent. mTOR inhibition by rapamycin in control mice induced autophagosome-like structures in axons and decreased synaptic vesicles to nearly the same level as the accompanying decrease in evoked dopamine release. In contrast, rapamycin had little or no effect on the number of synaptic vesicles or neurotransmitter release in autophagy-deficient neurons. Because these presynaptic effects were observed in dopaminergic presynaptic terminals in slices without their cell bodies, the induction of autophagy by rapamycin must have occurred locally in axons that typically lack mature lysosomes.[43] Autophagosomes may be synthesized locally in the synapses to sequester presynaptic components, such as synaptic vesicles, and to modulate presynaptic function. Both basal and acutely induced autophagy modulate presynaptic structure and function. Mice with chronic autophagy deficiency in dopamine neurons had abnormally large dopaminergic axonal profiles, released greater levels of dopamine in response to stimulation, and exhibited more rapid presynaptic recovery.

The role of autophagy in regulating synaptic plasticity was corroborated in study of neurotoxicity associated with nanoparticles known as quantum dots (QDs). Streptavidin-coated CdSe/ZnS QDs (QD525) induce the formation of autophagosomes in hippocampal neurons *in vivo*, accompanied by hippocampal CA1 synaptic dysfunction.[79,80] The PI3K-dependent autophagy inhibitors, wortmannin and 3-MA, can significantly reduce autophagic flux, ameliorate synaptic dysfunction by QDs *in vivo*, including the suppressed basal synaptic activities, depress *LTP*, and reduce synaptic density. These findings implicate induced autophagy in QDs-induced impairment of synaptic plasticity in the hippocampal CA1 area *in vivo*. These findings suggest that autophagy-blocking reagents may serve as a therapy for synaptic impairment.

11.9   NEURONAL AUTOPHAGY AND TURNOVER OF SYNAPTIC COMPONENTS

11.9.1   Synpatic Vesicles

While presynaptic autophagy is involved in synaptic transmission, the studies in the previous section have not addressed how particular presynaptic organelles, such as synaptic vesicles, may be specifically targeted by axonal autophagy. Synaptic vesicles undergo multiple rounds of recycling in the synapse via endocytosis and vesicle reformation.[81] It has been thought since the 1970s that newly reformed synaptic vesicles could either be actively reused as functional synaptic vesicles or be redirected to lysosomes as the final destination for degradation.[82]

Autophagy initiated in presynaptic boutons may act as a route to direct synaptic vesicle pools toward lysosomal degradation[83] in a manner distinct from the classical endosomal route.[84,85] Supporting this notion, several autophagy proteins have been found around the clusters of vesicles, suggesting the autophagic degradation of synaptic and secretory vesicles.[84] This process may involve selective sequestration of vesicle clusters into structures resembling early autophagosomes. The sequestration process is triggered by Rab26, a member of the Rab-GTPase superfamily related to the exocytotic Rab3/Rab27 subgroup.[86] Rab26 is enriched in large clusters of synaptic vesicles to which components of the pre-autophagosomal machinery, for example, autophagy proteins Atg16L1, LC3, and Rab33B, are recruited. Consistently, Rab26 was not found on all synaptic vesicles but was rather confined to vesicle aggregates that may be functionally impaired. LC3 was recruited to these dysfunctional vesicle clusters, which indicates the formation of autophagosomal membrane.[83] Atg16L1 and Rab33B-positive autophagosome structures appear to be filled almost exclusively with synaptic vesicles and proteins typically associated with large dense-core vesicles.[83] Therefore, Rab26 may provide a direct link between synaptic vesicles and the core autophagy machinery that initiates a selective autophagy pathway to regulate presynaptic turnover.

Synaptic vesicle clusters containing Rab26 and Atg16L1 can undergo exo-endocytotic cycling.[83] Intriguingly, clathrin has recently been shown to interact with Atg16L1, and targeting plasma membrane constituents toward autophagosome precursors via clathrin-mediated endocytosis.[87] Since clathrin-mediated endocytosis constitutes a main endocytotic pathway for synaptic vesicles, it is conceivable that Rab26- and clathrin-mediated autophagy and endocytosis may act in concert to target synaptic vesicles to pre-autophagosomal structures.

11.9.2 Synaptic Modulators

Neuronal activity modulates the strength and the number of synapses during development and adulthood. Neurons coordinate activity-dependent processes throughout the cell via the activation of gene transcription. The influx of calcium into neurons in response to synaptic activity leads to the activation of transcription factors. Members of the MEF2 family (MEF2A to MEF2D) of transcription factors are tightly regulated by several distinct calcium signaling pathways, and negatively regulate excitatory synapse number in an activity-dependent manner during synaptic development *in vitro* in neuronal cultures[23] and *in vivo* in cerebellar cortex.[24] Calcineurin, a calcium- and calmodulin-regulated phosphatase, dephosphorylates and activates MEF2, which in turn promotes the remodeling of synapses by inducing a program of gene transcription that may involve *arc* and *synGAP*. Recent evidence implicates MEF2 as important *in vivo* for such processes including synapse elimination, homeostatic control of synapse number, and mGluR5-dependent LTD. Activation of metabotropic glutamate receptor 5 (mGluR5) on the dendrites, but not cell soma, of hippocampal CA1 neurons is required for MEF2-induced functional and structural synapse elimination.[88] Activity-regulated cytoskeletal-associated protein (Arc) is required for MEF2-induced synapse elimination, where it plays an acute, cell-autonomous, and postsynaptic role.

Recent evidence supports a role for CMA in the selective degradation of MEF2D. This process involves binding of heat shock protein Hsc70 to substrate proteins via a KFERQ-like motif and their subsequent targeting to lysosomes via the lysosomal membrane receptor Lamp2a. Yang et al.[89] showed that MEF2D can be translocated from the nucleus to the cytoplasm under basal condition, where it interacts with CMA regulator Hsc70 via its N-terminal domain to be delivered to lysosomes for degradation. Inhibition of CMA causes accumulation of inactive MEF2D in the cytoplasm. Consistently, MEF2D levels were found to increase in the brains of α-synuclein transgenic mice and PD patients. Overexpression of wild-type and PD-related mutant α-synuclein both disrupted the MEF2D-Hsc70 binding and resulted in neuronal death. While CMA was suggested to modulate neuronal survival machinery by regulating MEF2D degradation, it is also conceivable that degradation of MEF2s by CMA plays diverse roles in neuronal development and synaptic plasticity.

11.9.3 Postsynaptic Components

11.9.3.1 AMPA Receptor

Glutamate receptors play central roles in synaptic plasticity. At many glutamatergic synapses, NMDAR activation triggers a signaling cascade in the postsynaptic density that induces recruitment of α-amino-3-hydroxy-5-methyl-4-isoxazolepropionic acid (AMPA) receptors into the postsynaptic membrane, leading to LTP and enhanced synaptic strength. A weaker and prolonged activation of NMDARs can lead to removal of postsynaptic AMPA receptors (AMPARs) and LTD. It is of key importance that the trafficking of synaptic AMPARs is controlled carefully to modify synaptic strength during plasticity.

Shehata et al.[90] (2012) investigated whether autophagy is a component of neuronal activity–regulated proteolysis that plays a role in synaptic plasticity. In primary cultured hippocampal neurons, using an NMDAR-dependent chemical LTD protocol in which neurons are challenged by brief low-dose NMDA, they identified increased autophagy following chemical LTD. The combined use of chemical LTD and the autophagy inhibitor bafilomycin, which blocks the vacuolar proton pump and inhibits auotophagosome-lysosomal fusion, appeared to increase autophagosome formation. Inhibiting the formation of autophagosomes by wortmannin completely blocked chemical LTD–induced autophagy. Chemical LTD induces autophagy by activating protein phosphatases, the negative upstream regulator of mTOR that contributes to synaptic plasticity and memory through the regulation of local translation.[91] It is thus plausible that mTOR provides a switch between protein synthesis and degradation in the context of synaptic plasticity and memory.

An increase in autophagosomes in dendritic shafts and spines after chemical LTD may account for the involvement of autophagy in the degradation of AMPARs. After LTD stimulation, AMPARs are internalized by endocytosis, after which they can be either redirected back to the synapse via recycling endosomes or accumulated by lysosomes for final degradation.[92,93] Two major subtypes of AMPARs are present in adult hippocampus: GluR1/2 and GluR2/3 heteromeric receptors, with GluR1 subunit implicated in the endocytosis of AMPARs. The time course of GluR1 degradation after chemical LTD coincides with the maximum increase in the LC3-II/LC3-I ratio. Inhibiting autophagy via knockdown of *Atg7* partially recovered GluR1 levels after chemical LTD, suggesting that chemical LTD–induced autophagy contributes to AMPAR degradation. Treatment with bafilomycin allowed detection of a considerable number of autophagosomes in the dendritic shafts of pyramidal neurons even in the absence of neuronal stimulation, suggesting that autophagosomes are present in dendritic shafts even under controlled conditions, but that their rapid fusion to lysosomes hinders their detection.

A mechanism for autophagic degradation of AMPARs in dendritic shafts and spines may involve a change in endosome recycling. The formation of more amphisomes due to the fusion of endosomes with autophagosome would reduce the recycling endosome population, and direct more AMPAR-containing endosomes to autophagy and lysosomal degradation. Autophagy may thus contribute to the NMDAR-dependent synaptic plasticity required to maintain LTD. To confirm this physiological role, however, requires further study in brain slice or *in vivo*.

11.9.3.2 GABA Receptor

Synaptic clustering of GABA$_A$ receptors (GABAARs) is important for the function of inhibitory synapses and influences the balance of excitation and inhibition in the brain. Presynaptic terminals are known to induce GABAAR clustering during synaptogenesis, but the mechanisms of cluster formation and maintenance are unknown. Rowland et al.[94] studied how presynaptic neurons direct the formation of clusters, by examining GABAAR localization in postsynaptic cells that fail to receive presynaptic contacts in *Caenorhabditis elegans*. Body-wall muscles on the dorsal side of the *C. elegans* normally receive synaptic inputs from both inhibitory GABA motor neurons and excitatory cholinergic motor neurons, and axon pathfinding can be genetically manipulated so that presynaptic contacts to body-wall muscles are selectively or completely eliminated. Selective loss of GABA inputs caused GABAARs to be diffusely distributed at or near the muscle cell surface, while selective loss of acetylcholine innervation showed no effect on GABAAR localization, suggesting that the direct contact with GABA presynaptic terminals induces GABAAR clustering. Autophagosomes were five-fold higher in non-innervated than innervated muscles, suggesting that lack of GABA innervation induces autophagy.

The mechanism of GABAAR trafficking to autophagosomes awaits elucidation. The autophagy pathway appears to reduce GABAAR surface expression, while presynaptic contacts block GABAAR trafficking from the cell surface to the autophagosomes.[94] Under identical conditions, GABAARs traffic to autophagosomes but acetylcholine receptors in the same cell do not, suggesting that autophagy selectively downregulates surface GABAARs.

An important implication of these studies is that autophagy provides a mechanism to control the balance of neuronal excitation and inhibition by regulating the strength of inhibitory and excitatory synapses. At inhibitory synapses, GABAARs are continually internalized by endocytosis; when the GABAARs recycle to the plasma membrane, synapse strength is maintained and when they traffic into a degradative pathway, synapse strength is reduced.[95] Thus, GABAAR surface levels are decreased via autophagy, and GABAARs in autophagosomes gradually disappear when the influx of new receptors from the plasma membrane is blocked. Importantly, autophagy at some synapses is selective for GABAARs and not acetylcholine receptors, and so can differentially modulate inhibitory and excitatory responses.

Autophagy proteins might further be involved in GABAAR export to the plasma membrane. GABARAP, as discussed above, is a homolog of LC3 and the yeast autophagy protein Atg8, and undergoes C-terminal lipidation, allowing association with autophagosome membrane.[96] GABARAP has been shown to facilitate trafficking of GABAARs from the Golgi apparatus to plasma membrane.[97] The involvement of autophagy proteins in both the biosynthesis and degradation of GABAARs raises the possibility that GABAAR trafficking and autophagy utilize overlapping biochemical mechanisms.

11.9.3.3   Nicotinic Acetylcholine Receptor

The nicotinic acetylcholine receptor (nAChR) is a pentameric ligand-gated transmembrane ion channel present in CNS synapses and in the postsynaptic membrane of the nerve-muscle synapse, the neuromuscular junction, where it mediates nerve-induced muscle contraction. Following its arrival via secretory vesicle fusion at the membrane, nAChR can either persist or be endocytosed and recycled back to the membrane or can be degraded in lysosomes.[98,99]

nAChR turnover is regulated by atrophic muscle stress conditions, such as immobilization and denervation, yet little is known about the mechanisms driving its turnover. The E3 ubiquitin ligase TRIM63 has been proposed as a major regulator of ubiquitin-proteasomal degradation of nAChR.[100] Endogenous TRIM63 is highly enriched in the perisynaptic region and interacts with CHRN. A direct interaction between TRIM63 and p62 suggests selective autophagy of nAChRvia TRIM63.[100] Notably, TRIM63 protein is more abundant upon total fasting but not upon amino acid deprivation, which primarily blocks protein synthesis through the mTOR complex 1 (mTORC1) via amino acid–sensing mechanisms. The full starvation protocol significantly increases the amount of endocytic vesicles containing nAChR per NMJ, which are largely positive for TRIM63 and autophagic/lysosomal markers including LC3, p62 and lysosomal associated membrane protein 1 (Lamp1), consistent with autophagy of nAChR.

nAChR is targeted to the LC3-II-positive autophagosomes in a process of selective autophagy that is regulated and mediated by TRIM63, SH3GLB1, and p62/SQSTM1.[98] Among these, SH3GLB1 protein is involved in the regulation of ATG9 trafficking and mediates fission of autophagic donor membranes to trigger starvation-induced autophagy, while TRIM63 plays a crucial role in atrophy-induced endocytic retrieval of nAChR and its subsequent autophagic processing. It is possible that SH3GLB1 is important in initial endocytic nAChR vesicle formation and then escorts the receptor through further steps. p62 may be involved in nAChR vesicle formation and serves in progressing these carriers to the autophagosome. nAChR turnover appears to be regulated only upon atrophy by TRIM63, while basal turnover might be controlled by other E3 ligases.

11.10   CONCLUSIONS

Autophagy provides an evolutionarily conserved catabolic process to degrade long-lived proteins and dysfunctional or superfluous organelles in the lysosomes of eukaryotic cells. It has been implicated in a wide range of physiological and pathological processes, including responses to nutrient deprivation, development, intracellular clearance, suppression of tumor formation, aging, cell death and survival, and immunity.[101] Early studies reported a lack of obvious autophagosomes in healthy neurons under nutrient-rich conditions. This observation corroborated a long-held belief that autophagy was absent in neurons, partly because neurons are quite resistant under typical starvation conditions, for example, postnatal neurons can be cultured for weeks in serum-free medium.[102] More recently, autophagy was found to occur constitutively in neurons under physiological conditions.[32,103] The autophagic machinery is so efficient in neurons that the turnover of neuronal autophagosomes is rapid, and autophagosomes do not accumulate in healthy neurons at detectable levels.[103]

In strong contrast to earlier claims that autophagy did not occur in neurons, a broad range of new studies has explored the role of autophagy in trafficking membrane proteins to and from the synapse. While there are still few relevant publications, recent evidence indicates that synapses utilize autophagy as a means of modulating neurotransmitter release, neurotransmitter receptor trafficking, and synaptic plasticity (Figure 11.3). At presynaptic sites, autophagy modulates the recycling of synaptic vesicles, a process that involves direct autophagy of synaptic vesicles or Rab26- and clathrin-mediated endocytosis of synaptic vesicle membranes after exocytosis. At the postsynaptic site, autophagy functions as a selective degradation pathway for GABA, AMPA, and acetylcholine receptors under specific physiological conditions, especially upon the loss

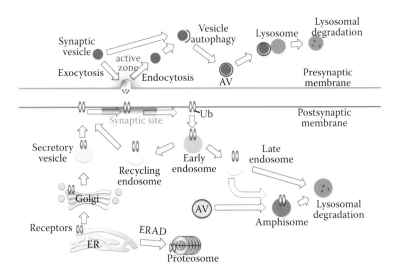

FIGURE 11.3 Autophagy of pre- and postsynaptic components. At presynaptic terminals, synaptic vesicles can be targeted to autophagosomes (AVs) directly or indirectly through the fusion of endosome-autophagosomes known as amphisomes after vesicle release. At postsynaptic terminals, neurotransmitter receptors are synthesized in ER and transported to synaptic membrane through *trans* Golgi network and secretory vesicles. Excessive receptor proteins retained in ER are targeted to proteasome degradation via ERAD pathway. Endocytosed receptors can be targeted to lysosomes directly by the fusion of late endosomes with lysosomes, or through amphisomes due to the fusion of late endosomes and AVs.

of presynaptic denervation. These findings also suggest a role for autophagy in controlling the balance of neuronal excitation and inhibition, consistent with recent findings for increased susceptibility to seizures in a neuronal-specific *Atg7*-knockout mouse line.[104,105] A plausible mechanism for the formation of autophagosomes and autophagic degradation of presynaptic vesicles and postsynaptic neurotransmitter receptors in response to presynaptic innervation may involve a change in endosome recycling. The formation of more amphisomes due to the fusion of endosomes with autophagosomes would reduce the recycling endosome population, and direct more receptor/vesicle-containing endosomes to autophagosomes for lysosomal degradation.

From this perspective, autophagy resembles other fundamental biological phenomena that have been adopted by the nervous system to provide synapses with a wide range of states. Impairment of any step of the autophagic pathway in neurons may disrupt neuronal homeostasis, which manifests itself with synaptic and neurite defects. It seems likely that different subtypes of neurons in the brain may feature variations in the autophagic pathway, and are differentially vulnerable to perturbations of the autophagic flux. Future studies are needed to investigate potential cellular differences of autophagy, and how this difference may relate to the selective vulnerabilities of specific populations of neurons in diseases, for example, the dopaminergic neurons in PD. It is remarkable that there are few lysosomes typically found in axons, but multivesicular bodies are very common, suggesting that these will be quite important in axons and synapses. In addition to the role of autophagy in synaptic structures, autophagic machinery appears to be essential also for highly specialized circuits and processes, such as the control of the visual cycle.[106] To better comprehend the involvement of autophagy in the maintenance and pathology of CNS synapses and circuits, effort should be invested in elucidating the upstream signals that regulate autophagic machinery in neurons and the selective types of autophagy that occur in neurons during development and synaptic plasticity. To do so, the development of tools to image and monitor selective types of autophagy in the CNS *in vivo* is essential, and identification of these molecular mechanisms is critical to the development of new forms of therapy.

REFERENCES

1. Nixon RA. The role of autophagy in neurodegenerative disease. *Nat Med.* 2013;19(8):983–997.
2. Son JH, Shim JH, Kim KH, Ha JY, Han JY. Neuronal autophagy and neurodegenerative diseases. *Exp Mol Med.* 2012;44(2):89–98.
3. Ragagnin A, Guillemain A, Grant NJ, Bailly Y. Neuronal autophagy and prion proteins. In Bailly Y. (Ed.), *Autophagy—A Double-Edged Sword—Cell Survival or Death?* 2013. InTech, Rijeka, 377–419.
4. Lee KM, Hwang SK, Lee JA. Neuronal autophagy and neurodevelopmental disorders. *Exp Neurobiol.* 2013;22(3):133–142.
5. Cecconi F, Di Bartolomeo S, Nardacci R, Fuoco C, Corazzari M, Giunta L, Romagnoli A, et al. A novel role for autophagy in neurodevelopment. *Autophagy.* 2007;3(5):506–508.
6. Tang G, Gudsnuk K, Kuo SH, Cotrina ML, Rosoklija G, Sosunov A, Sonders MS, et al. Loss of mTOR-dependent macroautophagy causes autistic-like synaptic pruning deficits. *Neuron.* 2014;83(5):1131–1143.

7. Colón-Ramos DA. Synapse formation in developing neural circuits. *Curr Top Dev Biol.* 2009;87:53–79.

8. Huttenlocher PR, Dabholkar AS. Regional differences in synaptogenesis in human cerebral cortex. *J Comp Neurol.* 1997;387:167–178.

9. Bailey CH, Kandel ER. Structural changes accompanying memory storage. *Annu Rev Physiol.* 1993;55:397–426.

10. Martin SJ, Grimwood PD, Morris RG. Synaptic plasticity and memory: An evaluation of the hypothesis. *Annu Rev Neurosci.* 2000;23:649–711.

11. Shinoda Y, Tanaka T, Tominaga-Yoshino K, Ogura A. Persistent synapse loss induced by repetitive LTD in developing rat hippocampal neurons. *PLoS One.* 2010;5:e10390.

12. Zuo Y, Yang G, Kwon E, Gan WB. Long-term sensory deprivation prevents dendritic spine loss in primary somatosensory cortex. *Nature.* 2005;436(7048):261–265.

13. Moody WJ. Control of spontaneous activity during development. *J Neurobiol.* 1998;37:97–109.

14. Tessier CR, Broadie K. Activity-dependent modulation of neural circuit synaptic connectivity. *Front Mol Neurosci.* 2009;2:8.

15. Penzes P, Cahill ME, Jones KA, VanLeeuwen JE, Woolfrey KM. Dendritic spine pathology in neuropsychiatric disorders. *Nat Neurosci.* 2011;14(3):285–293.

16. Zhou Q, Homma KJ, Poo MM. Shrinkage of dendritic spines associated with long-term depression of hippocampal synapses. *Neuron.* 2004;44(5):749–757.

17. Van Spronsen M, Hoogenraad CC. Synapse pathology in psychiatric and neurologic disease. *Curr Neurol Neurosci Rep.* 2010;10(3):207–214.

18. Lewis DA, Levitt P. Schizophrenia as a disorder of neurodevelopment. *Annu Rev Neurosci.* 2002;25:409–432.

19. Anderson MP. Arrested glutamatergic synapse development in human partial epilepsy. *Epilepsy Curr.* 2010;10(6):153–158.

20. Pardo CA, Eberhart CG. The neurobiology of autism. *Brain Pathol.* 2007;17(4):434–447.

21. Glausier JR, Lewis DA. Dendritic spine pathology in schizophrenia. *Neuroscience.* 2013;251:90–107.

22. Pfeiffer BE, Zang T, Wilkerson JR, Taniguchi M, Maksimova MA, Smith LN, Cowan CW, Huber KM. Fragile X mental retardation protein is required for synapse elimination by the activity-dependent transcription factor MEF2. *Neuron.* 2010;66(2):191–197.

23. Flavell SW, Cowan CW, Kim TK, Greer PL, Lin Y, Paradis S, Griffith EC, Hu LS, Chen C, Greenberg ME. Activity-dependent regulation of MEF2 transcription factors suppresses excitatory synapse number. *Science.* 2006;311(5763):1008–1012.

24. Shalizi A, Gaudillière B, Yuan Z, Stegmüller J, Shirogane T, Ge Q, Tan Y, Schulman B, Harper JW, Bonni A. A calcium-regulated MEF2 sumoylation switch controls postsynaptic differentiation. *Science.* 2006;311(5763):1012–1017.

25. Tsai NP, Wilkerson JR, Guo W, Maksimova MA, DeMartino GN, Cowan CW, Huber KM. Multiple autism-linked genes mediate synapse elimination via proteasomal degradation of a synaptic scaffold PSD-95. *Cell.* 2012;151(7):1581–1594.

26. Yamamoto A, Yue Z. Autophagy and its normal and pathogenic states in the brain. *Annu Rev Neurosci.* 2014;37:55–78.

27. Mizushima N. Autophagy: Process and function. *Genes Dev.* 2007;21(22):2861–2873.

28. Hara T, Nakamura K, Matsui M, Yamamoto A, Nakahara Y, Suzuki-Migishima R, Yokoyama M, et al. Suppression of basal autophagy in neural cells causes neurodegenerative disease in mice. *Nature.* 2006;441(7095):885–889.

29. Komatsu M, Waguri S, Chiba T, Murata S, Iwata J, Tanida I, Ueno T, et al. Loss of autophagy in the central nervous system causes neurodegeneration in mice. *Nature.* 2006;441:880–884.

30. Komatsu M, Waguri S, Koike M, Sou YS, Ueno T, Hara T, Mizushima N, et al. Homeostatic levels of p62 control cytoplasmic inclusion body formation in autophagy-deficient mice. *Cell.* 2007;131:1149–1163.
31. Nixon RA, Cataldo AM, Mathews PM. The endosomal-lysosomal system of neurons in Alzheimer's disease pathogenesis: A review. *Neurochem Res.* 2000;25:1161–1172.
32. Maday S, Holzbaur EL. Autophagosome biogenesis in primary neurons follows an ordered and spatially regulated pathway. *Dev Cell.* 2014;30:71–85.
33. Boland B, Nixon RA. Neuronal macroautophagy: From development to degeneration. *Mol Aspects Med.* 2006;27:503–519.
34. Jaeger PA, Wyss-Coray T. All-you-can-eat: Autophagy in neurodegeneration and neuroprotection. *Mol Neurodegener.* 2009;4:16.
35. Linke R, Faber-Zuschratter H, Seidenbecher T, Pape HC. Axonal connections from posterior paralaminar thalamic neurons to basomedial amygdaloid projection neurons to the lateral entorhinal cortex in rats. *Brain Res Bull.* 2004;63(6):461–469.
36. Bamber BA, Rowland AM. Shaping cellular form and function by autophagy. *Autophagy.* 2006;2(3):247–249.
37. Kirkin V, McEwan DG, Novak I, Dikic I. A role for ubiquitin in selective autophagy. *Mol Cell.* 2009;34(3):259–269.
38. Korolchuk VI, Menzies FM, Rubinsztein DC. Mechanisms of cross-talk between the ubiquitin-proteasome and autophagy-lysosome systems. *FEBS Lett.* 2010;584(7):1393–1398.
39. Fader CM, Colombo MI. Autophagy and multivesicular bodies: Two closely related partners. *Cell Death Differ.* 2009;16(1):70–78.
40. Rusten TE, Simonsen A. ESCRT functions in autophagy and associated disease. *Cell Cycle.* 2008;7(9):1166–1172.
41. Hanley JG. Endosomal sorting of AMPA receptors in hippocampal neurons. *Biochem Soc. Trans.* 2010;38(2):460–465.
42. Shen W, Ganetzky B. Autophagy promotes synapse development in Drosophila. *J Cell Biol.* 2009;187:71–79.
43. Hernandez D, Torres CA, Setlik W, Cebrián C, Mosharov EV, Tang G, Cheng HC, et al. Regulation of presynaptic neurotransmission by macroautophagy. *Neuron.* 2012;74(2):277–284.
44. Bunge MB. Fine structure of nerve fibers and growth cones of isolated sympathetic neurons in culture. *J Cell Biol.* 1973;56(3):713–735.
45. Hollenbeck PJ. Products of endocytosis and autophagy are retrieved from axons by regulated retrograde organelle transport. *J Cell Biol.* 1993;121:305–315.
46. Peça J, Feng G. Cellular and synaptic network defects in autism. *Curr Opin Neurobiol.* 2012;22(5):866–872.
47. Chu CT, Plowey ED, Dagda RK, Hickey RW, Cherra SJ 3rd, Clark RS. Autophagy in neurite injury and neurodegeneration: In vitro and in vivo models. *Methods Enzymol.* 2009;453:217–249.
48. Wishart TM, Parson SH, Gillingwater TH. Synaptic vulnerability in neurodegenerative disease. *J Neuropathol Exp Neurol.* 2006;65:733–739.
49. Spillantini MG, Schmidt ML, Lee VM, Trojanowski JQ, Jakes R, Goedert M. Alpha-synuclein in Lewy bodies. *Nature.* 1997;388(6645):839–840.
50. Nixon RA. Autophagy, amyloidogenesis and Alzheimer disease. *J Cell Sci.* 2007;120:4081–4091.
51. DiFiglia M, Sapp E, Chase KO, Davies SW, Bates GP, Vonsattel JP, Aronin N. Aggregation of huntingtin in neuronal intranuclear inclusions and dystrophic neurites in brain. *Science.* 1997;277:1990–1993.
52. Ironside JW. Prion diseases in man. *J Pathol.* 1998;186:227–234.

53. Sanchez-Varo R, Trujillo-Estrada L, Sanchez-Mejias E, Torres M, Baglietto-Vargas D, Moreno-Gonzalez I, De Castro V, et al. Abnormal accumulation of autophagic vesicles correlates with axonal and synaptic pathology in young Alzheimer's mice hippocampus. *Acta Neuropathol.* 2012;123:53–70.

54. DeKosky ST, Scheff SW. Synapse loss in frontal cortex biopsies in Alzheimer's disease: Correlation with cognitive severity. *Ann Neurol.* 1990;27:457–464.

55. Masliah E, Mallory M, Alford M, DeTeresa R, Hansen LA, McKeel DW Jr, Morris JC. Altered expression of synaptic proteins occurs early during progression of Alzheimer's disease. *Neurology.* 2001;56:127–129.

56. Scheff SW, Price DA, Schmitt FA, DeKosky ST, Mufson EJ. Synaptic alterations in CA1 in mild Alzheimer disease and mild cognitive impairment. *Neurology.* 2007;68:1501–1508.

57. Yu WH, Cuervo AM, Kumar A, Peterhoff CM, Schmidt SD, Lee JH, Mohan PS, et al. Macroautophagy—A novel beta-amyloid peptide-generating pathway activated in Alzheimer's disease. *J Cell Biol.* 2005;171:87–98.

58. Yang DS, Stavrides P, Mohan PS, Kaushik S, Kumar A, Ohno M, Schmidt SD, et al. Reversal of autophagy dysfunction in the TgCRND8 mouse model of Alzheimer's disease ameliorates amyloid pathologies and memory deficits. *Brain.* 2011;134:258–277.

59. Webb JL, Ravikumar B, Atkins J, Skepper JN, Rubinsztein DC. Alpha-synuclein is degraded by both autophagy and the proteasome. *J Biol Chem.* 2003;278:25009–25013.

60. Cuervo AM, Stefanis L, Fredenburg R, Lansbury PT, Sulzer D. Impaired degradation of mutant alpha-synuclein by chaperone-mediated autophagy. *Science.* 2004; 305:1292–1295.

61. Kabuta T, Furuta A, Aoki S, Furuta K, Wada K. Aberrant interaction between Parkinson disease-associated mutant UCH-L1 and the lysosomal receptor for chaperone-mediated autophagy. *J Biol Chem.* 2008;283:23731–23738.

62. Narendra D, Tanaka A, Suen DF, Youle RJ. Parkin is recruited selectively to impaired mitochondria and promotes their autophagy. *J Cell Biol.* 2008;183:795–803.

63. MacLeod D, Dowman J, Hammond R, Leete T, Inoue K, Abeliovich A. The familial Parkinsonism gene LRRK2 regulates neurite process morphology. *Neuron.* 2006;52:587–593.

64. Plowey ED, Cherra SJ 3rd, Liu YJ, Chu CT. Role of autophagy in G2019S-LRRK2-associated neurite shortening in differentiated SH-SY5Y cells. *J Neurochem.* 2008;105:1048–1056.

65. Cherra SJ 3rd, Kulich SM, Uechi G, Balasubramani M, Mountzouris J, Day BW, Chu CT. Regulation of the autophagy protein LC3 by phosphorylation. *J Cell Biol.* 2010;190:533–539.

66. Helton TD, Otsuka T, Lee MC, Mu Y, Ehlers MD. Pruning and loss of excitatory synapses by the parkin ubiquitin ligase. *Proc Natl Acad Sci U S A.* 2008;105:19492–19497.

67. Trempe JF, Chen CX, Grenier K, Camacho EM, Kozlov G, McPherson PS, Gehring K, Fon EA. SH3 domains from a subset of BAR proteins define a Ubl-binding domain and implicate parkin in synaptic ubiquitination. *Mol Cell.* 2009;36:1034–1047.

68. Doss-Pepe EW, Chen L, Madura K. Alpha-synuclein and parkin contribute to the assembly of ubiquitin lysine 63-linked multiubiquitin chains. *J Biol Chem.* 2005;280:16619–16624.

69. Clayton DF, George JM. Synucleins in synaptic plasticity and neurodegenerative disorders. *J Neurosci Res.* 1999;58:120–129.

70. Spencer B, Potkar R, Trejo M, Rockenstein E, Patrick C, Gindi R, Adame A, Wyss-Coray T, Masliah E. Beclin 1 gene transfer activates autophagy and ameliorates the neurodegenerative pathology in alpha-synuclein models of Parkinson's and Lewy body diseases. *J Neurosci.* 2009;29:13578–13588.

71. Winslow AR, Chen CW, Corrochano S, Acevedo-Arozena A, Gordon DE, Peden AA, Lichtenberg M, et al. α-Synuclein impairs macroautophagy: Implications for Parkinson's disease. *J Cell Biol.* 2010;190:1023–10337.

72. Borsello T, Croquelois K, Hornung JP, Clarke PG. N-methyl-d-aspartate-triggered neuronal death in organotypic hippocampal culturesis endocytic, autophagic and mediated by the c-Jun N-terminal kinase pathway. *Eur J Neurosci.* 2003;18:473–485.

73. Shacka JJ, Lu J, Xie ZL, Uchiyama Y, Roth KA, Zhang J. Kainic acid induces early and transient autophagic stress in mouse hippocampus. *Neurosci Lett.* 2007;414:57–60.

74. Puyal J, Ginet V, Grishchuk Y, Truttmann AC, Clarke PG. Neuronal autophagy as a mediator of life and death: Contrasting roles in chronic neurodegenerative and acute neural disorders. *Neuroscientist.* 2012;18:224–236.

75. Yue Z, Horton A, Bravin M, DeJager PL, Selimi F, Heintz N. A novel protein complex linking the delta 2 glutamate receptor and autophagy: Implications for neurodegeneration in lurcher mice. *Neuron.* 2002;35:921–933.

76. Nishiyama J, Matsuda K, Kakegawa W, Yamada N, Motohashi J, Mizushima N, Yuzaki M. Reevaluation of neurodegeneration in lurcher mice: Constitutive ion fluxes cause cell death with, not by, autophagy. *J Neurosci.* 2010;30:2177–2187.

77. Bigford GE, Alonso OF, Dietrich D, Keane RW. A novel protein complex in membrane rafts linking the NR2B glutamate receptor and autophagy is disrupted following traumatic brain injury. *J Neurotrauma.* 2009;26:703–720.

78. Wang Y, Wang W, Li D, Li M, Wang P, Wen J, Liang M, Su B, Yin Y. IGF-1 alleviates NMDA-induced excitotoxicity in cultured hippocampal neurons against autophagy via the NR2B/PI3K-AKT-mTOR pathway. *J Cell Physiol.* 2014;229:1618–1629.

79. Chen L, Miao Y, Chen L, Jin P, Zha Y, Chai Y, Zheng F. The role of elevated autophagy on the synaptic plasticity impaired by Cdse/ZnS quanta dots. *Biomaterials.* 2013;34:10172–10181.

80. Gao J, Zhang X, Yu M, Ren G, Yang Z. Cognitive deficits induced by multi-walled carbon nanotubes via the autophagic pathway. *Toxicology.* 2015;337:21–29.

81. Saheki Y, De Camilli P. Synaptic vesicle endocytosis. *Cold Spring Harb Perspect Biol.* 2012;4:a005645.

82. Ceccarelli B, Hurlbut WP, Mauro A. Turnover of transmitter and synaptic vesicles at the frog neuromuscular junction. *J Cell Biol.* 1973;57:499–524.

83. Binotti B, Pavlos NJ, Riedel D, Wenzel D, Vorbrüggen G, Schalk AM, Kühnel K, et al. The GTPase Rab26 links synaptic vesicles to the autophagy pathway. *Elife.* 2015;4:e05597.

84. Lee S, Sato Y, Nixon RA. Lysosomal proteolysis inhibition selectively disrupts axonal transport of degradative organelles and causes an Alzheimer's-like axonal dystrophy. *J Neurosci.* 2011;31:7817–7830.

85. Maday S, Holzbaur EL. Autophagosome assembly and cargo capture in the distal axon. *Autophagy.* 2012;8:858–860.

86. Pavlos NJ, Grønborg M, Riedel D, Chua JJ, Boyken J, Kloepper TH, Urlaub H, Rizzoli SO, Jahn R. Quantitative analysis of synaptic vesicle Rabs uncovers distinct yet overlapping roles for Rab3a and Rab27b in Ca2+-triggered exocytosis. *J Neurosci.* 2010;30:13441–13453.

87. Ravikumar B, Moreau K, Jahreiss L, Puri C, Rubinsztein DC. Plasma membrane contributes to the formation of pre-autophagosomal structures. *Nat Cell Biol.* 2010;12:747–757.

88. Wilkerson JR, Tsai NP, Maksimova MA, Wu H, Cabalo NP, Loerwald KW, Dictenberg JB, Gibson JR, Huber KM. A role for dendritic mGluR5-mediated local translation of Arc/Arg3.1 in MEF2-dependent synapse elimination. *Cell Rep.* 2014;7:1589–600.

89. Yang Q, She H, Gearing M, Colla E, Lee M, Shacka JJ, Mao Z. Regulation of neuronal survival factor MEF2D by chaperone-mediated autophagy. *Science.* 2009;323:124–127.

90. Shehata M, Matsumura H, Okubo-Suzuki R, Ohkawa N, Inokuchi K. Neuronal stimulation induces autophagy in hippocampal neurons that is involved in AMPA receptor degradation after chemical long-term depression. *J Neurosci.* 2012;32:10413–10422.
91. Hoeffer CA, Klann E. mTOR signaling: At the crossroads of plasticity, memory and disease. *Trends Neurosci.* 2010;33:67–75.
92. Collingridge GL, Isaac JT, Wang YT. Receptor trafficking and synaptic plasticity. *Nat Rev Neurosci.* 2004;5:952–962.
93. Hirling H. Endosomal trafficking of AMPA-type glutamate receptors. *Neuroscience.* 2009;158:36–44.
94. Rowland AM, Richmond JE, Olsen JG, Hall DH, Bamber BA. Presynaptic terminals independently regulate synaptic clustering and autophagy of GABAA receptors in *Caenorhabditis elegans. J Neurosci.* 2006;26:1711–1720.
95. Kittler JT, Moss SJ. Modulation of GABAA receptor activity by phosphorylation and receptor trafficking: Implications for the efficacy of synaptic inhibition. *Curr Opin Neurobiol.* 2003;13(3):341–347.
96. Kabeya Y, Mizushima N, Yamamoto A, Oshitani-Okamoto S, Ohsumi Y, Yoshimori T. LC3, GABARAP and GATE16 localize to autophagosomal membrane depending on form-II formation. *J Cell Sci.* 2004;117:2805–2812.
97. Leil TA, Chen ZW, Chang CS, Olsen RW. GABAA receptor-associated protein traffics GABAA receptors to the plasma membrane in neurons. *J Neurosci.* 2004;24:11429–11438.
98. Khan MM, Strack S, Wild F, Hanashima A, Gasch A, Brohm K, Reischl M, et al. Role of autophagy, SQSTM1, SH3GLB1, and TRIM63 in the turnover of nicotinic acetylcholine receptors. *Autophagy.* 2014;10:123–136.
99. Akaaboune M, Culican SM, Turney SG, Lichtman JW. Rapid and reversible effects of activity on acetylcholine receptor density at the neuromuscular junction in vivo. *Science.* 1999;286:503–507.
100. Bruneau E, Sutter D, Hume RI, Akaaboune M. Identification of nicotinic acetylcholine receptor recycling and its role in maintaining receptor density at the neuromuscular junction in vivo. *J Neurosci.* 2005;25:9949–9959.
101. Huang J, Klionsky DJ. Autophagy and human disease. *Cell Cycle.* 2007;6:1837–1849.
102. Brewer GJ. Serum-free B27/neurobasal medium supports differentiated growth of neurons from the striatum, substantia nigra, septum, cerebral cortex, cerebellum, and dentate gyrus. *J Neurosci Res.* 1995;42:674–683.
103. Tooze SA, Schiavo G. Liaisons dangereuses: Autophagy, neuronal survival and neurodegeneration. *Curr Opin Neurobiol.* 2008;18:504–515.
104. McMahon J, Huang X, Yang J, Komatsu M, Yue Z, Qian J, Zhu X, Huang Y. Impaired autophagy in neurons after disinhibition of mammalian target of rapamycin and its contribution to epileptogenesis. *J Neurosci.* 2012;32:15704–15714.
105. Giorgi FS, Biagioni F, Lenzi P, Frati A, Fornai F. The role of autophagy in epileptogenesis and in epilepsy-induced neuronal alterations. *J Neural Transm.* 2015;122:849–862.
106. Kim JY, Zhao H, Martinez J, Doggett TA, Kolesnikov AV, Tang PH, Ablonczy Z, et al. Noncanonical autophagy promotes the visual cycle. *Cell.* 2013;154:365–376.

Index